Shellfish Processing and Preservation

Nalan Gökoğlu

Shellfish Processing and Preservation

Springer

Nalan Gökoğlu
Fisheries Faculty
Akdeniz University
Antalya, Turkey

ISBN 978-3-030-60302-1 ISBN 978-3-030-60303-8 (eBook)
https://doi.org/10.1007/978-3-030-60303-8

This Springer imprint is published by the registered company Springer Nature Switzerland AG
The registered company address is: Gewerbestrasse 11, 6330 Cham, Switzerland

Dedicated to my dear husband, Mehmet Gökoğlu, and to my dear children, Kemal Gökoğlu and Murathan Gökoğlu

Preface

Shellfish is a broad term for various aquatic molluscs, crustaceans, and echinoderms that are used as food. They have economic and ecological importance. They have been consumed as food for many years. Shellfish also provide high-quality protein with all the dietary essential amino acids human nutrition. Shellfish is a major component of global seafood production. Shellfish aquaculture is rapidly growing. This production is evaluated by processing them in various methods overall the world. They are very perishable foods. Generally, they are preferred as alive. Otherwise, they must be preserved just after catching or harvesting. Preservation of seafood is very important issue in terms of quality and human health.

Generally, books on seafood processing and preservation are focused on finfish. There are a limited number of books specialized on shellfish. During my studies on the preparation of the book, I realized that there is a limited number of references on this subject while searching the literature. This situation excited me. I saw that there really is a need for such a book. For this reason, I hope that my book will meet this need and be useful for researchers, students, and industry working in this field.

I would like to thank Prof. Dr. Mehmet Gökoğlu for his support and help for taking photos.

Antalya, Turkey
Nalan Gökoğlu

Abbreviations

ADP	Adenosine diphosphate
AEW	Acidic electrolyzed water
AMP	Adenosine monophosphate
ASP	Amnesic shellfish poisoning
ATP	Adenosine triphosphate
Aw	Water activity
AZP	Azaspiracid poisoning
BA_s	Biogenic amines
BHA	Butylhydroxyanisole
BHT	Butylhydroxytoluene
BTX	Brevetoxin
Ca	Calcium
CCP	Critical control point
CLE	Chamuang leaf extract
CNS	Central nervous system
CO_2	Carbon dioxide
CSW	Chilled sea water
Cu	Copper
DHA	Docosahexaenoic acid
EAA	Essential amino acids
EOs	Essential oils
EPA	Eicosapentaenoic acid
FD	Freeze-drying
FDA	Food and Drug Administration
DHA	Docosahexaenoic acid
DMA	Dimethylamine
DPA	Docosapentaenoic acid
DSP	Diarrhetic Shellfish Poisoning
DTX	Dynophysistoxins
EAA	Essential amino acids
EDTA	Ethylenediaminetetraacetate

EFD	Electro freeze drying
EHD	Electro hydrodynamic drying
EPA	Eicosapentaenoic acid
FA	Ferulic acid
FAs	Fatty acids
FAO	Food and Agriculture Organisation of the United Nations
FDA	Food and Drug Administration
Fe	Iron
GMP	Good manufacturing practices
HACCP	Hazard Analysis Critical Control Points
HAV	Hepatitis A virus
H_2S	Hydrogen disulphide
HHP	High hydrostatic pressure
HUFA	Highly unsaturated fatty acids
HPF	High pressure freezing
4-HR	4-Hexyresorcinol
Hx	Hypoxanthine
HxR	Inosine
IQF	Individual quick freezing
IR	Infrared radiation
K	Potassium
LDPE	Low density polyethylene
L-DOPA	3, 4-Hydroxyphenylalanine
MAP	Modified atmosphere packaging
Mg	Magnesium
$MgCl_2$	Magnesium chloride
Mn	Manganese
MPa	Megapascal
Na	Sodium
NaClO	Sodium hypochlorite
NH	Ammonia
NoV	Nova virus
NSP	Neurotoxic shellfish poisoning
OA	Okadaic acid
OCP	Organochlorine pesticides
OFA	Oxygenated ferulic acid
P	Phosphore
PAHs	Polycyclic aromatic hydrocarbons
pAV	Paraanisidine value
PCB	Polychlorinated biphenyl
PEST-PE	Polyester polyethylene laminate
PPO	Polyphenoloxidase
PSP	Paralytic shellfish poisoning
PTX	Pectenotoxin
PUFAs	Polyunsaturated fatty acids

PV	Peroxide value
Rh	Relative humidity
SCUBA	Self-contained underwater breathing apparatus
SO_2	Sulphur dioxide
STP	Sodium tripolyphosphate
TBARs	Thiobarbituric acid reactive substances
TMA	Trimethylamine
TMAO	Trimethylamineoxide
TVB	Total volatile bases
TVB-N	Total volatile basic nitrogen
TVC	Total viable count
TW	Tap water
US	United States
USA	United States of America
UV	Ultraviolet
VP	Vacuum packaging
YTX	Yessotoxins
Zn	Zinc

Contents

About the Author

Nalan Gokoglu is a professor at the University of Akdeniz, Faculty of Fisheries. She graduated from the Ankara University, Food Engineering Department. She received her PhD in seafood processing technology. Her research activities and subjects of investigation are on fish processing technology, seafood quality, seafood safety, and seafood preservation. She has published various articles and books on seafood issues. She has served as the Dean of Fisheries Faculty (2009-2016). She is currently head of Fish Processing Department. She has various international activities. She was a visiting researcher in 1996 as a DAAD scholar in Germany and in 1998 as a British Council scholar in the UK. She also participated to various international conferences as speaker and chair. She teaches courses at the undergraduate and graduate level. She also supervised several master and PhD students.

List of Figures

List of Tables

Chapter 1
Introduction to Shellfish

1.1 Shellfish

Shellfish are defined as creatures which have shell and live in the sea. They are aquatic invertebrates. These animals do not have a backbone but instead have an outer shell. Most of the shellfish species are obtained from the sea. However, some species are also available in fresh water. Their name is shellfish, but they are not fish. They are edible creatures and used ad human food.

Shellfish are mainly categorized into molluscs and arthropods.

Molluscs
Shellfish
Arthropods

Molluscs are categorized into gastropods, cephalopods and bivalves.

Gastropods (Abalone, snail, whelk)
Molluscs → Cephalopods (Squid, octopus, cuttlefish)
Bivalves (clam, mussel, oyster, scallop)

Arthropods are categorized into crustaceans.

Arthropods → Crustaceans (Shrimp, crab, lobster, crayfish)

N. Gökoğlu, *Shellfish Processing and Preservation*,
https://doi.org/10.1007/978-3-030-60303-8_1

Global production of fish, crustaceans, molluscs, and other aquatic animals has been reported as 170.9 million tonnes in 2016. Crustaceans from both fresh water and sea contributed about 14.6 million metric tonnes to world fisheries and aquaculture production. Approximately 6.7 million tonnes of this total production is obtained from catch fisheries, while 7.8 million tons of this production was produced by aquaculture (FAO 2016).

Total molluscs production in 2016 was reported as about 23.5 million metric tons by FAO. This represented 13.75% contribution to the global aquaculture production. The production of bivalve (oysters, arches, clams, scallops, and mussels) included in the molluscs is reported to be approximately 14.8 million tons in 2016 (FAO 2016, 2018). A significant increase in bivalve production has been observed since 1996. In this 10-year period, bivalve production increased annually about 6.4% (FAO-Globefish 2007; FAO 2016). This increase occurred due to the rapid grow of aquaculture. China is the largest bivalve producer and it is reported that it meets 70% of global production. Although the demand for bivalves is increasing worldwide, the increase in production is not sufficient to meet this demand. Japan, America, South Korea, and Thailand follow China in terms of production volume in bivalve production. Scallops are leading in the world bivalve export in 2016 (Pawiro 2010; FAO 2016). Although most of the shellfish production is not involved in international trade, there are some species that have been transported from New Zealand to Chile over long distances, such as the green mussel. Although mussel culture is widespread, it cannot meet the growing demand. Despite the increasing prices, consumers' demand for mussels is not diminishing (FAO-Globefish 2018).

Shellfish has been considered as a food source for people in various countries since prehistoric times. In recent years, increasing population and purchasing power in the world has increased demand for seafood. Consumption of seafood per person increased from 18.5 kg per person to 20.3 kg per person per year from 2011 to 2016 (FAO 2018). The per capita shellfish consumption in 2013 was 4.9 kg, subdivided into 1.8 kg of crustaceans, 0.5 kg of cephalopods, and 2.6 kg of other molluscs (FAO 2016).

Shellfish, in general, contain appreciable quantities of digestible proteins, essential amino acids, bioactive peptides, long-chain polyunsaturated fatty acids, astaxanthin and other carotenoids, vitamin B_{12} and other vitamins, minerals, including copper, zinc, inorganic phosphate, sodium, potassium, selenium, iodine, and also other nutrients (Venugopal and Gopakumar 2017).

Shellfish contain quality protein with essential amino acids for the protection and growth of the human body. It has been reported that the crude protein content of seafood varies between 17 and 22% and this level can vary between 7 and 23% in crustaceans and molluscs (Gokoglu and Yerlikaya 2015). In the studies, varying amounts of protein contents have been reported for various shellfish products (Table 1.1). Holland et al. (1993) reported protein contents of 10.8 g, 12.1 g, 19.5 g, and 17.6 g (per 100 g raw edible portion) for oyster, mussel, crab, and prawn, respectively. Nurnadia et al. (2011) reported the protein content as 13.94, 19.2, 15.99 and 13.31% for cuttlefish, prawn, cockles, and oyster, respectively. Lawson et al. (1998) stated that they found protein contents of Northern shrimp and squid as

Table 1.1 Proximate composition of some shellfish species (g/100 g)

		Protein	Fat	Ash	Moisture	
Clam	*Mya areneria*	9.7–15.6	1.4–2.5	1.7	78.5–87.8	Wheaton and Lawson (1985)
	Venirupis semidecusata	12.2–13.6	0.7–0.9	–	84.9	Souci et al. (1981)
	Mixed species	12.77	0.97	1.87	81.82	Silva and Chamul (2000)
Oyster	Ostreadiae spp.	5.0–14.3	0.7–2.6	1.1–2.7	76–93	Wheaton and Lawson (1985)
		7.80	1.50	1.80	84.80	Wheaton and Lawson (1985)
	Ostrea edulis	6.0–10.7	0.79–2.1	2.0	80.5–87.5	Souci et al. (1981) Silva and Chamul (2000)
	Crasostrea gigas	9.45	2.30	1.23	82.06	Silva and Chamul (2000)
Mussel	*Mytilus edulis*	8.0–11.7	0.8–1.9	1.3–2.4	80.4–86.7	Souci et al. (1981)
		11.90	2.24	1.59	80.58	Silva and Chamul (2000)
Shrimp	Mixed species	16.2–22.7	0.1–3.2	1.3–6.8	69.6–84.8	Wheaton and Lawson (1985)
	Metapenaeus monoceros	20	0.7	2.1	77.4	Venugopal and Gopakumar (2017)
	Crangon crangon	16.4–21.3	0.8–2.3	1.2–1.7	76.7–79.7	Souci et al. (1981)
Lobster	*Panilirus* spp.	16.2–21.6	0.6–1.9	1.2–3.4	71.5–81.2	Wheaton and Lawson (1985)
		20.60	1.51	1.39	4.07	Silva and Chamul (2000) Silva and Chamul (2000)
	Homarus vulgaris	14–18.8	1.8–1.9	1.7–2.2	77.5–85.2	Souci et al. (1981)
	Thenus orentalis	21.6	0.6	2.3	75.6	Venugopal and Gopakumar (2017)
Crab	*Callinectes sapidus*	11.9–19.2	0.4–1.5	1.3–1.8	77.4–86.7	Wheaton and Lawson (1985)
		18.06	1.08	1.81	79.02	Silva and Chamul (2000) Silva and Chamul (2000)
	Mixed species	7.2–22.4	0.1–12.5	1.4–6.2	61.84.7	Wheaton and Lawson (1985)
	Scylla serata	17.5	0.2	1.6	79.2	Venugopal and Gopakumar (2017)
Squid	*Loligo* spp.	19.9	0.9	0.5	75	Venugopal and Gopakumar (2017)
	Loliginidae	15.58	1.38	1.41	78.55	Silva and Chamul (2000)
	Pectinidae spp.	15.2–20.1	0.3–1.6	1.3–1.8	74.6–85.6	Wheaton and Lawson (1985)
		16.78	0.76	1.53	78.57	Silva and Chamul (2000)

17.2 and 17% respectively. Ersoy and Sereflisan (2010) reported protein contents of 11.87 and 11.97% for two freshwater mussels (*U. terminalis and P. littoralis*). In another study, protein contents of breast, claw meat and hepatopancreas of blue crab were found to be 19.5, 18.8 and 18.8 respectively (Kucukgulmez et al. 2006). Gokoglu and Yerlikaya (2003) found protein levels of claw and body meats as 15 and 14.71% of blue crab and 21.54 and 22.64 of swim crab.

The shellfish has low fat and low calories and is therefore considered a healthy diet. The lipid contents were reported by Holland et al. (1993) as 1.3 g, 1.8 g, 5.5 g, and 0.6 g (per 100 g raw edible portion) for oyster, mussel, crab, and prawn, respectively. Nurnadia et al. (2011) reported the fat content as 1.35, 1.06, 1.93 and 1.24 g/100 g for cuttlefish, prawn, cockles, and oyster, respectively. Ersoy and Sereflisan (2010) reported lipid contents of 2.55 and 1.05% for two freshwater mussels (*U. terminalis and P. littoralis*).

On the other hand, the presence of unsaturated fatty acids in high level puts the shellfish in an important place in human nutrition. Long-chain omega-3 fatty acids such as eicosapentaenoic (EPA), docosapentaenoic (DPA), and docosahexanoic (DHA), which are polyunsaturated fatty acids (PUFA), play an important role in human health and nutrition. Molluscs contain a wide range of PUFAs which some are known as essential fatty acids (Ab Lah et al. 2017). It was reported that 32% of the shrimp lipid was composed of PUFAs and 64% of this was ω-3 PUFA and 33% was ω-6 PUFA (Dayal et al. 2013). Yerlikaya et al. (2013) reported that the fatty acid composition of the shrimps was affected by the shrimp species and the depth of the catch. They found the PUFA content of shallow water shrimps higher than the PUFA content of deep-water shrimps. They also determined the highest DHA content among the three shrimp species in *Aristomorpha foliacea* and the highest EPA content in *Penaeus kerathurus*. EPA and DHA contents of white shrimp (*Litopenaeus schimitti*) were found to be 80.55–83.79 and 97.43–101.25 mg/100 g, respectively (Pires et al. 2018).

Cholesterol is the main sterol of fish such haddock, pollock, salmon and crustaceans such as shrimp and lobster. Cholesterol is present in crustaceans such as shrimp, crab, and lobster, and in molluscs such as squid, octopus, and cuttlefish. However, as shellfish products contain low fat and unsaturated fatty acids and have low calories, they are not risky for human health. Because of these characteristics, they are considered as heart friendly foods. The intake of unsaturated fatty acids together with the consumption of shellfish reduces blood cholesterol levels. Cholesterol content of fish, crustaceans and molluscs vary significantly according to sex, season, environmental conditions, nutrition, and growing conditions (Ozogul et al. 2015).

Shellfish are rich in terms of mineral content (Table 1.2). Several kinds of shellfish are rich sources of iron. Clams have enough iron. Oyster, mussels, abalone, and shrimp also contain significant amounts of iron. Although red meats are rich iron sources due to the presence of heme iron, shellfish should also be considered to have easily absorbable iron. On the other hand, oysters are rich sources of zinc, and it is reported that consuming a 100 g portion of the oyster is sufficient to meet the daily need of the human (Dong 2001; Food and Nutrition Board 2004). Squid, lobster, oysters, and other shellfish are reported to be excellent sources of copper (Dong 2001).

Table 1.2 Copper, zinc, and iron contents of some shellfish species (mg/100 g)

		Copper	Zinc	Iron	
Clam	Mixed species	0.34	1.37	13.98	USDA (n.d.)
	Mactridae spp.	0.07–0.1	–	–	Wheaton and Lawson (1985)
	Veneridae spp.	0–1.92	0.5–7.7	1.6–13.0	Wheaton and Lawson (1985)
Oyster	Pacific	1.58	16.62	5.11	USDA (n.d.)
	Ostreadia spp.	0.003–60.6	1.7–200	0.89–10.1	Wheaton and Lawson (1985)
	Ostrea edulis	1.2–3.7	6.5–160		Souci et al. (1981)
Mussel	*Mytilus edulis*	0.09	1.60	3.95	USDA (n.d.)
Shrimp	Mixed species	0.26	1.11	2.41	USDA (n.d.)
		0.01–1.1	2.2	0.95–13.5	Souci et al. (1981)
	Crangon crangon	0.08–0.14	1.75–3.1	2.15–1.3	Souci et al. (1981)
Lobster	Mixed species	0.38	5.67	1.22	USDA (n.d.)
	Homarus vulgaris	7.0	1.6	0.95–1.3	Souci et al. (1981)
Crab	Blue	0.67	3.54	0.74	USDA (n.d.)
	Mixed species	0.01–5.07	1.4–9.36	0.25–6.1	Wheaton and Lawson (1985)
	Dungeness	0.67	4.27	0.37	USDA (n.d.)
Squid	Mixed species	1.89	1.53	0.68	USDA (n.d.)
		0.3–1.5	0.76–8.4	0.5–18.5	Wheaton and Lawson (1985) Wheaton and Lawson (1985)
Scallop	Mixed species	0.05	0.95	0.29	USDA (n.d.)
	Pectinidae spp.	0.01–0.31	1.7–2.95	1.0–3.0	Wheaton and Lawson (1985)

In addition to this superior nutritional value, the shellfish are highly sensitive foods and can be easily degraded and underwent quality changes. Immediately after harvesting, the shellfish can quickly deteriorate and therefore require attention during processing and storage. It is reported that the shellfish should be subjected to pre-treatments such as cleaning, washing, depuration, peeling, etc. according to the nature of the animal immediately after harvesting (Venugopal and Gopakumar 2017). In addition to the most common applications such as chilling and freezing, the shellfish is kept exposed to a wide variety of processes such as heat treatment, dehydration, drying, high pressure application, modified atmosphere packaging. Occasionally, combination of a few of these applications are tried to be used to increase the storage efficiency.

References

Ab Lah R, Smith J, Benkendorff K (2017) Investigation of nutritional properties of three species of marine turban snails for human consumption. Food Sci Nutr 5(1):14–30

Dayal JS, Ponniah AG, Khan HI, Madhu Babu EP, Ambasankar K, Vasagam KPK (2013) Shrimps—a nutritional perspective. Curr Sci 104(11):1487–1491

Dong FM (2001) The nutritional value of shellfish. Washington Sea Grant: WSG-MR 09-03. U.S. Natl. Oceanic and Atmospheric Administration, Seattle, pp 4–8

Ersoy B, Sereflisan H (2010) The proximate composition and fatty acid profiles of edible parts of two freshwater mussels. Turk J Fish Aquat Sci 10:71–74

FAO (2016) Yearbook of fishery statistics summary tables. Fisheries and Aquaculture Department, Rome, 200 p. http://www.fao.org/3/i9942t/I9942T.pdf

FAO (2018) The State of World Fisheries and Aquaculture 2018—meeting the sustainable development goals. Rome

FAO-Globefish (2007) Bivalves-commodity update, October 2007. Published by Globefish, (FAO), Rome

FAO-Globefish (2018) Globefish highlights. A quarterly update on world seafood markets. FAO, Rome, p 6

Food and Nutrition Board (2004) Dietary Reference Intakes (DRIs): Tolerable Upper Intake Levels (UL), elements. Food and Nutrition Board, Institute of Medicine, National Academies Press, Washington, DC

Gokoglu N, Yerlikaya P (2003) Determinaton of proximate composition and mineral contents of blue crab (*Callinectes sapidus*) and swim crab (*Portunus pelagicus*) caught off the Gulf of Antalya. Food Chem 80:495–498

Gokoglu N, Yerlikaya P (2015) Chemical composition of fish. In: Seafood chilling, refrigeration and freezing. Wiley, Chishester, pp 5–13

Holland B, Brown J, Buss DH (1993) Fish and fish products, the third supplement to McCance & Widdowson's The Composition of Foods, 5th edn. HMSO, London

Kucukgulmez A, Celik M, Yanar Y, Ersoy B, Cikrikci MC (2006) Proximate composition and mineral contents of the blue crab (*Callinectes sapidus*) breast meat, claw meat and hepatopancreas. Int J Food Sci Technol 41:1023–1026

Lawson JW, Magalhaes AM, Miller EH (1998) Important prey species of marine vertebrate predators in the northwest Atlantic: proximate composition and energy density. Mar Ecol Prog Ser 164:13–20

Nurnadia AA, Azrina A, Amin I (2011) Proximate composition and energetic value of selected marine fish and shellfish from the West coast of Peninsular Malaysia. Int Food Res J 18:137–148

Ozogul F, Kuley E, Ozogul Y (2015) Sterol content of fish, crustacea and mollusc: effects of cooking methods. Int J Food Prop 18:2026–2041

Pawiro S (2010) Bivalves: global production and trade trends. In: Rees G, Pond K, Kay D, Bartram J, Santo Domingo J (eds) Safe management of shellfish and harvest waters. IWA Publishing, London, pp 11–21

Pires DR, de Morais ACN, Coelho CCS, Marinho AF, Góes LCDSA, Augusta IM, Ferreira FS, Saldanha T (2018) Nutritional composition, fatty acids and cholesterol levels in Atlantic white shrimp (*Litopenaeus schimitti*). Int Food Res J 25(1):151–157

Silva JL, Chamul RS (2000) Composition of marine and freshwater finfish and shellfish species and their products. In: Martin RE, Carter EP, Flick GJ, Davis LM (eds) Marine and freshwater products handbook. Technomic Publishing Co Inc., Lancaster, pp 31–45

Souci SW, Fachmann W, Kraut H (1981) Food composition and nutrition tables 1981/82. Wissenschaftliche Verlagesellschaft mbH, Stuttgart

USDA National Nutrient Database for Standard Reference, United States Department of Agriculture, Agricultural Research Service (n.d.). nal.usda.gov/fnic/foodcomp/search/

Venugopal V, Gopakumar K (2017) Shellfish: nutritive value, health benefits, andconsumer safety. Compr Rev Food Sci Food Saf 16:1219–1242

Wheaton FW, Lawson TB (1985) Properties of aquatic materials. Processing aquatic food products. Wiley, pp 21–75

Yerlikaya P, Topuz OK, Buyukbenli HA, Gokoglu N (2013) Fatty acid profiles of different shrimp species: effects of depth of catching. J Aquat Food Prod Technol 22(3):290–297

Chapter 2
Crustacean Shellfish

2.1 Shrimps/Prawns

2.1.1 General Information About Shrimps/Prawns

2.1.1.1 Biology of Shrimps/Prawns

Shrimps/prawns are member of crustaceans' subphylum, they are creatures having an extended abdomen, in lengths ranging from microscopic size to 35 mm. Shrimps/prawns taxonomically belong to the swimming group of decapod crustaceans in suborder Macrura Natantia (Chan 1998a). The body of the shrimps consists of a combined head and thorax (cephalothorax) and abdominal (abdomen) region of the ring-shaped segments (Artuz 2005). At the anterior end of the crust covering cephalothorax (carapace), there is a sharp spike-shaped protruding rostrum, which plays a role in the differentiation and recognition of species. The shrimp has two pairs of antennas. The part of abdomen is longer and stronger than cephalothorax. This part forms the edible portion of shrimps. At the back end of the abdomen there is a fan-shaped tail (telson) consisting of five lobes. The legs on the thorax are used to hold and walk (pereiopods), whereas the legs on the abdomen are paddle-shaped, swimming legs (pleopods) that allow the animal to swim. These swimming legs are shaped to help the eggs to be kept in the female and the care of the fry (Kumlu 2001; Artuz 2005; Sathiya and Valarmathi 2018).

Shrimps and prawns generally live well on the bottom of the seabed although they can swim well. They rise to the higher water body from the ground time to time. Shrimps, which usually bury themselves partly in the sand during the day, rise upwards to hunt at night. Many shrimp species, especially Penaeids, are both carnivorous and herbivorous, and can generally take small-shelled aquatic organisms, maggots, aquatic plants, and organic particles as nutrients. Nutrition-related migrations are performed depending on the day-night period. In other words, shrimps that live at the bottom and are fed by organisms during the day migrate vertically or

N. Gökoğlu, *Shellfish Processing and Preservation*,
https://doi.org/10.1007/978-3-030-60303-8_2

horizontally in the water to find food at night. Most of the shrimps are fed by filtering the water containing the nutrient particles (plankton) or the decay material in the ground sludge. Among the shrimp species, there are also those who have consistently followed the pelagic life (Artuz 2005). As in the other Decapods, most of the shrimps have separate sexes, but certain species first undergo a male phase and then to a female (Farfante 1988).

Sometimes shrimps are locally named differently. In Turkish, small species are commonly referred to as "Teke" and large species are called "shrimp". Shrimps, which are called "Crevette" in French, are divided into two subgroups as "Shrimp" and "Prawn" in English. Usually "shrimp" is called for small species; "prawn" is used for large species. There is no significant difference between the two terms and even the use is sometimes confused. Their use is often mixed or even reversed in different countries. Therefore, no attempt has been made to restrict or define their meaning. Naming according to their size causes confusion in international trade. Sometimes colour is used in naming. Pink species are called prawn brown species. However, this leads to significant confusion (Anonymous 1972; Chan 1998a). Mostly the term "shrimp" will be used in this book, but the term used in a scientific article will be given as in same form when referring to scientific research.

2.1.1.2 Distribution of Shrimps/Prawns

The shrimps have a widespread area from the equator to the poles. It is commonly found in fresh water, brackish water, and seas. About 2000–3000 species are known, of which only about 300 are of commercial importance, and especially about 100 species constitute a significant part of world fishing (Farfante 1988; Kocataş et al. 1991; Chan 1998a; Gillett 2008). Although the marine species range from the shore to a depth of 5700 m, most of the shrimp species of commercial importance have spread to the depths of 100 m above the continental shelf. Despite its few representatives living in the pelagic region, most of them live in benthic area and especially in muddy, sandy-muddy, or rocky bottoms (Farfante 1988).

2.1.1.3 Trade and Marketing of Shrimps/Prawns

Shrimp and prawns are heavily traded products and represent the second main group of species exported. So far, the biggest share of production was in Latin America and Eastern and Southeast Asian countries, but most of the consumption is in developed markets. Although shrimp captured from nature in the production of total shrimp has an important contribution, most of the shrimp today is obtained by aquaculture. Total shrimp, prawn production as 8.7 million tonnes is reported by FAO in 2016 (FAO 2018). The part of 3.5 million tonnes of this production comes from capture fisheries and 5.2 million tonnes from aquaculture. With the increase in demand in developing countries, the prices of shrimps have increased in line with the general trend in the last 2–3 years. It is one of the most important products in terms of value in international trade. In terms of value in exports, shrimps account for about 16% of all exported fishery products (FAO 2018).

Although commercially important shrimp species vary by country, some species are important worldwide. *Penaeus indicus, Penaeus monodon, Penaeus semisulcatus, Penaeus merguiensis, Penaeus canaliculatus, Metapenaeus dobsoni, Metapenaeus affinis, Metapenaeus monoceros, Metapenaeus brevicornis, Metapenaeus kutchensis, Parapenaeopsis stylifera, Parapenaeopsis sculptilis, Parapenaeopsis hardwickii, Solenocera indica* as commercially important species in India have been reported. It is reported that the most important part of the catch and the economically important shrimps are penaeids belonging to *Penaeus*, *Metapenaeus*, *Parapenaeopsis* and *Solenocera* genera (Sathiya and Valarmathi 2018). So far, shrimp species found in the waters of Turkey is 61, the number of those who have the economic value is limited. 7 of these 61 species are evaluated commercially. From these species, *Penaeus japonicus, P. semisulcatus, Metapenaeus monoceros, Penaeus kerathurus and Parapenaeus longirostris* are common in all seas except the Black Sea. Shrimp species of economic value belong to the families of Penaeidae and Caridea (Bascinar 2004). Hernandez-Padilla et al. (2018) reported that white shrimp (*Penaeus vannamei*), blue shrimp (*P. stylirostris*), brown shrimp (*P. californiensis*) and crystal shrimp (*Penaeus brevirostris*) are commercial species in the southeastern Gulf of California, Mexico. *Penaeus indicus*, *Penaeus monodon*, *Penaeus meruguiensus*, *Penaeus semisulcatus* and *Metapenaeus dobsoni* are reported that the most important commercial species from Sri Lanka (Piratheepa et al. 2012), *Penaeus semisulcatus*, *Metapenaeus affinis* and *Parapenaeopsis stylifera* from North West Arabian Gulf (Al-Maliki and Al-Khafaji 2018).

2.1.1.4 Some Important Shrimp Species

There are approximately 3047 species known as shrimp and prawn. These are divided into four main groups: Sergestoidea (about 94 species), Penaeoidea (about 376 species), Stenopodidea (at least 60 species), Caridea (ena less 2 517 species) (Chan 1998a).

The studies on penaeoids are very comprehensive and are currently known to comprise 4 families including 191 species in the Western Central Pacific. Among these families Penaeidae is the most important family. Penaeidae is generally medium to large and is often found in large quantities in shallow waters throughout the continental shelf, commonly catching with trawls, purse seines, and traps (Chan 1998a). Stenopodoid, a second large group of shrimps, form Infraorder Stenopodidea, which does not have any commercial significance. Some of these are reported to have commercial value due to seen as sporadic in aquarium trade (Farfante 1988; Chan 1998a). Carideans, with many members of economic importance, constitute the third largest group of shrimps, consisting of about ten super families (Farfante 1988). Sergestoid shrimp are small and no interest to fisheries except for the genus acetes.

Of the more than 2500 known shrimp species, 343 are of economic importance. Of these, 110 species belong to the Penaeidae family and constitute 80% of the shrimps captured from nature (Kumlu 2001). Member of this family are usually marine origin. Some members, mainly those of the genera *Parapenaeus penaeopsis* occur at deep water (more than 750 m). Penaeids are mostly benthic and mainly found on soft bottom of sand and mud. Their size ranges from 2.5 to 35 cm body length. *Penaeus* species are caught extensively by trawls, seines, set nets and traps. *Penaeus merguiensis* and *Penaeus monodon* are the most important species. *Penaeus* aquaculture, mostly *Penaeus monodon*, is very popular in many countries. *Metapenaeus* is the secondary genus in aquaculture. The third commercially important species is *Parapenaeopsis* (Chan 1998a). In Caridae prawns, the first two of the walking legs are clamps. In the Stenopodidae shrimp, the first three of the walking legs are clamped, but III. double walking legs are much longer and longer than other (Kumlu 2001).

Information about some important shrimp species is given below.

Penaeus semisulcatus **(De Haan, 1844)**
Penaeus semisulcatus is called as green tiger shrimp. In green tiger shrimp, cephalothorax is bulging from the sides and a groove in the middle region of the back. They are found in all countries (Australia, Philippines, Indonesia, Japan, China, and Korea), which are coastal to the Indian Ocean. After entering the Mediterranean through the Suez Canal, they are settled on the coasts of Israel, Egypt, and Turkey. This species, which grows to a maximum of 228 cm, prefers to live in mud or sandy areas at a depth of up to 130 m. It is caught offshore by trawls, sometimes in coastal areas by fish corrals (Atay 1997; Chan 1998a; Kumlu 2001) (Fig. 2.1).

Metapenaeus monoceros **(Fabricus, 1798)**
It is called as spotted shrimp. Its body is covered with small spots. There is no tooth in the rostrum. This shrimp species, which showed a widespread from India to the Eastern shores of Africa (Madagascar, Mozambique, Kenya, Somalia etc.) entered the Mediterranean through the Suez Canal. It lives in sandy muddy areas at depths up to 60 m, more commonly at depths of 10–30 m. It prefers coasts with salinity of 0.5–0.30% (Kumlu 2001) (Fig. 2.2).

Penaeus kerathurus **(Forskal, 1775)**
It is called as grooved shrimp. In grooved shrimps, there are grooves on the top and sides of the carapace. The ridges on the sides of the backbone reach up to the

Fig. 2.1 *Penaeus semisulcatus*

Fig. 2.2 *Metapenaeus monoceros*

Fig. 2.3 *Penaeus kerathurus*

rostrum. It is found all over the Mediterranean and on the East coast of the Atlantic Ocean. This species grows up to a maximum of 22.5 cm. Young individuals are found intensively in river mouths and coasts where food density is present. Its consumption is most preferred because it is delicious (Kumlu 2001) (Fig. 2.3).

Penaeus japonicus **(Batı, 1888)**
P. japonicus is like grooved. There are 9–11 teeth on the upper edge of the rostrum and 1 tooth on the lower edge. Habitats are up to 90 m from the coastline, usually less than 50 m. They live on sandy and sandy-mud bottoms. Natural habitats are the Eastern countries (Japan, China, Indonesia, and Australia) and the Indian coasts and the Red Sea. Habitats are up to 90 m from the coastline. But they usually live in sandy and sandy-muddy bottoms less than 50 m. This shrimp, which enters the Mediterranean Sea through the Suez Canal, can usually reach 22.5 cm. Some individuals can reach up to 30 cm. It is a marine species and is found in muddy and sandy areas and prefers depths up to 90 m (Chan 1998a; Kumlu 2001) (Fig. 2.4).

Metapenaeus stebbingi **(Nobii, 1904)**
This shrimp species, which has a widespread range from the coasts of India and Pakistan to the coasts of East Africa, entered the Mediterranean Sea from the Red Sea. It is accepted that it is not suitable for culture due to its small size and slow growth. The maximum size is 13.9 cm. In the depths up to 90 m, muddy or muddy-sandy bottoms, the coasts where there are brackish waters, lagoons and river mouths

Fig. 2.4 *Penaeus japonicus*

Fig. 2.5 *Parapenaeus longirostris*

are abundant. Especially in the months when the temperature is high, they are dense at 0–10 m depths (Kumlu 2001).

Parapenaeus longirostris **(Lucas, 1846)**
Parapenaeus longirostris is a deep-water shrimp species that lives in the depths of 20 m up to 700 m and can live in all areas of the Mediterranean Sea. It also shows a wide distribution from Portugal to Angola coast in the east of the Atlantic Ocean, and from USA to France in the west of the Atlantic Ocean. Maximum size is reported as 16 cm in males and 19 cm in females. It lives in deep waters (Chan 1998a) (Fig. 2.5).

Penaeus monodon **(Fabricus, 1798)**
It is known as the largest of the Panaeids. Usually the maximum length is 35 cm in females and 26.8 cm in males. Habitats are up to 150 m from the coastline, usually less ten 30 m. They live on bottoms of sand, mud, or slits. Habitats are distributed from the eastern coast of Africa to the Red Sea, Japan, Australia, and Fiji. It has great economic importance. They are caught by trawls, gill nets, seine nets, stake nets, traps, and artisanal gear. It is a very important species for aquaculture. In many Southeast Asia countries (Endonesia, Malaysia, Thailand, Philippines) and Australia is culture in large-scale ponds (Chan 1998a).

Farfantepenaeus aztecus **(Ives, 1891)**
Maximum length in females is 236 mm, in males is 195 mm. Their habitats are estuarine and ocean littoral. They live up to depths of 110 m from the coastline. They prefer mud or sandy mud bottoms. They are mainly caught by shrimp trawls. Other gears such as seines, channel nets, cast nets, push nets, lift nets, and set gear are also used to take these shrimps (Tavares 2002) (Fig. 2.6).

Fig. 2.6 *Farfantepenaeus aztecus*

Aristaemorpha foliacea **(Risso, 1827)**
This species live in marine deep waters from 250 to 1300 m on the muddy bottoms. Habitats are Mediterranean coasts of Spain, France, Italy, Algeria, and Israel (Tavares 2002) (Fig. 2.7).

Aristeus antennatus **(Risso, 1816)**
Aristeus antennatus is an ecologically and economically important deep-water species in the Mediterranean Sea. This shrimp, known as red shrimp, live in a muddy environment and at a depth of 200–250 m of the west of the Mediterranean. *A. antennatus* is widely distributed in the Central area of the East Atlantic, in the Mediterranean Sea, apart from the North Adriatic Sea. It is caught by the bottom trawl (Marra et al. 2015) (Fig. 2.8).

Crangon crangon **(Linnaeus 1758)**
Crangon crangon, also known as brown shrimp, is found in shallow waters from 0 to 12 m and grows to around 60 mm in size. It has greyish-brown colour. The absence of a rostrum separates it from Pandalus shrimps. Brown shrimps are often found in soft sandy and muddy bottoms, bays, and gulfs. They are resistant to water temperature and salinity changes and their lives are 3–4 years (Anonymous 1972).

2.1.1.5 Fishing of Shrimps/Prawns

A major characteristic of most large-scale shrimp fishing is the use of trawl gear. Many types of trawls are used to catch shrimp, but because the otter trawl is the most important commercial gear in many countries, it deserves special mention. Otter trawl gear is also used in many fisheries apart from shrimp fisheries. Although the otter trawl is the most common form of shrimp trawl, several other similar types of gear are used to catch shrimp on a large scale; two common types are the beam trawl and the pair trawl. Most of the industrial shrimp fishing in the world is done by trawling, yet there are some large-scale shrimp fisheries that use other gear. In

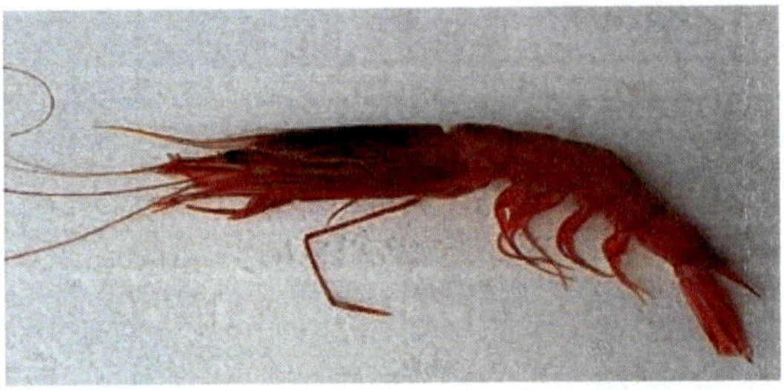

Fig. 2.7 *Aristaemorpha foliacea*

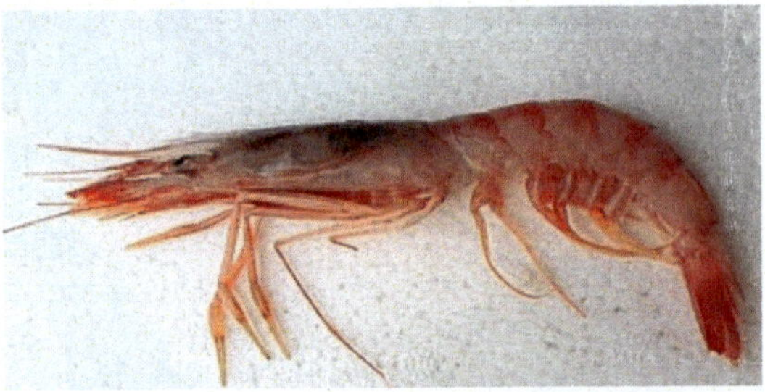

Fig. 2.8 *Aristeus antennatus*

large scale fisheries operations, shrimp boats in some countries use pots, traps and gillnets for reasons such as bottom topography and troll use such as rocky bottoms (Gillett 2008).

2.1.1.6 Nutritional Composition of Shrimps/Prawns

Shrimps are high in protein, low in saturated fat and calories, and have a neutral flavour (Table 2.1). Due to these characteristics, shrimps form a natural additive in salads, pastas, curry, soups, and stir-fried dishes. Shrimps have also been identified as a rich source of vitamin B_{12}, selenium, ω-3 highly unsaturated fatty acids (HUFA) and astaxanthin, a potent natural antioxidant. Despite the several nutritional parameters of shrimp based on which it can be considered as a healthy food, there is reluctance among dieticians and health professionals as well as consumers because of its relatively higher cholesterol (Krzynowek and Panunzio 1989; Essien 1995; Dayal et al. 2013).

Shrimp meat contains low levels of saturated fats and high levels of polyunsaturated fatty acids such as eicosapentaenoic (EPA) and docosahexaenoic (DHA) acids (Krzeczkowski 1970; Bottino et al. 1979; King et al. 1990). Thirty-two percent of shrimp lipid is composed of polyunsaturated fatty acids (PUFA) (Table 2.2). Within this, about 64% is contributed by ω-3 PUFA and the remaining 33% by ω-6 PUFA. Eating 100 g edible portion of shrimp in a day will provide >180 mg of EPA + DHA (Dayal et al. 2013).

Species, habitat, season, and feed can influence the proximate composition, particularly the lipid profile. Bono et al. (2012) examined the shrimp caught by trawlers from muddy bottoms at 500–700 m depth in five different fishing areas in the central

Table 2.1 Proximate composition of some shrimp species (g/100 g)

Species	Moisture	Protein	Fat	Ash	Carbohydrate	Reference
Penaeus monodon	80.47	17.1	1.23	0.95		Sriket et al. (2007)
Penaeus vannamei	77.21	18.8	1.30	1.47		
Litopenaeus schimitti	75.65–77.95	17.74–18.99	0.86–1.40	1.24–1.93	1.65–2.57	Saldanha et al. (2018)
Penaeus semisulcatus	474.87–75.77	20.44–21.70		1.55–1.65		Yanar and Celik (2006)
Metapenaeus monoceros	74.70–75.57	21.06–22.46	–	1.59–1.62	–	

Table 2.2 Fatty acid composition of some shrimp species (% of total FA)

	SFA	MUFA	PUFA	n-3	EPA + DHA	Reference
Penaeus kerathurus	30.95	23.54	45.35	34.30	30.78	Tsape et al. (2010)
Aristeus antennatus	32.73	37.59	29.68	24.59	23.87	Yerlikaya et al. (2013)
Aristeomorpha foliacea	31.19	34.92	33.88	29.41	28.67	
Plesionica martia	29.93	37.87	32.20	26.03	25.78	
Parapenaeus longirostris	35.73	30.32	33.95	27.02	26.72	

Mediterranean and determined that trophic and geographic factors significantly affect the shrimp (*Aristaeomorpha foliacea*) composition. Yerlikaya et al. (2013) reported that the depth of water caught by shrimps affects the composition of fatty acids. The levels of PUFAs of shallow water shrimps were found to be higher than those of deep-water shrimps. Venugopal (2020) also indicated that shallow water shrimp species generally have richer n-3 PUFA contents than their deep-water counterparts, coming through their feeding on PUFA-rich microalgae. In a study investigating the seasonal variation of the nutritional content of speckled shrimp (*Metapenaeus monoceros*), it was reported that the moisture, protein, and ash content of shrimp meat ranged between 74.70 and 75.57%, 21.06 and 22.46% and 1.59 and 1.62%, respectively (Yanar and Celik 2006). Seasonal changes in the composition of shrimp fatty acids have been reported by other researchers (Bragagnolo and Rodriguez-Amaya 2001; Kasai and Sakai 2004; Mika et al. 2013). It has also been reported that the composition of fatty acids varies depending on the sex of the shrimp. The highest PUFAs content was found in female red shrimp (Aristaeomorpha foliacea) (Olgunoglu et al. 2015) and jinga shrimp (*Metapenaeus affinis*) (Dincer and Aydin 2014) than in those of males.

Shrimps are also a rich source of astaxanthin, a lipid-soluble caratenoid. Crustaceans have carotenoproteins in some tissues and organs and their chromophores are generally composed of free carotenoids, especially astaxanthin. The colour of shrimps is due to the carotenoid content, which provides typical red-orange tissue pigmentation and changes according to their natural environment (Okada et al. 1994). Carotenoids are compounds that have significant benefits on health as antioxidants that protect cells and tissues from the harmful effects of free radicals and single oxygen. Carotenoids are either hydrocarbons, such as β-carotene or xanthophylls, or the oxygenated derivatives of carotenes such as astaxanthin, astacene, canthaxanthin, cryptoxanthin, lutein, neoxanthin, violaxanthin, and zeaxanthin. These pigments are present in esterified form or as protein complexes in the shellfish. Raw shrimp are brown or green due to red orange colour of free and esterified carotenoids. Astaxanthin is the major red-orange coloured carotenoid in many shrimp species which remains bound to proteins or chitin in shrimp (Venugopal 2020). The astaxanthin level of wild shrimps has been reported to vary between 740 and 1400 μg/100 g in edible meat portions, which again supports the argument for including them as part of the daily diet (Sachindra et al. 2005). Yanar et al. (2004) examined seasonal changes in the mussel tissue total caratenoid contents of the most commercially important shrimp species (*Penaeus semisulcatus* and *Metapenaeus monoceros*) in the north-eastern Mediterranean Sea caught in different seasons. They found considerably higher caratenoid contents for both species during spring and summer than other seasons. It is reported that the total caretenoid content varies according to the shrimp species and tissue (Sachindra et al. 2005; Bono et al. 2012). They are high in carapace, followed by head and minimum in the meat, varying from 17 to 200 mg per g (Venugopal 2020).

Raw shellfish species including shrimp have ash contents up to 2% that contains minerals, which include sodium, potassium, calcium, magnesium and phosphorus and microelements such as selenium, fluorine, iodine, cobalt, manganese, molybdenum, and others (Table 2.3). Shellfish contain nearly twice the amounts of minerals as compared with finfish (Venugopal 2020). The two major sources of minerals for marine organisms are seawater and feed. The mineral contents of shrimp species can vary seasonally. Ca, K, P, Na, and Fe contents of two shrimp species (*Penaeus semisulcatus* and *Metapenaeus monoceros*) from Eastern Mediterranean changed seasonally (Yanar and Celik 2006).

2.1.2 Quality Changes in Shrimps/Prawns

In practice, shrimps are collected by trawl nets. Sometimes bagged trawl nets are used. The nets are lowered to the bottom and withdrawn from the back of the ships. Shrimp and other sea creatures are collected in the nets when the nets dragged at the bottom. The net is pulled to the ship and shrimp and other sea creatures in the net are discharged to the deck of the ship. Shrimp fishing is practically carried out at night on sandy, muddy grounds. The type of the ground where the shrimp is caught determines its quality. Shrimps caught from muddy ground can contain up to 30 million bacteria, while those caught in sandy bottoms contain less.

The quality of the shrimp begins to change when the trawl is withdrawn. Shrimp trawls are drawn from the bottom of the sea from 1.5 to 5 h depending on the shrimp concentration. In the case of trawling, the earliest caught shrimps die before they are dumped to the deck and may begin to deteriorate. Especially during the shrimp catching season, high temperature causes the shrimp to deteriorate while on the trawler. After the trawl is unloaded to the ship's deck, the shrimps are separated

Table 2.3 Macro and micro elements of some shrimp species (mg/100 g)

	Macro elements					References
	Na	K	Ca	Mg	P	
Penaeus semisulcatus	324.6	365.6	10.7	6.91	244.4	Yanar et al. (2011)
	128.6–160.8	214.1–226.1	60.3–61.1		157.8–183.1	Yanar and Celik (2006)
Metapenaeus monoceros	123.4–151.1	215.9–237.2	59.1–62.4		163.6–171.8	Yanar and Celik (2006)
Metapenaeus affinis	125.2–125.3	303.1–355.8	23.6–24.5	37.2–41.4	–	Dincer and Aydin (2014)

	Micro elements				
	Mn	Cu	Zn	Fe	
Penaeus semisulcatus	0.13	0.41–0.42	2.36	2.02	Yanar et al. (2011)
			0.97	1.36–1.58	Yanar and Celik (2006)
Metapenaeus monoceros				1.41–1.66	Yanar and Celik (2006)
Metapenaeus affinis	0.01	–	–	0.08–0.13 –	Dincer and Aydin (2014)

from other sea creatures. In the meantime, crab, fish, sponges, and shellfish can be found in the nets. The shrimp may take a few hours to separate, so that the shrimp may begin to deteriorate. In addition, exposure to wind and sun on the deck accelerates the deterioration of the shrimp. Therefore, the processing of shrimp starts on board.

In some fishing boats, the heads of the shrimps are removed. However, this process is mostly applied to large shrimps. If a large amount of small shrimp is caught or if the port is to be returned in a short time, this is not done. Shrimps are taken into metal baskets after the heads are removed and washed with sea water flowing through a hose. They are then stored in the tank of the boat in crushed ice. Temperatures are very high in southern regions where shrimp is most caught. The water temperature rises to 30–35 °C at a temperature of approximately 30 °C. On some of the ships' metal decks, the temperature rises even higher. As a result, shrimps are damaged if not processed quickly. Bacterial and autolytic deterioration together with the so-called "black spot" begins.

2.1.2.1 Spoilage of Shrimps/Prawns

In general, same deteriorative changes are seemed in shellfish as those of fish. Therefore, degradation of fresh fish or shellfish passes through the following stages: rigor mortis, dissolution of rigor mortis, autolysis, and bacterial deterioration (Abu Bakar et al. 2008).

The shelf life of shrimps is variable, and its shelf life depends on the nature of the shrimp starter flora and the deterioration bacteria that may develop during storage (Gonçalves et al. 2003). On the other hand, storage conditions are also an effective factor on shelf life. Therefore, the researchers reported different shelf life for different shrimp species. For example, at 0 °C, shelf life was reported as 6 days for *Parapenaeus. longirostris* (Mendes et al. 2005), whereas it was reported to be 10 days for *Macrobrachium rosenbergii* (Leitao and Rios 2000) and 13–16 days for tropical shrimps stored on ice. On the other hand, the shelf life of *M. rosenbergii*, which was iced and kept at 0–4 °C, was reported as 12–16 days (Lalitha and Surendran 2006). Abu Bakar et al. (2008) recommend that *Macrobrachium rosenbergii* should not be kept for more than 14 days in ice, more than 8 days at 10 °C and more than 12 h at room temperature.

Shrimps undergo various chemical and physical changes after harvesting and during storage. Degradation occurs by the combined action of microbial chemical and enzymatic activities. Degradation can easily be determined by observing changes in physical properties. Changes in colour, odour, and texture are some of the characteristics observed in spoiled fish.

Enzymatic Spoilage

The time from shrimp death to icing is critical. Immediately after death, autolytic and bacterial enzymes begin to break down in proteins, lipids, and carbohydrates. Autolysis is mean the breakdown or decomposition of larger molecules such as proteins, lipids, and carbohydrates by enzymes after the death (Wakjira 2011).

Enzymes are protein-like substances found in the flesh and stomach of shellfish which initiate or accelerate chemical reactions. In a live organism, enzymes are in balance via the help of. Enzymes which are kept under control by digestive or blood systems are still active in post-mortem period and cause flavour changes before bacterial spoilage during cold storage. Textural and appearance changes are also seen due to enzymatic activity just after death (Gokoglu and Yerlikaya 2015).

Enzymes are the main cause of autolytic spoilage. Higher enzymatic activities increase autolytic spoilage due to ionic strength. Enzymes and bacteria do not cause degradation due to their natural defence systems during the life of all aquatic species. However, after harvesting, enzymes cause autolytic changes, while microorganisms on the surface of the organism enter the tissues and cause loss. After death, the biological regulation of the enzymes is lost, and the enzymes hydrolyse muscle proteins and resolve rigor-mortis. The digestive system of the organism, which is overfed before death, is deteriorated very quickly because of the large number of enzymes (Paul 2015). At harvesting, fish and shellfish contain food in their intestines and have enzymes. After the death of animals, enzymes penetrate the muscle, weaken, and soften the meat wall (Wheaton and Lawson 1985; Hultmann 2003). After enzymatically breakdown of proteins lipids and carbohydrates, flavour, texture, and appearance of organism change.

Nucleotide Degradation

Adenosine-5′-triphosphate (ATP) is available in muscle of shrimp in postmortem period acts as an energy source for muscle contraction. During storage, adenosine diphosphate (ADP) forms by depleting of ATP via endogenous enzymes. Adenosine triphosphate (ATP) degrades to adenosine diphosphate (ADP), adenosine monophosphate (AMP), inosine monophosphate (IMP), Inosine (Ino) and hypoxanthine (Hx). Hypoxanthine is a quality indicator in cold storage and is characterized by bitter taste in the product (Huss 1995). Creatine phosphate is a reservoir of energy-rich phosphate in the muscle cells and ATP is formed by the reaction between creatine phosphate and ADP in the live organism. After depletion of creatine phosphate, ATP is produced from ADP via phosphorylation. When regeneration is over after death, ATP rapidly degrades, and rigor-mortis develops (Huss 1988).

Enzymatic Browning

After catching/harvesting, colour changes occur in the shell segments of shrimp, especially at the place where the head part is removed, by the effect of environmental factors (sun, temperature, etc.). In this formation, as well as environmental factors, the late removing of the head after in harvesting, the application of no or insufficient cooling accelerates the colour change. This colour change is called "melanosis", "darkening", "browning" and "black spot" (Erkan et al. 2007). Shortly after catching, darkening is the main problem of those dealing with shrimp trade. This darkening is one of the industry's biggest problems, because the negative changes occur in the sensory characteristics of shellfish creatures. It results in shorter shelf life and lower product quality, thus reducing the acceptability of consumers and thus causing financial losses (Kim et al. 2000; Nirmal and Benjakul 2009a). The melanosis event in crustaceans is often seen before the onset of deterioration during storage after death. It is reported that melanosis begins within a few hours after catching. Especially in autumn and winter, they are more sensitive due to moulting and nutritional deficiencies (Montero et al. 2004). This change has a significant effect on quality and is caused by enzymatic degradation due to the effect of oxygen and light in air. Melanosis develops very rapidly in shrimps stored in cold without any preservatives and can lose its market value within 24 h (Rotllant et al. 2002). Melanosis is a biochemical process and occurs by the enzyme polyphenoloxidase. In living crustaceans polyphenoloxidase enzyme is stored as proenzyme in the blood and lymphas and in the cuticle (epiderm). Stimulation of the immune system, the scarification of the cuticle and the process of wound healing is responsible for activation of the enzyme polyphenoloxidase (Martínez-Alvarez et al. 2005a, 2008a). After harvest and death, PPO systems are still active and can promote the development of black pigments around the shell and on the meat surface (Thepnuan 2007). The enzymatic browning reaction is a reaction that catalyses the phenoloxidase, monophenoloxidase, diphenoloxidase and tyrosinase, called the polyphenoloxidase enzyme. The phenoloxidase enzyme converts the phenolic compounds into quinones and the quinones are then transformed into coloured melanoids (Kim et al. 2000; Marshall et al. 2000; Villamiel et al. 2006).

Melanosis is a natural post-mortem process originating from the polymerization of phenols in water-insoluble black pigments (melanin). Polymerization is catalysed by polyphenoloxidase, an enzyme complex (tyrosinases and catecholoxidases) found in almost all organisms. As soon as the shrimp is taken out of the water, the melanosis begins with the reaction of the atmosphere with oxygen or by the enzyme reaction during storage. The enzymatic browning reaction is catalysed by the phenoloxidase, monophenoloxidase, diphenoloxidase and tyrosinase-catalysed enzyme. The development of physiological functions of aquatic organisms is based on the important contributions of polyphenoloxidases. Polyphenoloxidases are important in the hardening of the outer shell after shell replacement in shellfish. It is also responsible for wound healing (Corzo-Martinez et al. 2012). The substrate of the polyphenoloxidase enzyme is phenolic compounds. Phenolic compounds (substrates) are secondary metabolites synthesized during the metabolism of aromatic amino acids in tissues (Hernandez-Romero et al. 2006). These compounds as food components are important in many aspects such as their functions in human health, taste and odour formation, colour formation and change, antimicrobial and antioxidative effects, roles in enzyme inhibition, (Marshall et al. 2000).

Polyphenoloxidase enzyme catalyses two basic reactions: hydroxylation (cresolase activity) and oxidation (catecholase activity). By using phenolic substances and oxygen as substrates, they hydroxylate the existing hydroxyl groups by zooming to the *o*-position. It then oxidizes diphenols to *o*-benzoquinones, which are then oxidized by non-enzymatic mechanism and oxidized to dark-coloured products, melanins (Nicolas et al. 2003; Kim et al. 2000; Martínez-Alvarez et al. 2005b). This initial reaction is called monophenolase activity and transforms tyrosine into 3, 4-hydroxyphenylalanine (L-DOPA). L-DOPA is then oxidized to *o*-quinones which are called PO diphenase activity (Aladaileh et al. 2007). Difenoloksidaz is activity which catalyzes oxidation of *o*-diphenolic compounds to *o*-quinone and initiates the formation of brown coloured compounds (Cemeroglu et al. 2004; Villamiel et al. 2006).

The distribution of PPOs in different body regions of crustaceans is different. They are usually present in the cuticle, in particular on the inner surface of the chromatophores; however, these enzymes have been reported in the hepatopancreas, muscle and hemolymph of *Penaeus monodon* (Kim et al. 2000). It is generally reported that melanosis occurs in the carapace of cephalothorax, in the caudal region and mainly in the abdomen regions where cuticle segments are combined and cuticle joins with pleopods (Montero et al. 2001a; Nirmal and Benjakul 2012; Ogawa et al. 1984). Darkening starts from the head-thorax region where the polyphenoloxidase activity is highest, and gradually spreads to the abdomen and tail (Zamorano et al. 2009).

The severity of melanosis formation in crustaceans has been reported to vary with species due to differences in substrate and enzyme concentration (Benjakul et al. 2005; Nirmal and Benjakul 2012). In some species PPO activity are faster than others. PPO activity in deep water pink shrimp was found to be faster than white shrimp and melanosis spread was slower in black tiger shrimp (Montero et al. 2001a). In a study in which the biochemical properties of the PPO enzyme were investigated, it was determined that PPO has different molecular weight, optimum pH, thermal stability, and kinetic parameters in different shellfish products (Nirmal and Benjakul 2012).

Inhibition of Melanosis

Shrimps need to be processed immediately or maintained with an effective method to maintain nutritional value, slow, or stop microbial growth, and increase shelf life. Numerous studies have been carried out to prevent melanosis that comes at the beginning of the factors limiting the shelf life. These studies mostly focused on the inhibition or prevention of polyphenoloxidase activity. Various techniques and mechanisms have been developed over many years to control undesired enzyme activity. These techniques aim to eliminate one or more of the essential compounds (oxygen, enzyme, copper, and substrate) in the browning reaction (Gokoglu and Yerlikaya 2008). Before referring to the ways of prevention of melanosis, factors affecting the development of melanosis should be known. These factors are pH of muscle tissue, temperature, dehydration, high pressure, enzyme concentration, the presence of oxygen, seasons, species and geographical origin of crustaceans and technological factors (Kim et al. 2000; Mendes 2006).

Melanosis is considered a process that needs to be controlled or eliminated. Controlling this reaction starts with a good understanding of the mechanisms of marine products, the properties, substrates and inhibitors of the enzyme and the chemical, biological and physical factors that affect each of them (Gonçalves and de Oliveira 2016).

The method of catching, age and sex affects melanosis formation. Applications during and after catching trigger a defence mechanism that activates PPO in crustaceans, resulting in increased melanosis (Bartolo and Birk 2002; McEvily et al. 1991).

Melanosis Inhibition Methods

1. Chilling
   Cooling temperatures slow the enzyme activity. Enzymatic reactions can be controlled by temperature. Every 10 °C increase in temperature increases the reaction. Conversely, a reduction of 10 °C also slows down the biological activity. Due to the close relationship between deterioration and storage temperature, crustacean products should be rapidly cooled immediately after harvesting (Pardio et al. 2011). Melanosis is slowed down by ice cooling but cannot be completely prevented (Martínez-Alvarez et al. 2007).
2. Heating
   Heating is one of the most common methods applied to foods to destroy microorganisms and inactivate enzymes. The thermal stability of polyphenoloxidase varies according to the source of the enzyme. Its catalytic activity is generally

destroyed at 70–90 °C (Manheem et al. 2012). This method has some disadvantages. These are loss of vitamin, aroma, texture and colour, as well as loss of water-soluble compounds when water or steam is used. It also has a lot of energy requirements and waste problems.
Pre-cooking is another process that inactivates enzymes and inhibits melanosis. Keeping for 2 min at the boiling temperature is reported to be sufficient for PPO inactivation (Martínez-Alvarez et al. 2009).

3. Dehydration
The water content affects the enzyme activity as it forms a solvent in which the reactants can dissolve, move, and react. Water activity in enzymatic reactions is important. Reduced water activity also reduces enzyme activity.
4. Freezing
Dissolved salts, sugars and other carbohydrates are potent inhibitors in high concentrations. Increased solute concentration during freezing increases the enzyme inhibition. As with high temperature inactivation, low temperature also has some negative effects. Freezing can change texture and other properties. Rotllant et al. (2002) reported that the rapid freezing of shrimps was a good way to prevent melanosis and that melanosis was not observed in shrimps kept frozen for 3 months.
5. High pressure
High pressure is also recommended to prevent melanosis (Encarnacion et al. 2010). The effect of high pressure on enzymes is related to alternating and nonreversible changes in protein structure. Losses in catalytic activity may be different depending on the type of enzyme, the nature of the substrate, the temperature, and the length of the process. A suitable combination of pressure temperature should be applied to increase enzyme inactivation.
6. Use of inhibitors
The inhibitors used in the prevention of darkening are limited by their effects on toxicity, health, taste aroma and texture. Inhibitors are categorized according to activity. Inhibitors used to prevent melanosis as follows:

Reducing Agents

The main role of reducing agents in the prevention of darkening is to reduce the orthoquinones to colourless diphenols or to react irreversibly to form colourless products with orthoquinones. The use of reducing compounds is the most effective method of preventing darkening (Kim et al. 2000; Gokoglu and Yerlikaya 2008).

Sulphides and derivatives (especially sodium met bisulphite) are polyphenoloxidase inhibitors, which are commonly used to prevent darkening in practice. Treatment with sulphide-based bleach inhibitors may delay the darkening for a long time depending on the concentration used (Nirmal and Benjakul 2009a). Bisulphites show competitive inhibition by binding sulfhydryl groups of the active part of the enzyme polyphenoloxidase. On the other hand, bisulfide inhibition depends on the reaction of the sulphides with the quinones and results in the formation of irreversibly inhibited sulfokinone forms of polyphenoloxidase (Kim et al. 2000). Sulphide addition to control the development of black spot (melanosis) in shrimps has been used worldwide for many years. In addition, sulphides are commonly used as antimicrobial agents to prevent bacterial growth in stored shrimps (Maldhavi et al. 1995). Although sulphides are effective in enzymatic browning, they are used in limited amounts due to their negative health effects. The accumulation levels of sulphites cause health problems in some consumer individuals, especially asthma patients (Martínez-Alvarez et al. 2007; Gokoglu 2004; Rotllant et al. 2002). In the literature, there are reports describing allergic reactions of sensitive individuals after consumption of sulphide-containing foods (FDA 2000). Therefore, sulphite concentrations are limited by the regulations. However, the regulation that determines the maximum tolerated SO_2/kg level in crustacean tissues varies between countries and organizations. Following oral intake of food treated with sulphide, allergic asthma has been shown to cause reactions in the body (Martínez-Alvarez et al. 2005b). It is also stated that sulphide residues initiate acute poisoning when they reach soft tissues, and in some cases, they can lead to death with difficulty of breath, cyanosis, and pulmonary oedema (Atkinson et al. 1993).

Alternative methods have been investigated to prevent enzymatic browning in shrimp due to the health problems caused by sulphites. 4-Hexylresorcinol (Hardisson et al. 2002; Benner et al. 1994; Martínez-Alvarez et al. 2005a; Mendes 2006) is one of the effective alternatives proposed and effective to prevent enzymatic browning. 4-hexylresorcinol (4-HR) is stable in salt water, does not lose its effectiveness in the presence of organic materials and is effective even at much lower concentrations than sulphides. In this respect, it has been suggested that 4-hexylresorcinol may be highly effective in inhibiting decay in crustaceans in many studies. The low level of residues in the edible part of the shrimp and the lack of a toxicological feature in terms of consumer health are among the advantages of 4-hexylresorcinol over sulphides. Many researchers have suggested the use of 4-HR as an alternative to sulphite. The use of this product prevents potential health risks, such as hypersensitivity reactions due to sulphide. There are several studies involving the addition of this compound to different types of crustaceans. The effects of different doses were demonstrated in these studies. These investigations reported the effective doses to be 50 mg/kg for brown shrimp (*Penaeus aztecus*) and pink shrimp (*Penaeus duodarum*) (McEvily et al. 1991; Otwell et al. 1992) and 100 kg/mg for deep water pink shrimp (*Parapenaeus longirostris*) (Guandalini et al. 1998). Other researchers reported 8% inhibition with 1000 mg/kg concentration in tiger shrimp (*Penaeus japonicus*) (Montero et al. 2001a). These different findings may be due to interspecies differences, cyclic changes in physiological sensitivity, or melanosis inhibitor concentration and the method of administration used.

In a study, shrimp (*Parapenaeus longirostris*) was treated with 4-hexylresorcinol on the fishing vessel at different times of the year and it was reported that this inhibitor prevented the development of melanosis and at the same time increased shelf life with increasing concentration (Montero et al. 2006). Montero et al. (2001a) reported that 4-hexylresorcinol was effective in preventing melanosis at a concentration of 0.5% applied to shrimps immediately after death. Some studies have been carried out on pink shrimp (*Penaeus duodarum*) stored in ice, and it has been found that

0.005% 4-hexylresorcinol treatment for 1 min is more effective than 1.25% sulphide for longer storage times (Otwell et al. 1992). The effects of 4-hexylresorcinol on different shrimp species *Penaeus aztecus* (Benner et al. 1994), *Penaeus esculentus, Penaeus plebejus*, *Metapenaeus enchavoury* and *Metapenaeus benettae* (Slattery et al. 1995) were also reported.

Another reducing agent, ascorbic acid, acts as an oxygen scavenger to reduce molecular oxygen. Inhibition mechanism of ascorbic acid is the reduction of ortho-quinones to diphenols. In addition, dehydroascorbic acid is oxidized until oxidation is delayed. Ascorbic acid is often used in combination with citric acid to produce acidic pH (Golan-Goldhirsh and Whitaker 1984). Erythorbic acid is the stereoisomer of L-ascorbic acid and is used as an antioxidant in various processed foods (Clark et al. 2009). Erythorbic acid is a free radical scavenger, changes the redox potential, and reduces quinones to diphenols (Golan-Goldhirsh and Whitaker 1984). Erythorbic acid was found to be effective in preventing the darkening of apple slices when used with 1% citric acid (Sapers and Ziolkowski 1987). The use of erythorbic acid with citric acid is referred to as sulphide substitution. In most studies, it is reported that L-ascorbic acid and erythorbic acid have equal antioxidant properties (Haard and Simpson 2000). The role of ascorbic acid and erythorbic acid in inhibiting enzymatic degradation is its ability to reduce quinones to diphenols. Hsu et al. (1988) reported that ascorbic acid inhibits the enzyme polyphenoloxidase in the mushrooms. Usually ascorbic acid is used in combination with other agents such as citric acid, calcium chloride. Ascorbic acid and its isomer are the best reducing as sulphide alternative. It is generally used as an anti-melanotic agent in fruit juices, purees, sliced, fruits, canned fruits, and vegetables. The blackening usually proceeds after the depletion of ascorbic acid (Sapers 1993; Osuga et al. 1994; Ashie et al. 1996a). It has been reported that the melanosis was better prevented in case of the use of ascorbic acid in combination with other inhibitors. For example, it was determined that the use of ascorbic acid with citric acid was more effective than the use alone (Eskin et al. 1971; Sapers 1993). This is probably due to the stability of ascorbic acid in the acidic environment and the inhibition effect of the acidic medium on the catalytic activity of the enzyme. There are studies on the use of ascorbic acid and erythorbic acid in fruits and vegetables to prevent darkening (Ponting et al. 1972; Sapers and Ziolkowski 1987; Sapers et al. 1989, 1990; Santerre et al. 1988; Gil et al. 1998). There has not been much study on the use of ascorbic acid and erythorbic acid in the prevention of melanosis in shrimps. In one study, it was determined that ascorbic acid and erythorbic acid were effective in delaying the darkening of shrimps, especially in combination with sodium metabisulphite (Toktas 2018; Gokoglu and Toktas 2018).

Antioxidants

Antioxidants protect the food by preventing the rancidity or colour degradation caused by oxygen. Synthetic and natural antioxidants are used for this purpose. Synthetic antioxidants include butylhydroxy toluene (BHT) and butylhydroxy anisole (BHA) natural antioxidants, such as tocopherols, flavonoids, cinnamic acid derivatives, and phenolic compounds such as coumarins. In recent years, interest has been focused on natural antioxidants due to the toxicity of synthetic antioxidants. Natural antioxidants are the ingredients that people have consumed for hundreds of years or mixed into foods. Therefore, it is seen as safe by consumers. Fruits, vegetables, cereals, teas, wines, and some spices are natural antioxidant sources (Buricova and Reblova 2008). Natural antioxidants are primarily plant phenolics, which occur in all parts of plants such as fruits, vegetables, seeds, leaves, nuts (nuts, walnuts), roots and shells. Plant phenolics serve as reducing agent, free radical scavenger, metal chelating agent and singlet oxygen scavenger (Jayathilakan et al. 2007). In a study investigating the effect of natural antioxidants on melanosis in shrimps, it has been found that the extract of the grape seed is effective in delaying the melanosis in the shrimp for a short time (Gokoglu and Yerlikaya 2008). Green tea and rosemary extracts were found effective to delay melanosis in shrimp (*Aristerus antennatus*) (Yatmaz and Gokoglu 2016). Nirmal and Benjakul (2009a) also reported that green tea extract prevented melanosis. The extract obtained from edible enokitake mushroom tried as melanosis inhibiting agent and was found to be successful (Jang et al. 2003).

Chelating and Complexing Agents

Enzymes generally have metal ions in their active parts. Removal of these ions with chelating agents may inactivate the enzyme. Chelating agents can form compounds via non-dispersive electron pairs in their molecular structure by pro-oxidative agents such as copper and iron ions (Kim et al. 2000). The chelators used in the food industry include sorbic acid, polycarboxylic acids (citric, malic, tartaric, and succinic acids), polyphosphates (ATP and pyrophosphates), macromolecules (proteins) and EDTA (Ethylenediaminetetraacetate). It is stated that the enzyme activity of PPO, which needs copper ions for its activation, can be prevented or reduced by chelating agents or by the formation of compounds. Kojic acid are reported as a potential antimelanotic agent (Chen et al. 1991). Kojic acid was found to inhibit PPO activity in shrimp (Benjakul et al. 2006).

Acidulants

Acids are generally used to maintain the optimum pH for the catalytic activity of the enzyme. Polyphenoloxidase enzyme is inactivated by decreasing pH. Acids are often used in combination with other antimelanotic agents (Gomez-Guillen et al. 2005). Acidifiers, such as citric, malic, and ascorbic acids, can lower the pH of a system, and thus may inactivate PPO (Montero et al. 2001a).

Citric acid is one of the acidifiers commonly used in the food industry. The citric acid has inhibitory effect on melanosis by lowering the pH as well as chelation of copper in the active part of the enzyme. However, it has no antimelanotic effect alone. Citric acid is often used in combination with other agents used to prevent melanosis. Combination of L-lactic acid and 4-hexylresorcinol in brown shrimp (*Penaeus aztecus*) has been reported to be effective in preventing melanosis (Benner et al. 1994). In a study investigating the effect of organic acids on melanosis and shelf life in shrimp (*Penaeus japonicus*), 1% lactic, citric, and acetic acids and 0.3% sodium metabisulphite solutions and their combinations were used. Sodium metabisulphite combinations were found to be effective in delaying melanosis (Gokoglu 2004). In a study in which the effect of ferulic acid on the inhibition of polyphenoxidase and quality changes in pacific white shrimp (*Litopenaeus vannamei*) stored in ice was investigated, different concentrations of ferulic acid (FA) and oxygenated ferulic acid (OFA) were tried and inhibition of the concentration of polyphenoloxidase activity was observed (Nirmal and Benjakul 2009a). In another study, ascorbic acid, citric acid, sodium benzoate, kojic acid and 4-Hexylresorcinol were tried separately and combinations to prevent melanosis and microbial degradation in shrimp (*Penaeus japonicus*). It has been determined that sodium benzoate and kojic acid are effective in the inhibition of melanosis in shrimps, and that 4-Hexylresorcinol is effective with both alone and combination with ascorbic acid or citric (Montero et al. 2001b).

7. Alternative methods

   Today, consumers prefer natural ones instead of synthetic additives. Therefore, researchers are studying on the use of natural additives to prevent melanosis. Plant phenolics as potential natural additives have received more attention due to their antioxidant and antimicrobial activities. These compounds are predominantly tocopherols, flavonoids, cinnamic acid derivatives and coumarins (Banerjee 2006). They are present in fruits, vegetables, leaves, nuts, seeds, flowers, and barks. They are found naturally in plants such as grapes, green tea and others and have antioxidant effect, which gives them inhibitory activity against PPOs (Kim et al. 2000). The extract obtained from edible enokitake mushroom tried as a preventive agent and was found to be successful (Jang et al. 2003). Grape seed extract (Gokoglu and Yerlikaya 2008; Sun et al. 2014) and rosemary extract were used in shrimps to prevent melanosis and were found to be effective (Yatmaz and Gokoglu 2016). Gokoglu and Yerlikaya (2008) stored the shrimp (*Parapenaeus longirostris*) at 4 °C after dipping into ethanol extracts of grape seed which was prepared with distilled water, and they observed inhibition effects of grape seed extract on melanosis. Sun et al. (2014) also immersed white shrimp (*Litopenaeus vannamei*) into grape seed extracts in different concentrations (0, 7.5 and 15 g/L). They suggested that grape seed extract could be used as an effective natural antimelanotic agent as alternative to synthetic ones. The rosemary extract contains antioxidant compounds such as carnosol, carnosic acid, rosmanol, epirosmanol isorosmanol, methyl carnosate and phenolic acids (O'Grady et al. 2006). Certain components in rosemary extracts also show antimicrobial activity. In one study, it was determined that rosemary extract was effective in preventing the darkening of *Aristeus antennatus* and its combination with sodium metabisulphite was more effective. In the same study, green tea extract showed similar effect (Yatmaz and Gokoglu 2016). Other researchers have also reported similar results with green tea extract (Nirmal and Benjakul 2011a) Treatment of pacific white shrimp (*Litopenaeus vannamei*) with polyphenolic glycosides-rich Chamuang leaf extract (CLE) has been found to be more effective in the prevention of melanosis when compared with sodium metabisulphite (Shiekh et al. 2019). Pomegranate extract was found effective in delaying melanosis in shrimp (*Litopenaeus vannamei*) (Fang et al. 2013). Lower melanosis scores were found in Pacific white shrimp treated with lead (*Leucaena leucocephala*) seed extract throughout the storage of 12 days compared with the control and sodium metabisulphite (1.25%) treated samples (Nirmal and Benjakul 2011a, b). The effects of Thailand native leaf extracts (Liang, Mun-poo, cashew and lead) on shrimp polyphenoloxidase (PPO) activity were compared. The highest inhibitory activity was observed with cashew leaf extract. It has been reported that immersion of the shrimp in the solution of the cashew leaf extract (1%) is effective in preventing melanosis during refrigerated storage (Sae-Leaw and Benjakul 2019).

Microbial Spoilage

Shrimp is a perishable product due to their protein and free amino acid content and post-mortem changes occur faster than fish. Rapid microbial deterioration during post-mortem storage is a serious problem in shrimp processing (Ramesh Babu et al. 2017). The muscular tissue of the shrimp deteriorates faster than the mammalian muscles, as in the fish. The faster breakdown of the crustaceans is due to higher water content, higher free amino acid content, and faster autolysis by existing enzymes and less connective tissue compared to other fleshy foods (Abu Bakar et al. 2008). Free amino acid and non-nitrogenous substance contents of shrimps constitute a suitable medium for microbial growth (Zeng et al. 2005). Moreover, neutral pH, high water activity (Aw) and high content of low molecular weight compounds make shrimp matrix an ideal substrate for bacteria. The growth of these bacteria and their metabolic activities cause undesired flavours, colours, and texture changes (Gram and Huss 1996; Jaffres et al. 2011). Shrimps spoil because of improper handling, and this spoilage is irreversible, can never bring back with further processing. Shrimps can contaminate with microorganisms in case of poor hygienic condition, inappropriate processing, and preservation and storage conditions. Protecting the quality of the shrimp or preventing its deterioration is one of the most important problems facing the industry worldwide (Abu Bakar et al. 2008).

Dominant Microflora

Shrimps can have high levels of microorganism load even when caught. One of the factors affecting the microbial quality of the shrimps is the status of microbial contamination of the waters they live in. It is likely that the microorganism load of shrimps caught from coastal areas close to polluted regions is high. On the other hand, fishing gears that drag the bottom of the water for shrimp catching are used. During the application of this technique, it is possible to increase the level of contamination of shrimps by activating the sediments rich in microorganisms in the sea bottom. However, during processes such as peeling, packaging, washing, baking, and freezing, there is a rapid increase in the burden of microorganism.

The shrimp-related bacterial flora is primarily included in the genera *Flavobacter*, *Achromobacter*, *Bacillus* and *Micrococcus*. The dominance of each genus varies considerably during storage. While Gram-negative bacteria constituted 73% of the total flora in fresh prawn (*Macrobrachium rosenbergii*) and dominant flora was *Enterobacteriaceae* and *Aeromonadaceae*, after 19 days of iced storage, Gram-negative bacteria constituted more than 80% of the bacterial flora and also *Pseudomonas*, *Aromonas hydrophila*, *A. veronii boivar sobria* and *Shewanella putrefaciens* were identified as the dominant spoilage organisms (Lalitha and Surendran 2006). Akintola and Bakare (2011) reported that the total number of aerobic bacteria in freshly caught freshwater shrimps ranged from 3 to 5 log cfu/g and that the isolates consisted mainly of Gram-negative bacteria and *E. coli* was dominant (61%). *Acinetobacter*, *Enterobacter*, and *Flavobacterium* species were reported as predominant spoilage microflora of fresh shrimp (*Penaeus aztecus*) (Heinsz et al. 1988). Dominant spoilers in *Litopenaeus vannamei* have been reported as *Pseudomonas* followed by *Enterobacteriacea* (Don et al. 2018). Tsironi et al. (2009) found that 91% of the bacteria isolated from frozen shrimps were *Psychrobacter phenylpyruvicus*. Jeyasekaran et al. (2006) reported that the bacterial flora of fresh raw shrimp (*Penaeus indicus*) consisted of the genera *Aeromonas*, *Pseudomonas*, *Vibrio*, *Flavobacterium* and *Serratia*, and 38% of the flora was *Aeromonas*. Fatima et al. (1988) found that the shrimp (*P. merguiensis*) had a bacterial load of 10^9 cfu/g on the 20th day of iced storage. However, Jeyaweera and Subasinghe (1988) observed a total viable count of 10^7 cfu/g in *P. indicus* on the 17th day of iced storage. Some researchers reported that *Vibrio*, *Aeromonas*, *Pseudomonas*, *Acinetobacter* and *Moraxella* were the dominant bacterial genera associated with the shrimp, Penaeus (Vanderzant et al. 1973; Cobb et al. 1976). Çolakoğlu et al. (2006) found that the numbers of total aerobic bacteria and *Pseudomonas* were 6.78 log cfu/g and 6, 95 Log cfu/g in shrimps (*Parapenaeus longirostris*) stored at 7 °C ± 1 after 6 days.

Mechanism of Microbial Spoilage

After death, shrimps become contaminated with a wide variety of microorganisms. These microorganisms multiply faster in the product and cause degradation. Amines, biogenic amines, organic acids, sulphides, alcohols, aldehydes, and ketones are formed by microbial growth and metabolism and produce unpleasant odours and flavours. Bacteria convert the fish odour and flavour into sour, heavy, fruity, ammonia and faeces odours through their enzymes, (Quang 2005; Ghaly et al. 2010). Apart from odour and flavour changes, bacteria are also responsible for the appearance and physical characteristics of meat.

Many degradation products are formed from seafood components via bacterial activity. Levels of these products are used as objective degradation indicators in seafood. Microorganisms and their enzymes catalyse autolytic and proteolytic changes in muscle tissue of fish during storage. Peptides, amino acids, ammonia, and some other low molecular N-substances are formed by protein degradation. Some microorganisms cause the formation of biogenic amines (Gokoglu and Yerlikaya 2015). Certain compounds gradually accumulated in the flesh of fish and shellfish due to the deteriorative changes. Progress in deterioration can be determined by measuring of these compounds (Connell 1995).

Biogenic amines (BAs) are low molecular weight organic bases with biological activity. By removing the carbon dioxide from the amino acids, the amine of this amino acid is formed. This phenomenon can be formed by organ-specific enzymes as well as microbial. Removing of carbon dioxide from amino acids is called decarboxylation, and the formed amine is called biogenic amine. Microorganisms with decarboxylase enzyme activity convert amino acids into their biogenic amines. Biogenic amines in seafood have seen as an important cause of foodborne diseases (Biji et al. 2016). The important biogenic amines in seafood are histamine, tyramine, tryptamine, putrescine and cadaverine which are formed by decarboxylation of amino acids histidine, tyrosine, tryptophan, ornithine, and lysine. Most bacterial species can convert histidine into histamine. *Morganella morganii, Klebsiella pneumoniae* and *Hafnia alvei* are reported to be the most potent histamine producers (Biji et al. 2016). *Morganella psychrotolerans, Photobacterium phosphoreum, Photobacterium psychrotolerans Clostridium* spp., *Vibrio alginolyticus, Acinetobacter lowffi, Plesiomonas shigelloides, Pseudomonas putida, Pseudomonas fluorescens, Aeromonas* spp. are reported as other histamine capable species (Hwang et al. 2010). On the other hand, *Proteus vulgaris, Proteus mirabilis, Enterobacter aerogenes, Enterobacter cloacae, Serratia fonticola, Serratia liquefaciens* and *Citrobacter freundii* are reported as enteric species that produce biogenic amines (Kung et al. 2009). Putrescine was found to be the dominant biogenic amine in Taihu white prawn (*Exopalaemon modestus*) during ice storage (Du et al. 2017). Putrescine has been suggested as an index of spoilage by other researchers (Mietz and Karmas 1978; Shakila et al. 1995). The microbial activity of prawns (*Macrobrachium rosenbergii*) during storage caused a decrease in amino acids arginine, lysine, and histidine. This decrease was correlated with formation of biogenic amine such as putrescine, cadaverine and histamine (Abu Bakar et al. 2008). In the shrimp (*Penaeus semisulcatus*) stored in ice, histamine-producing bacteria were not detected but putrecine and cadaverine-forming bacteria were detected (Lakshmanan et al. 2002).

Total volatile bases (TVB) and trimethylamine (TMA) content are the most common chemical parameters used to assess the quality. Trimethylamine (TMA) is an indicator for seafood spoilage and plays a large role in the loss of quality of seafood products. The formation of TMA from the reduction of trimethylamine oxide (TMAO) occurs by bacterial degradation. *Shewanella putrefaciens* and *Photobacterium phosphoreum* have been reported as specific degradation organisms which reduce TMAO to TMA (Shaw and Shewan 1968; Dalgaard et al. 1993). *Shewanella putrefaciens* and *Shewanella baltica* showed strong activity for reduction of TMAO in *Litopenaeus vannamei* during the storage at 4 °C (Zhu et al. 2017). In post-mortem period, TMAO degradation results in accumulation of volatile amines which responsible for off-odours (Gokoglu and Yerlikaya 2015). Shrimp and prawns are rich in non-protein compounds such as free amino acids and trimethylamine oxide (TMAO) (Lopez-Caballero et al. 2002; Abu Bakar et al. 2008). The acceptability limit of shrimp TMA was reported to be 5 mg/100 g (Cobb et al. 1973; Shamshad et al. 1990; Zeng et al. 2005; Okpala 2014a). However, some researchers have suggested different limits for different shrimp species. It is suggested as 9–12 mg/100 g for *Penaeus notialis* (Dabade et al. 2015), as <3 for *Littepenaeus vannamei* (Don et al. 2018). Initial TMA content was reported to be 2.5 mg/100 g for *Penaeus notialis* (Dabade 2015).

Volatile compounds from degradation indicators are produced by specific degradation organisms. Total volatile basic nitrogen (TVB-N), consisting of trimethylamine (TMA), ammonia (NH3) and dimethylamine (DMA), has been reported to be the most widely used parameter for microbiological degradation of seafood (Chan et al. 2006; Pacquit et al. 2007). Total volatile base nitrogen (TVB-N) is a combination of total amount of ammonia (NH3), dimethylamine (DMA) and trimethylamine (TMA) and is used as a freshness index (Wu and Bechtel 2008). There is also an increase in the amount of TVB along with the increase in TMA during seafood spoilage. While some researchers (Connell 1995; Dalgaard 2000) reported the acceptability limit value for TVB-N to be 30 mg/100 g in seafood products, some of them reported to be 35 mg/100 g (Amegovu et al. 2012; Huss 1995; Shamshad et al. 1990). However, this value may not be suitable for fresh shrimps and TVB-N values are generally higher in shrimps (Oehlenschläger 1997). The fact that the baseline TVB values were reported to be high for the shrimp stored in various methods confirms this statement. The initial TVB value was reported as 20.57 mg/100 g (Azam et al. 2013) and 17 mg/100 g (Abu Bakar et al. 2008) for freshwater shrimp (*Macrobranchium rosenbergii*), as 29 mg/100 g for deep water pink shrimp (*Parapenaeus longirostris*) and narwal shrimp (*Parapandalus narval*) (Condurso et al. 2016), 33.5 mg/100 g for Northern shrimp (*Pandalus borealis*) (Qingzhu 2003), 18.29 mg/100 g for *Penaeus japonicas* (Gokoglu 2004), 15.24 mg/100 g for cold water shrimp (*Pandalus borealis*) (Paul 2015).

Leitao and Rios (2000) reported that TVB increased in *Macrobranchium rosenbergii* at 5 °C and reached to maximum allowable level after 5 days. They also found a positive correlation between TVB and total psychrotrophic bacteria counts. It was reported that the same shrimp species reached to acceptable limit after 12 weeks during frozen storage (Azam et al. 2013). TVBN and TMA values increased in shrimp (*Penaeus notialis*) stored at 0, 7 and 28 °C throughout the storage (Dabade 2015) and in Pacific white shrimp (*Litopenaeus vannamei*) (Okpala et al. 2014), *Penaeus merguiensis* (Fatima et al. 1988) during ice storage.

Tryptophan which is an amino acid is metabolised to indole by some microorganisms. These microorganisms, which are often gram-negative, are found in shrimps and grow at relatively high temperatures (15–35 °C). To produce indole, a bacterial enzyme called tryptophanase, which oxidizes the free amino acid tryptophan is required. Enterobacteriaceae provides a significant indole production. It is reported that indole is used by the US Food and Drug Administration to confirm the results of sensory analysis in shrimps. The acceptable indole level determined by the FDA is <250 μg/kg. Deterioration can be differentiated with a strong odour when indole level was over 250 µg/kg (Food and Drug Administration (FDA) 1982; Mendes et al. 2005). Lower indole levels in deep water pink shrimp (*Parapenaeus longirostris*) were found at low temperature compared to higher storage temperatures. It is determined that high temperature accelerated indole formation (Mendes et al. 2005). Similar results were obtained with *Penaeus merguiensis* by Shamshad et al. (1990). Indole has been reported as the best chemical indicator to confirm chemical degradation in shrimps (Tripathy 2013). It is reported that indole test is used to determine if raw shrimp is exposed to high temperature because storage temperature is taken into consideration rather than storage time for indole formation (Botta 1995). The amount of indole formed in shrimps depends on the temperature and processing and storage applications and bacterial population. In a study on the effect of temperature on the level of indole, it was found that indole may be a good indicator for shrimp stored at high temperature and handled in poor hygienic the conditions (Mendes et al. 2000).

Oxidative Spoilage

Fat ratio in shrimp is quite low compared to other aquatic products. Although their low-fat content, shrimps are susceptible to oil oxidation during storage because their lipids contain high levels of polyunsaturated fatty acids. Oksuz et al. (2009) have determined the average total lipid ratio in deep water pink shrimp (*P. longirostris*) to be 1.1% and the ratio of polyunsaturated fatty acids in total fatty acids to be 42.13%. Li et al. (2011) reported that the rate of polyunsaturated fatty acids varied between 32.8 and 47.5% in seven shrimp species. Oxidation causes physiochemical and taste changes in shrimp. Since the shrimp has low fat content, fat oxidation is more important in frozen products (Bak et al. 1999; Tsironi et al. 2009).

The peroxide value (PV) of the pacific white shrimp (*L. vannamei*) stored in ice exceeded the limit value after 8 days. Para anisidine value (pAV) which indicates secondary lipid oxidation products in seafood reached to 224.04 value at 12 day in these shrimp samples (Okpala et al. 2014).

Sensorial Spoilage

Degradation is a phenomenon that can be detected by smell, taste, touch, and sight. These changes detected by our sensory organs are caused by the effects of microbial, chemical, and enzymatic activities. The chemical compounds which formed from the metabolism of spoilage microorganisms produce unacceptable flavours and odours associated with sensory spoilage. Progression of lipid oxidation leads to undesirable odour and taste formation.

It is reported that shrimp freshness can be determined by measuring the changes in sensory properties such as appearance, odour, colour, and texture. Changes in these attributes can be measured by sensory or instrumental methods (Olafsdottir et al. 2004). Progression of lipid oxidation leads to undesirable odour and taste formation. Frozen prawn (*Macrobrachium rosenbergii*) was sensorial acceptable during storage for 8 weeks at −20 °C. Odour and flavour was in excellent up to 4th weeks. After 12th week, there were losses in bright appearance, natural odour, and flavour (Siddiqui et al. 2011). However, Rahman et al. (2000) determined that two shrimp species, giant freshwater prawn (*Macrobrachium rosenbergii*) and marine tiger shrimp (*Panaeus monodon*), were in acceptable condition for colour, odour, appearance, and texture until the 5th day of ice storage.

2.1.3 Shrimp/Prawn Processing and Preservation

2.1.3.1 Alive Transportation of Shrimps

With the increasing purchasing power, consumers started to be more selective about quality when buying shrimps. Most consumers want to provide shrimps alive. Live shrimps show a better quality and flavour. Live shipping has become one of the major applications in the shrimp industry to increase the value and to keep the freshness of the final product. Since the autolysis begins rapidly after the death of shrimps, the quality of the food made with shrimp depends on the freshness.

If shrimps are to be marketed live, special care must be taken to reduce stress and minimize harvest damage as far as possible to ensure long-term post-harvest survival. There are several ways to keep shrimps alive. These; still water, balance system, closed circulation system, open circulation system, plastic bags with or without oxygen and use of iced/iceless sawdust. Size and species of shrimp and length of travel are important factors to consider before choosing the ideal transportation method. The most recommended of these methods is the use of sawdust. For this purpose, chilled sawdust is used. This method is easy and inexpensive (Iranto and Giyatmi 1997).

Temperature and dissolved oxygen are the two main parameters which must be checked constantly and improved prior and during transportation. The basic principle of live transport of shrimps is to reduce the temperature to slow down metabolic activity. The water temperature is reduced to a limit where the animal's metabolic rate is minimized, so that storage and transport in this case does not cause an increase in metabolic rate. This creates an anaesthetic effect on shrimp. Cold anesthetized shrimp movements are minimal, no stress occurs, weight loss is virtually absent and does not produce faeces as there is no food intake (Schoemaker 1991). After the shrimps are moved to their destination, the temperature is brought to ambient temperature and the anaesthesia effect removes. Generally, temperatures below 14 °C and above 35 °C are fatal for shrimps. Some species can tolerate 12 °C. Richards-Rajadurai (1989) reported that *P. monodon* could not tolerate 12–14 °C but was suitable for *M. japonicus*. During experimental live storage under cold anaesthesia, it was found that farmed *P. monodon* lived better at 16 °C than at

12 °C (Goodrick et al. 1995). It was found that *M. rosenbergii* was able to live for 4–8 h in chilled sawdust.

Penaeus japonicus, called kuruma shrimp, is a shrimp species which traditionally is transported live. Kuruma shrimp *Penaeus japonicus* is reported to be the only shrimp species commercially transported in chilled sawdust as it lies in hibernation (Salin 2005). It is also reported that *Penaeus japonicus* was exported live from Korea, China, and Japan. On the other hand, Salin and Vadhyar (2001a, b) studied live transportation of cold-anaesthetised of *Penaeus monodon*. They reported that the demand for live shrimps has increased worldwide and the competitive advantage of fish farming on fishery has led to an extremely favourable view of the live shrimp market. Live transport of freshwater shrimps was also tried to increase their market value.

2.1.3.2 Preliminary Processing of Shrimps/Prawns

The processing of seafood products involves a series of processing steps, until the raw material is converted into the desired processed product form. The first and the most important ones of these steps are the pre-treatments to make the raw material workable. While the pre-treatments to be applied according to the type and form of the material, the technology to be applied and the technical potential of the manufacturer vary, the common goal is to separate the edible part of the product from the non-edible part. Thus, the separated edible part facilitates further processing and allows the non-edible parts to be used more effectively in various areas such as animal feed production. It also provides economy by reducing product losses. Today, most of the pre-treatments applied to aquatic products are made with machines. However, due to the structural nature of some materials, it is necessary to perform some pre-treatments manually (Bykowski 1990; Gokoglu 2002).

Sorting/Grading

During the harvest, shrimps of various sizes can be found in the ponds. Shrimp size affects its marketability. In some markets mixed shrimps are sold at the same price, while in others, the price varies depending on the size. Therefore, shrimps may need to be classified according to their size. Grade grading of shrimps according to their size is done manually or by using mechanical classifiers working according to various principles.

Peeling

The skeleton of the crustaceans is composed of a shell, known as the exoskeleton or cuticle, which is located on the outer surface of the body and serves as a protective covering against predators. The shell is hard but segmented into sections that allow the animal to move in water. Due to biological growth, the animal should change its shell several times in its life. This phenomenon is called moulting (Dang et al. 2018a). Shrimps should replace their shells to larger ones for growing. The physiology, morphology, and biochemistry of shrimp change during moulting process (Dall et al. 1990). The connection of the shell to the muscle occurs via a collagen-like connective tissue protein. The continuity of this attachment makes peeling difficult.

Shrimps can be peeled mechanically and manually. A typical shrimp peeling machine has rubber coated adjacent piston cylinders. This machine operates according to the friction principle. A cylinder is strengthened and uses friction to rotate other adjacent rollers. When the shrimp falls between the two rolls rotating in opposite directions, the rollers grasp the shells and pull them out of the meat (Lapeyre et al. 2004). The machines are efficient in terms of time and labour, but sometimes meat can break or not being fully peeled. Loss in tail segments due to strong adherence to the muscles during the peeling is an important problem of mechanical peeling (Linton and Gordon 1992). Mechanical shrimp peelers press down on shell during peeling process and therefore it can break the meat. Due to these problems of mechanical peelers, hand peeling is used as an alternative method. Compared to mechanical peeling, hand peeling has advantageous such as keeping meat integrity and reducing shell residues in the meat. However, hand peeling is costly, laborious, time-consuming, and hazardous for food safety due to contamination.

Because of the negative effects of mechanical shell peelers on shrimp meat, researchers have developed several methods to loosen shrimp shells before going to the peeling machine. One of these methods is pre-cooking. In the peeling process with the machine, it is reported that in pre-cooked shrimps more meat yield is obtained and even the loss of the tail meat is reduced, and the shape and colour are preserved. This is attributed to the expansion of the shell or the contraction and compression of the meat or by a combination of both (Dang et al. 2018a).

However, thermal process is not suitable for every situation. Sometimes cookers need raw shrimp meat. Another method applied is maturation when the cooking process does not fully weaken the connection between the shell and the meat. Ice or brine curing process was tried by the researchers before cooking. The maturation, which is a traditional method, also has the characteristics of easy application and cheapness. It has been reported that loosening in maturation method occurs via enzymatic degradation of the connective tissue between the meat and the shell (Taylor 1993).

Recent studies have focused on emerging methods to loosen shrimp shell. Thus, it is aimed to perform mechanical peeling with less loss in shrimp meat. With the idea that the connection between the shell and the meat can be broken by protein denaturation, the studies are directed towards high hydrostatic pressure (HHP) applications. Hansen and Nielsen (1988) developed a patented method. In this method, the sudden pressure drop caused the water in the shell to warm up, causing the loosening of the shell. Then, Yang et al. (2010) found that the pressure application of 200 MPa for 3 min was sufficient for peeling in shrimps (*Penaeus vannamei*). In another patented study, pressure application of between 224 and 327 MPa for 15–120 s was reported to have excellent results in peeling of shrimps (Jabbour and Hognason 2011).

Although there are studies about enzyme, microwave, and ultrasound applications for peeling of plant-based foods, there is not enough study about the use these methods for shrimps. These are the issues that need to be considered and studied. We have found a very recent study on this subject. In this research, individual and combined ultrasound and enzyme methods have been used to facilitate the shrimp peeling by loosening the shell from the meat. The ultrasound pre-treatments increased peelability of shrimp with increasing ultrasonic amplitude and time. The parallel ultrasound-enzyme combination substantially improved the peelability of shrimp due to erosion and diffusion effects. Cavitation bubbles generated from sound waves pitted the surface of shrimp shell, generating pathways for enzyme diffusion into the muscle-shell attachment. The ultrasound either alone or in parallel combination with enzyme modified the shrimp shell not only the morphology but also the structural and functional properties (Dang et al. 2018b).

In another study in which protease solutions were used to promote the relaxation of shrimp shells prior to peeling, enzymatic maturation was found to effectively increase the peelability of shrimps. It has been reported that this enzymatic maturing process offers a better shrimp product in terms of texture and colour compared to an industrially used brine matured reference (Dang et al. 2018c).

Deveining

The vein is the dark line which appears in the upper dorsal region of prawns or shrimps' flesh. It can be a source of contamination with sand and bacteria for shrimps. Deveining is the removal of the vein by cutting longitudinally along the dorsal region of the prawns or shrimps and pulling the vein.

2.1.3.3 Chilling of Shrimps/Prawns

Shrimps have limited shelf life. For this reason, they should be stored immediately after harvest or catching. The most common preservation method used for this purpose is chilling. Chilling is the process of lowering the temperature of the fish to a temperature just above the freezing point. Chemical and biochemical reactions responsible for quality deterioration during storage will slow down with decrease in storage temperature, and the storage life will be prolonged at lower temperatures (Riaz and Qadri 1990).

The main purpose of chilling is to reduce or stop the activities of microorganisms by reducing the temperature and to prevent the physical, chemical, and biochemical phenomena occurring under normal conditions. Chilling cannot stop the spoilage completely, but it can delay a certain period. The first measures to preserve the freshness of the shrimp start on the fishing boat. Since the degradation is largely dependent on the temperature, the temperature of the caught fish should be reduced to 0 to −1 °C as soon as possible (Gokoglu 2002).

Chilling On-Board

Shrimps must be taken care of when it comes to the deck of the trawler. Shrimp catching should be done in a reasonable time. Unnecessarily extending time can damage shrimps in the nets. Shrimps should be processed quickly after coming to the deck. Exposure to sun and wind on deck should be avoided. Otherwise, the spoilage starts immediately. Firstly, shrimps must be separated from the other caught. Then shrimps should be washed with sea water and the sand, mud and bacteria should be removed.

Icing

The most common method for preserving the freshness of shrimps is icing. For optimum results, the ice method and ice quality should be considered. The shelf life of iced shrimp can be extended for a few days. The time from shrimp death to icing is critical. Immediately after death, autolytic and bacterial enzymes begin to break down in proteins, lipids, and carbohydrates. To reduce the bacterial load of the shrimp, it is inevitable to wash it thoroughly and reduce the ambient temperature. After separation and washing, the raw shrimps are drained and packed in ice in boxes. The time between catching and chilling must be short. A 1-h delay can cause deterioration, especially on a warm day (Anonymous 1972).

Ice-chilling of seafood is mainly due to the flow of cold melting water over the product. When the product to be cooled is hotter, the melting rate will be higher, so cooling is applied automatically and quickly without the risk of freezing. In addition to good cooling conditions, melting water also helps to remove blood, bacteria, etc. The chilling of the product to the temperature, which is the melting point of the ice, can be achieved by surrounding of the product completely with ice (Gokoglu 2002).

Boxes should not be deeper than 200 mm to prevent crushing of shrimps on the bottom. A thin layer of ice or broken ice block should be placed on the bottom of the box. A layer of shrimp should be placed on a layer of ice not higher than 50 mm and should be covered with a little more ice. Shrimp and ice should be added in this way until the box is filled. Boxes should not be overfilled, otherwise shrimps may be crushed. Shrimp should be covered with ice until landed. In summer, 1 kg of ice may be required for 1 kg of shrimp. Storage temperature should be kept at 1–3 °C so ice melts slowly (Anonymous 1972). The maximum storage time of the shrimp in ice varies up to 3 weeks depending on the washing process and the effectiveness of the icing after washing. Shrimp tissue is still biochemically active, although shrimps lose their vitality in a short time after their capture. Therefore, they enter an extremely rapid deterioration process due to the activity of both bacteria and original enzymes (autolysis). Shrimps stored in crushed ice will remain in good condition for 4 days. But for better results, shrimps should be processed within 2 days. Shrimps stored on ice for 6 days, the typical aroma disappear, and the meat softens; in this case it becomes difficult to remove the shell. After 8 days the fishy odour and aroma develops.

The quality of the water to be used in the icing is also important. The water must be of drinking water quality. Even if the ice is produced from clean drinking water, some bacteria may grow depending on the storage time and temperature. Thus, the long-standing ice is contaminated with deteriorating bacteria, and because of the use of old and dirty ice, the product will deteriorate more quickly (Gokoglu 2002).

Giant freshwater prawn (*Macrobrachium rosenbergii*) and marine tiger shrimp (*Panaeus monodon*) were stored in ice in two groups as head-on and headless. According to the sensory analysis results freshwater prawn were found acceptable up to 5 days in head-on shrimps and 6 days in headless shrimps during the storage, while marine tiger shrimps were found acceptable up to 8 and 9 days in head-on and headless shrimps respectively (Rahman et al. 2000). While a 10-day shelf life has been reported for *Macrobrachium rosenbergii* stored at 0 °C (Leitao and Rios 2000), it was reported as 12–16 days for the same fish species stored in ice (Lalitha and Surendran 2006). Pacific white shrimp (*Littopenaeus vannamei*) had shelf life of 8 days during the storage in ice (Odilichukwu et al. 2014).

Slurry Ice/Liquid Ice

The traditional icing method may have some undesirable effects on the product. Due to crushed ice, the product can be damaged, or some aroma losses can be seen. In addition, when large amounts catch is obtained labour and the cost increases. Liquid ice has recently been noted as a successful method for the rapid cooling of food products. Liquid ice provides better heat conduction due to physical properties. In addition, the liquid ice consists of microscopic ice crystals suspended in water, also prevents physical damage. On the other hand, as the liquid ice cuts off the contact with air, the biochemical events in the presence of oxygen are delayed (Huidobro et al. 2002). The ice slurry consists of a homogeneous mixture of small ice particles and carrier fluid. The liquid may be pure fresh water or a double solution containing water and freezing point lowering such as sodium chloride, ethanol, ethylene glycol and propylene glycol. Since the slurry maintains a constant low temperature during the cooling process, it provides a higher heat transfer coefficient than water or other single-phase liquids. Due to these properties of the ice slurry it is useful in many applications (Kauffeld et al. 2010). In a study in which shrimps (*Pandalus borealis*) were preserved using different cooling techniques, the use of liquid ice was found to be more effective compared to the flake ice or sea water flake ice mixture (Zeng et al. 2005).

Chilled Seawater (CSW)

Chilled sea water can be used as an alternative to ice to store raw shrimp at sea. It has the advantage of cooling shrimp rapidly. Some shrimp vessels are equipped to hold shrimps in sea water at 1–4 °C. CSW is very efficient in cooling because the fish are surrounded completely by the cooling medium. In its simplest form, CSW is made by adding fresh seawater to ice held in tubs or in subdivided waterproof compartments or in tanks in the fish hold. In the previous study, prawn (*Metapenaeus dobsoni*) was chilled in two ways, chilling with icing and chilled sea water. In the icing, washed shrimps were packed with crushed ice. The second group of prawns were kept in a closed sea water system cooled to 1.7–3.8 °C. The prawn held in chilled sea water have been found to be superior in general appearance, firmness of the meat and absence of external damage in shell and flesh as compared with those in ice. On the other hand, melanosis has been considerably inhibited. Better results with chilled sea-water have been obtained in terms of microbiological (Nayar et al. 1962).

Acidic Electrolyzed Water (AEW) Ice

Another alternative cooling method is acidic electrolyzed water (AEW) ice. In a study investigating the effect of acidic electrolyzed water ice on preserving the quality of shrimp (*Litopenaeus vannamei*), it has been reported that AEW ice is more effective in inhibiting the formation of total volatile basic nitrogen (TVBN) and bacterial growth compared to tap water ice (Lin et al. 2013; Wang et al. 2014). On the other hand, it has also been reported that acidic electrolyzed water ice can inhibit the shrinkage in muscle fibres of shrimps (Wang et al. 2015).

Superchilling

Super chilling is a process in which the temperature of a food product is reduced to 1–2 °C below the initial freezing point of the product. The initial freezing point of the food is between −0.5 and −2.8 °C. Generally, super chilling is positioned between freezing and refrigeration, where the ambient temperature is set below the initial freezing point. Super chilling produces ice on the surface, which absorbs the heat in the interior. This means that there is no need for external icing around the product in transport or storage for shorter periods of time. For long-term storage of the super-chilled product, storage at super-chilling temperatures will be necessary. Microbial growth is the most important factor limiting the shelf life and quality of foods. Microbial activity decreases at super cooling temperatures and most bacteria cannot grow (Kaale et al. 2011). Super-chilling (2 °C) and refrigeration (5 °C) were applied to shrimp (*Penaeus japonicus*) and their effects on the quality of shrimp were compared. Better results were obtained with super-chilled samples, compared to refrigerated ones. Colour, hardness was better maintained, and K value suppressed in case of super-chilling (Ando et al. 2004).

Combination of Chilling with Other Preservation Methods

Chilling was applied together with other preservation methods to increase the quality and the shelf life of shrimps. One of these applications is the cold storage of shrimps packaged in a modified atmosphere. Lannelongue et al. (1982) reported that microbial growth and total volatile nitrogen values were slowed down in fresh shrimps (*Penaeus aztecus*) packaged under modified atmosphere compared to shrimps stored in normal air atmosphere. Shrimps pre-treated with bisulphite wash solution and packed under modified atmosphere had shelf life up to 10 days compared to shrimp packed in air with or without sulphite wash (Kalleda et al. 2013).

Before chilling, treatment of shrimps with various antimicrobials may influence shelf life. Organic acids were used as antimicrobial preservative to protect the quality of shrimps. It was determined that total psychrotrophic bacteria, hypoxantine and trimethylamine levels were lower compared to the control group of the shrimp (*Panaeus aztecus*) dipped in ascorbic acid, citric acid, potassium sorbate and 4-hexyl resorcinol solutions and stored at −1 °C (Pardio et al. 2011). Nirmal and Benjakul (2009b) reported that in the shrimps treated with ferulic acid (1 and 2%), the increase in the number of mesophilic and psychophilic bacteria was prevented as well as the prevention of melanosis during storage for 10 days. Organic acid treatments improved the sensory quality and extended the shelf life of shrimp samples stored at 0 °C (Attala 2012).

Natural plant extracts are also used as preservative for protecting quality and extend the shelf life of shrimps before chilling. It was reported that the increase in psychrophilic bacteria and Enterobacteriaceae numbers was delayed during the cold storage of shrimps treated with green tea extract (0.1%) and ascorbic acid (0.005 and 0.01%) (Nirmal and Benjakul 2010). Grape seed extract treatment before storage in iced and refrigerated storage was found effective on keeping quality of *Litopenaeus vannamei* (Sun et al. 2014) and *Parapenaeus longirostris* (Gokoglu and Yerlikaya 2008). Orange peel extract having antioxidant and antimicrobial activity improved quality of shrimps during refrigerated storage (Vakili and Ardakani 2018). Application of turmeric in peeled shrimp (*Penaeus semisulcatus*) stored in ice extended its shelf-life up to 8 days (Prabhu et al. 2016).

Chitosan, a derivative of chitin, is obtained from waste shells of the crab, shrimp, and crawfish industries. It has superior characteristics. It is biodegradable, nontoxic, biocompatible, broadly antimicrobial, nutritional, and antioxidative. There are several studies on use of chitosan in chilled and refrigerated shrimps. TVB-N and TMA-N formation reduced in deep water pink shrimps (*Parapenaeus longirostris*) immersed into solutions containing chitosan (Bingol et al. 2015). Shrimp coated with a chitosan coating, shrimp coated in a chitosan-protein-lipid concentrate coating, shrimp dipped in a lactic acid solution and shrimp un-coated used as control were stored 5 °C for 17 days. Shrimp coated in a chitosan-protein-lipid concentrate inhibited the growth of microorganisms (Arancibia et al. 2015). Whole and headless shrimp (*Pandalus borealis*) were dipped in various concentrations of chitosan solution and stored at 4–7 °C for 20 days. The microbial growth, total volatile bases, nucleotide breakdown, and melanosis were monitored during the storage. At the concentrations between 0.0075 and 0.01% chitosan inhibited several microorganisms due to its strong antimicrobial properties (Simpson et al. 1997). The pretreatment of shrimp (*Litopenaeus vannamei*) with carboxymethyl chitosan and chitosan solution (1.0 or 1.5%) retarded growth of psychrophilic bacteria compared to control during the storage at 0 °C (Huang et al. 2012).

Some carboxylic acids and their salts have been treated. Dipping into sodium propionate and sodium acetate solutions before chilling of shrimp (*Penaeus monodon*) was found effective to extend shelf life compared to untreated control samples (Tam et al. 2018).

Essential oils (EOs) are natural antimicrobial in food preservation. They are volatile liquids extracted from plant material and contain antimicrobial compounds. They are commonly used to protect the quality of foods due to their antimicrobial properties. Shrimp (*Penaeus monodon*) was dipped into essential oils (cinnamon oil, garlic oil and lime oil) and organic acids (lactic acid, tartaric acid, and sodium diacetate) at 1:2 shrimp/treatment rates for 30 min. According to microbiological and physicochemical analyses essential oils and organic acid treatments were successful to extend the shelf life. Mixtures of tartaric acid and garlic oil and lactic acid and cinnamon oil were the best effective treatments to delay the spoilage of fresh shrimps (Noordin et al. 2018).

Ozone has been used for many food products to reduce the microbial load. Ozone destroys algae, viruses, bacteria, and fungi on contact. The biocidal effect of ozone is caused by its high oxidation potential, reacting with organic material up to 3000 times faster than chlorine. Washing with ozonated water of shrimps reduced the number of bacteria and improved microbial quality during cold storage (Tantratian et al. 2011). Quality attributes of sequential minimal ozone-treated ice stored shrimp (*Litopenaeus vannamei*) were investigated. Minimal ozone treatment sequentially applied at days 1, 3, 5, 8 and 11. Sequential minimal ozone treatment affected colour, titratable acidity, total volatile bases, trimethylamine, peroxide and para anisidine values of ice stored shrimp. After minimal ozone treatment, aerobic plate counts significantly decreased. During iced storage shrimps showed lower microbial count compared to control (Okpala 2014a, 2015). In another study, Okpala (2014b) reported that sequential minimal ozone treatment affected some physicochemical properties of Pacific white shrimp (*Litopenaeus vannamei*) during iced storage. Increasing ozone exposures and iced storage changed proximate composition, colour, and textural properties of Pacific white shrimp (*Litopenaeus vannamei*) (Okpala 2017).

Quality Changes of Chilled Shrimps

Chemical and biochemical reactions responsible for quality loss during cold storage can be slowed and shelf life can be increased (Lalitha and Surendran 2006). However, shrimps stored in cold loss their quality after a period due to microbial growth. There is an increase in the number of microorganism (especially psychrotrophic and psychrofilic) in the cold during storage. Jeyasekaran et al. (2006) reported that in white shrimps (*Penaeus indicus*) stored in ice, the total bacterial count increased from 10^6 to 10^9 cfu/g at 24 h, and total psychophilic bacteria increased from 10^3 to 10^6 cfu/g. Huang et al. (1996) found that the number of psychrotrophic bacteria which initially is 3.86 log cfu/g, reached 9.46 log cfu/day on the tenth day of chilled storage in the fresh shrimps (*Penaeus* spp.). Erdem and

Bilgin (2004) reported that the average total number of microorganisms in shrimp (*Palaemon adspersus*) was 3.55 log cfu/g, and this number increased to 6.71 log cfu/day during cold storage. Varlık et al. (2000) examined the changes in the quality of shrimps stored in cold storage period and determined that the samples were spoiled in terms of sensory, physical, and chemical on the second day.

2.1.3.4 Freezing

The storage of aquatic products for a long time without spoiling has been achieved by the development of freezing technique. In freezing technology which is among the methods of physical storage and the best method for preserving the quality of products, the water in the structure of the products converts to ice crystals and as a result of this the chemical, biochemical and microbiological activity slow down by reducing of the water activity and temperature (Gokoglu 2002).

As a result of the removal of energy from a liquid substance transition from the liquid phase to the solid phase is called freezing. By removing energy, the free movement of molecules in the liquid phase is gradually slowed down and the molecules tend to spontaneously aggregate into a regular structure. This is the first sign of the beginning of phase change. The phase change of a pure substance can only take place at a specific temperature. For water, this temperature is 0 °C. However, when a liquid reaches its freezing temperature, it does not cause the phase change to start. To start phase change, there must be a structure called "nucleus" in the medium. If there is no nucleus in the medium, the liquid must first cool down to its original temperature below the critical temperature called the freezing point. This is called super cooling. There is a second condition other than the super cooling for the phase change to occur. Accordingly, it is also necessary to have a structure that can lead to the transformation of molecules into solid phase. This can be either a small nucleus resulting from the crystalline structure, which is spontaneously formed by aggregating its molecules in the liquid, or it can be a foreign material similar to the properties of the crystal structure of that substance. The nucleation formed by the liquid itself is called homogenous nucleation. Homogeneous nucleation can only occur in liquids which do not contain any foreign matter, cooling the liquid to below its freezing level. The energy released because of freezing phenomena occurring by transition of water from the liquid phase to the solid phase causes heating of froze mass. However, if this energy is removed by the cooling medium, the temperature of the freezing mass remains constant at the freezing point until all material is frozen (Cemeroglu et al. 2003).

As is known, water must be present in the liquid phase to be utilizable by microorganisms. According to this, microorganisms cannot benefit from frozen water. One of the results obtained by freezing is to make the environment unfit for microorganisms in terms of water. The second effect of freezing is based on the strict stopping of microorganism activities below certain temperatures. Activity of both food poisoning and psycrophilic microorganisms stops under −10 °C. To prevent microbiological deterioration in frozen storage, the maximum temperature that can be applied is −10 °C. The freezing of seafood is the process for lowering the temperature to below freezing point. At these temperatures, most of the water in the material is converted into ice. The freezing point depends on the concentration of different substances dissolved in the tissue fluid. The dissolved and colloidal substances in the water of the product reduce the freezing point below 0 °C (Gokoglu 2002).

The freezing of seafood occurs theoretically in three stages. The first stage is the time until the product temperature is just below 0 °C. The second stage is the passing time until approximately 75% of the water is frozen, during which time the temperature does not drop, it remains at almost −1 °C. This stage is called thermal stop period. This period should last 2 h at most. In the third stage, when the freezing is continued, the product temperature decreases rapidly, and the remaining water is frozen (Gokoglu 2002).

To produce good quality frozen products, fast freezing is essential. Thus, since small ice crystals are formed in the cell, the cell is not overly damaged and as a result the intermixing of the intracellular fluid is prevented. In addition, with rapid freezing, water is transformed into ice crystals where it is located, and the cell water is prevented from passing through the intercellular space. By creating small ice crystals in intercellular spaces, the physical structure of the cell is protected. The critical point, between −0.5/−5 °C, is rapidly passed. Since the temperatures at which the activities of microorganisms are completely stopped are reached rapidly, the possibility of microbiological deterioration during freezing is eliminated (Gokoglu 2002; Gokoglu and Yerlikaya 2015).

Freezing Methods

Freezing of food is carried out using different freezing systems. A single freezing system cannot meet all freezing needs due to a wide range in kinds of food products and process properties. Product type, easy cleaning capability, hygienic design, desired product quality and reliable and economical operation are available among the selection criteria for freezing method. Plate contact, immersion, air blast, fluidized-bed, and cryogenic freezing are common freezing methods in food industry (Rahman and Velez-Ruiz 2007).

Plate Freezing

The basis of this method is that the packaged products placed between the two cooled plates are frozen by contact with the plate. The product to be frozen by this method must be have a regular shape or in blocks. It is not possible to freeze a packaged but amorphous mass with this method. It is very important that the package is in full contact with the plate on a smooth surface. Accordingly, packages of uniform shape and the same thickness can be placed side by side on the plate and a fast freezing in two directions can be achieved when sitting on the other plate from the top. This type of freezers is mostly used for packaged products (Gokoglu 2002).

The most common plate freezers are those with a multi-plate freezer type. In a well-insulated cabinet, several shelf-shaped slabs are arranged in a row. The effectiveness of plate freezers depends on the degree of contact between the product and the plate. The time of freezing depends on the type and thickness of the packaging material, the type of the product being frozen, the initial temperature and the thickness of the product (Cemeroglu et al. 2003).

Air Blast Freezing

In this method, the temperature of the food is reduced by the cold air flowing at a relatively high speed. Air blast freezers are systems that allow the cold air to circulate over the product which will be frozen with the help of fans. With the help of powerful fans, the moving air cools down as it passes through the evaporator and then rapidly passes through the frozen product. The air temperature in the freezer is −30 to −40 °C and the air velocity is 4–6 m/s (Gokoglu 2002). Lower air velocities cause freezing of the product slowly and high speeds significantly increase unit freezing costs (Rahman and Velez-Ruiz 2007). These types of freezers have a wide range of use in the freezing of packaged or unpackaged products of different shapes and sizes, in pans, trays and trolleys, in cabinets or on conveyor belts, individually or in blocks. Therefore, it is a commonly used freezing method. This method can also be divided into tunnel freezing, belt freezing and fluidized bed freezing depending on how the air interacts with the product.

The most common of these is tunnel type freezers. In the tunnel freezers, the product to be frozen is transported by a belt or by the movement of trays or racks in the tunnel. The movements of the product to be frozen and the cold air in the tunnel may be parallel or opposite. In opposite current tunnels, the product is given from one side of the tunnel and cold air from the other side.

In some tunnels, cold air is blown upwards from the bottom of the belt. With the fast air blowing, the substances on the belt are kept in a position as floating in the air. In this system, called as fluid bed freezers, a very fast freezing is performed on all sides of the particles that rise and fall in the air and full contact with the cold air. In this way, the individual freezing of each part is called "individual quick freezing" (IQF). This is a system that can be applied easily to products that are small enough to form a homogeneous and fluidized bed. In addition to the advantages of this system, the electrical energy consumed by the fans is very high. For a product to be frozen in the fluidised bed system, the product must be in small parts to be able to flow in a certain air stream. The individual freezing process (IQF) allows the shellfish to freeze quickly and in consumable quantities. IQF is a suitable method for products such as shrimp, scampi tail meat, squid fillet, squid rings and pectin (Venugopal 2006).

A spiral belt freezer consists of a long belt, wrapped in cylindrical form in two layers, thus requiring less space. The spiral freezer uses a laterally twisting conveyor belt. It is suitable for products with long freezing time (usually 10 min to 3 h) and for sensitive products. A belt located in an insulated cabin follows the spiral path from bottom to top, while cold air is supplied from the sides. The spiral shape of the belt causes the system to occupy little space. In addition, spiral belt freezers are used for the freezing of packaged amorphous products (Gokoglu 2002; Cemeroglu et al. 2003; Rahman and Velez-Ruiz 2007).

Immersion Freezing

In this method, the product which is packaged or un-packed is immersed in a suitable liquid cooled to low degrees, or the liquid is sprayed onto the product. When freezing the product without packaging, a rapid heat transfer between the product and the cooler is ensured. In addition, many products which have not in a distinct shape and as in particulate form can be successfully frozen by this method. The number of refrigerants that can be used in this method is limited. The reason for this is some of the features required in the freezing fluid. Freezing liquid should not freeze at low temperatures, not have toxic effect, not contain foreign colour, taste, and odour, not change the colour of the product, and have a sanitary quality. Most importantly, the liquid to be used must be compatible with the sensory properties of the product to be frozen. The fluids usually used are salt solutions (sodium chloride), sugar solutions and glycol and glycerol solutions. The most suitable liquid used for freezing seafood is salt water (Gokoglu 2002).

Cryogenic Freezing

Liquefied gases with a very low boiling point are called cryogenic liquids. The most used cryogenic liquids are liquid nitrogen and liquid carbon dioxide. Of these, the liquid nitrogen boils at −196 °C and the liquid carbon dioxide at −145 °C. In this method, direct contact between the product and the cryogenic liquid is achieved and a very rapid freezing takes place. In cryogenic freezing, liquefied gases are in direct contact with foods. The product is exposed to atmospheres below −60 °C by direct contact with liquid nitrogen or liquid carbon dioxide or vapours. This is a very quick-freezing method. Rapid formation of small ice crystals reduces damage caused by cell rupture, preserves colour, texture, aroma, and nutritional value. Fast freezing also reduces weight loss via evaporation, provides high product yield and needs less space. The most used liquid nitrogen from cryogenic liquids is produced by compressing the air into a liquid and then distilling the nitrogen through a special valve to remove the nitrogen gas from the oxygen. The nitrogen gas produced is re-compressed and liquid nitrogen gas is obtained. Liquid nitrogen gas remains liquid at atmospheric pressure and only a small portion of it converts into nitrogen gas at −196 °C. Liquid nitrogen is not toxic, does not react with the product, prevent oxidative reactions, does not need a separate cooling equipment and provides very fast freezing and for these reasons it is preferred priority. Liquid carbon dioxide is obtained by compressing the carbon dioxide gas under high pressure. The boiling point is −145 °C. The product is applied by spraying on liquid carbon dioxide product as it passes through a tunnel. In some systems, solid carbon dioxide (dry ice) is placed on the conveyor belt before the product is placed on it and then sprayed with liquid carbon dioxide. It is possible to find carbon dioxide in any form at any time.

It is not toxic unless it is in very high concentrations. It is a relatively inert gas, no danger of fire or explosion. In freezers with carbon dioxide, freezing is very fast and drip losses are reduced (Gokoglu 2002).

In a study in which two different freezing methods were used, forced convection freezing and cryogenic freezing, the researchers determined that cryogenic freezing is the best method in terms of cell and tissue structure, but they suggested forced convection freezing method for freezing of farmed whiteleg shrimp (*Litopenaeus vannamei*), because the cryogenic method is expensive (Diaz-Tenorio et al. 2007).

Innovative Freezing Methods

Temperature below −40 °C is defined as ultra-low freezing temperature. The freezing temperature greatly influences the structure and quality of a product. Ultra-low-temperature freezing increases the freezing rate because the surface heat transfer coefficient between the food and the freezing medium is very high and the temperature is very low (Wu et al. 2017). Tiger shrimps were frozen at different temperatures (−70, −80, −90 and −100 °C) and it was determined that the freezing time and freezing rate increased with the temperature decline (Boonsumrej et al. 2007). Similarly, anchovy was frozen at 3 different temperatures (−20, −40, and −80 °C) and the highest freezing rate and shortest freezing time were determined in the product frozen at −80 °C (Aydın and Gokoglu 2014).

High pressure freezing (HPF) is a novel freezing technique, which has the potential to manage the formation and distribution of ice crystals and thus improve the quality of frozen food (Cheng et al. 2017a). It is reported that the application of high pressure reduces the freezing and melting points of the water to a minimum of −22 °C at 207.5 MPa, as it opposes the volume increase in the formation of ice crystals. In addition, during high-pressure freezing, it is possible to form small ice crystals due to super cooling. In comparison with commercial freezing methods, freezing with high pressure shortens the thawing time in the fish muscle and reduces drip loss during thawing (Cheftel and Culioli 1997; Chevalier et al. 2000; Zhu et al. 2004). Slow freezing often results in larger ice crystal sizes that can cause extensive mechanical damage. In a study, pressure assisted, and pressure shift freezing were applied to actomyosin extracted from prawn (*Metapenaeus ensis*) to determine the effect of high pressure freezing on denaturation of actomyosin. At the end of study, it was found that pressure shift freezing at 200 MPa/20 °C/30 min conditions reduced the denaturation of actomyosin and a pressure of 300 MPa was the critical point to induce such a denaturation (Cheng et al. 2017b). In another study, fresh shrimp was frozen by pressure shift freezing at 100 MPa (−8.4 °C), 150 MPa (−14 °C) and 200 MPa (−20 °C), as well as by conventional air freezing at −20 °C and liquid immersion freezing at −20 °C. It was found that pressure shift freezing was the best method when investigated in terms of size, shape, and distribution of ice crystals (Su et al. 2014).

Ultrasound is used as a non-thermal food preservation method. Ultrasound has advantages compared to heat treatment. These are much less loss of sensory and chemical properties of food, saving the time used and a higher rate of inactivation of enzymes and microorganisms. Ultrasonic is a technology that can meet the needs such as determining the physical and chemical properties of foods, modifying, observing during the process and helping them to control the quality. The ultrasound applied material particles vibrate under the influence of ultrasonic waves and the total number of vibrations per minute is called frequency (Knorr et al. 2004). In addition, high-energy ultrasound application has been proven to accelerate crystal formation in the crystallization process. It has been determined that ultrasonic freezing processes shorten the time required for the product to freeze and improve the quality of the frozen product. It has been found that high-energy ultrasonic application affects the crystallisation process in many ways, such as supporting the formation of crystal nuclei, ensuring the formation of small and regular crystals and preventing the uneven surface structure (Kiani et al. 2013; Saclier et al. 2010). There is no study on use of ultrasound assisted freezing of shrimp. This is one of the issues to be studied in the future.

Changes in Frozen Storage

Freezing is more effective in long-term preservation, physical, chemical, and biochemical events continue during storage. The low temperature in freezing, although slowing down the formation of enzymatic and chemical changes, cannot completely prevent it (Fennema et al. 1973). The most important changes during frozen storage occur in proteins and lipids. These changes affect sensory properties such as odour, colour, and texture. Frozen storage causes changes in the proteins of fish meat known as freezing denaturation. These changes occur in the form of opening of protein molecules, secondary reactions between the different reactive groups of proteins and other components of the fish muscle. Proteins lose some of their solubility and have lower enzyme activity. As a result of these changes, significant changes in the functional properties of fish meat are observed. These occur in the form of water retention capacity; gel forming capability and decreases in emulsion capacity and negative changes in texture (Sikorski and Kolakowski 1990). The freezing process promotes changes in muscle tissue with the formation and accretion of ice crystals, dehydration and increase of solute (Shenouda 1980).

Protein denaturation and textural defects may occur depending on many factors such as species, freezing and thawing rates, storage temperatures, storage conditions between harvesting and freezing (Diaz-Tenorio et al. 2007). It has been reported that sea animals' muscles are more sensitive to protein denaturation caused by freezing than that of mammalian muscle (Matsumoto 1979).

During frozen storage, drying or dehydration may develop on the surface of frozen seafood. Excessive drying causes "freezer burn". To avoid this, the seafood is covered with ice glass called "glazing". Glazing means that the use a protective ice coating on frozen seafood. The glazing process, which is to wrap the product with a thin layer of ice, increases the shelf life of the product by protecting it from oxygen and dehydration. Glazing is usually accomplished by dipping or spraying frozen seafood in potable water (Solval et al. 2014).

During storage of frozen shrimps, lipid oxidation, protein denaturation, sublimation and recrystallization of ice crystals may occur. This causes undesirable taste and odour formation, drip loss, water loss, weight loss, dehydration, and toughening (Londahl 1997). Glazing process is applied between 6 and 12%. Ice percentage over >20% to be added to the product in coating is considered as over-coating which is accepted as fraud. It has been reported for shrimps that glazing as thick as 25–45% was defined as excessive glazing. For this reason, the amount of ice in glazed product should be inspected (Manso et al. 2013). The amount of glaze applied depends on the different factors such as product temperature, water temperature used for glazing, product size, glazing time, and product shape (Gonçalves et al. 2009). Several researches were conducted to determine the effects of these factors on glaze uptake of frozen shrimps. In one of these studies, effects of initial temperature of frozen shrimp and glazing time on glaze uptake were studied. In that study, different glaze percentages and their effects on physical and chemical changes of frozen shrimp during storage were also investigated. It was found that glazing time and shrimp temperature showed significant effect on glazing uptake. The increase of glazing time caused an increase in glazing uptake. Glazing time of 15 and 20 s showed best protection. On the other hand, the most reasonable glazing percentages were found as 15 and 20% in terms of the quality of final product (Gonçalves et al. 2003).

Freezing of Shrimps/Prawns

On Board Freezing

The shelf life of shrimps is considerably increased as the storage temperature is lowered. Fatima et al. (1988) examined the quality changes of ice-chilled (0 °C) and partially frozen (−3 °C) shrimps and found that they kept their quality for 8 days on ice and for 16 days in frozen storage. Tsironi et al. (2009) found that shrimps kept at −12 °C and −15 °C were still acceptable in the eighth and eleventh months, respectively. Shrimps, especially in catching in distant seas, need to be quickly frozen on ships to increase shelf life. Frozen shrimps may have a shelf life of 1 year or more. Mechanical freezing equipment on vessels allows the shrimp to be frozen immediately. This is the ideal system and the shrimp is always of first-class quality. Freezing of shrimp at sea is not economical, needs expensive freezing equipment and experienced staff. Shrimps can be frozen by dipping into a cold brine or a sugar and salt solution by air flow or by plate freezers. It is reported that the freezing process with dipping and air flow methods has been successfully applied in the fishing boats especially in North and Central America. Glazing is formed on the shrimp when is frozen in sugar and salt solution, this makes shrimps easier to separate when it thaws. Deep water shrimps can be frozen by immersing them into the brine at −20 °C for 10–15 min. Very long immersion times can result in an undesirable product formation due to excess salt intake. Shrimp can also be frozen in a vertical plate freezer in 50 mm blocks. Shrimps are placed in a polyethylene bag and placed between plates. The freezing time for a 50 mm block in a freezer plate at −35 °C is 90 min. Shrimps in a frozen polyethylene bag are placed in cardboard boxes and stored in this way (Anonymous 1972).

Freezing on Shore

All raw shrimp should be processed in factories near to they are landed. Shrimps are processed in land operations in cases where the ship does not have freezing facilities. All operations of shrimps are made better on land. If shrimps are iced at the sea immediately after capture, they can be processed with less loss on land under more hygienic conditions.

There are several definitions for frozen shrimp. Frozen shrimps are classified according to their market.

2.1.3.5 Canning

Canning is a method of protection the food by killing microorganisms via heat. It is generally well known that foods carry microorganisms that cause degradation. The advantages of thermally processed and ready-to-aet products over conventional methods are that they have a long shelf life and do not require any further processing prior to consumption (Sreenath et al. 2008). Canned shrimp is defined as a product consisted of only one specie or together different shrimp which have not lost their freshness, graded in terms of size and weight, peeled or not peeled, washed with potable water at temperature less than 10 °C, added non-iodized salt, cooked, sealed, sterilized by heating and can be eaten without requiring any further treatment (Anonymous 1993).

The first rule of obtaining a good product is the use of high quality and fresh raw materials. The quality of the finished product is never better than the quality of the raw material. The methods of catching, preservation and processing significantly affect the quality of the raw material. The success of the canning technology depends on the quality of the raw material. Care taken to preserve the original freshness until the raw material is transported from the hunting area to the processing plant is the first condition of success in canning technology.

Canned shrimps are divided into classes according to the shape and size of shrimps, the type of brine used in the production and whether flavouring contain or not.

According to Brining Method

Dry Salted Canned Shrimp

Dry salted shrimp is a product that its head and shells have been removed by hand or in a machine, deveined or not, washed, blanched in 10% salted water for at least 4 min and salted homogenously with a salt of 21 kg per 100 kg shrimp.

Wet Salted Canned Shrimp

Wet salted shrimp is a product that its head and shells have been removed by hand or in a machine, deveined or not, washed, blanched in 10% salted water for at least 4 min and placed into the cans and salted by adding salt solution.

According to Whether the Flavouring Contains

Seasoned Canned Shrimp

Seasoned canned shrimp is a product added seasonings to shrimps that its head and shells have been removed by hand or in a machine, deveined or not, washed, blanched in 10% salted water for at least 4 min and wet or dry salted. Seasonings and additives that can be used in canned shrimp: Tomato sauce, curry sauce, mustard seed meal, edible starch, hydrolysed herbal protein, ravioli, onion, red and green pepper, powdered red pepper, stalk and root celery, celery seed flour, darkened milk, garlic powder, tomato, butter, new spring, carafe, bay leaf, sugar, edible mushroom, distilled vinegar, black pepper.

Canning of Shrimps/Prawns

Shrimps coming to the canning factory are emptied into a washer. This is done on a perforated belt. When the shrimp goes on the belt, the workers on both sides of the belt separate the ones that are decayed and discoloured, the shrimps are then weighed. Shrimps are usually brought to the factory with head. According to the demand, some of the processors buy headless.

Sometimes, frozen headless shrimp can also be taken in rare cases. The process of removing the head and shells or only the shells is called "Picking". This can be done manually or by machine. When the head and shell are removed, the shrimp loses approximately 50–55% of its weight.

The picking process, shrimps are subjected to another examination for quality. In some factories, shrimps are separated according to their size. Large shrimps are deveined. Deveining improves the quality of the canned product. This process is done immediately with the machine.

The next process is pre-cooking. Most processors look at this as one of the most critical processes. A suitable pre-cooking shrimp is important for a good colour and easy bending during packaging. This process provides a quality product by preventing the water separation and meat shrinkage in the can because of reducing the amount of water in the product. As a result of heat treatment proteins are denaturized and a part of the water is released. This situation causes water separation, dilution of the sauce and oil and the meat shrinkage in the can (Gokoglu 2002). The blanching solution is hot salt water. Its concentration is 25° of salinometer, and its duration is 1–3 min, depending on the size of the shrimp. Shrimps are passed through the boiling brine solution. In some factories, shrimps may come directly from the pre-cooker (blancher) into cold water. The shrimps pass from the cold-water boiler onto a drying belt and the excess water is removed. Steam pre-cooking is done at 95–100 °C for 8–10 min depending on shrimp size. While the shrimp colour changes from pink to white during pre-cooking, the texture changes into firm and the shrimp gets its typical curl appearance. They then come to the grader, which can be separated into five categories (tiny, small, medium, large, and jumbo). Then they are visually inspected to correct grader's mistakes and foreign matter and crushed shrimps are removed (Featherstone 2016a).

Shrimps are filled into suitable containers. These containers can be tin cans and glass jars. However, the containers to be used must have properties such as hermetic sealing, heat resistance, and non-reactivity with the product. It is more common to use tin cans because of their advantages such as sterile preservation of the product, resistance to sterilization temperatures, ease of distribution, storage, and stacking. However, in the case of use of cans, there is a risk of corrosion (Horner 1992; Gokoglu 2002). In fact, flexible bags, which have been started to be used in 1960s, have been used more widely in recent years. These are robust polymer laminates consisting of a heavy-duty outer polymer layer which then binds to the heat sealable inner polymer layer. Of course, flexible bags also have some disadvantages. They are more expensive than tin cans; they require specially designed retorts and leak detection is more difficult (Horner 1992). In addition, retort pouches have several advantages, resulting in more quality, such as shelf stability, low weight and less storage space, ease of opening and preparation, and low temperature exposure. Because of the faster penetration of food into the food, there are also advantages such as cooking time and decreasing energy consumption. In a study compared the cooking process of shrimp in retortable pouches and cans, retort pouches showed 37% less thermal processing time than the canned product (Chiar et al. 2006). In another study, 35% reduction in processing time in kuruma shrimp processed in retortable pouches compared to aluminium can was observed (Mohan et al. 2006). Pouched foods are reported to have a better texture and flavour. Most retort pouches are made from a four-layer laminate consisting of a polyester outer layer, a nylon second layer, a third layer of aluminium foil and a polypropylene inner layer. Polypropylene polymers have a melting point of about 138 °C, which is higher than the commercial sterilization temperature of 121 °C. It has been reported that improved texture and flavour are observed in freshwater shrimp (*Macrobrachium rosenbergii*) packed with laminated flexible pouches in a curry medium (Majumdar et al. 2017).

Shrimps are placed by hand in the cans and carefully weighed. The can weight depends on the type and size of the shrimp. Because of the cans are contaminated during the transportation and storage stages before filling the canning containers must be washed. The cleaning of the cans is usually done by pressurized water or steam (Gokoglu 2002). The filled cans come to the sealing machine via conveyor belts. Hot saline solution is added to the shrimp cans. The product produced in this way is called "wet-pack".

The filled cans are placed in the autoclave with the metal baskets. Heat treatment is carried out under pressure. The processing time depends on the size of the can. At the end of the process, the cans from the autoclave are cooled, drained, and placed in boxes (Gokoglu 2002).

It has been reported that heating at 115 °C for 24 min affects the texture of canned shrimp (*Macbrobrachium rosenbergii*) due to collagen denaturation (Noomhorm and Vongsawasdi 1998). Processing at 124 °C caused the toughening at the first stage and the softening at the later stages of heating in shrimp muscle (Ma et al. 1983).

2.1.3.6 Drying of Shrimps/Prawns

Drying is one of the oldest food preservation methods. It means removing water from the product. Drying is a general definition and the process of removing water from the product by evaporation. The high moisture content of shrimps causes rapid deterioration. If they are not exposed to cold storage, they start to deteriorate shortly after being caught. However, cold storage is expensive and may not be present in some areas where the electricity becomes difficult. In many parts of the world, drying continues to be one of the best options for pre-processing of seafood. It is one of the oldest food preservation methods and can be applied to a wide range of food products, including shrimps. The principle of drying is to reduce moisture to levels that are too low to prevent microbial growth, and slow down enzymatic and other biological reactions (Akonor et al. 2016).

Principal of Drying

In the drying process, removing of water takes place by evaporation of the water as vapour from the surface and transferring of the water from the inner layers to the surface layer. The removal of water from the surface layer depends on the drying temperature, the humidity and velocity of the air, as well as the surface and pressure to be exposed.

Initially drying is the removal of water from the surface and near the surface by evaporation. This period is called "constant rate drying" period. In this period drying continues at a constant rate. The drying rate depends on the air velocity, humidity, temperature, the surface area of the product, the amount of heat transferred from the air to the product per unit time, and other conditions. Since the evaporation takes place on the surface, the water in the fish muscle must first be transferred to the surface. This transfer is largely carried out by diffusion. This period is characterized by a gradual decrease in the rate of drying and this period is called "falling rate drying" period (Doe and Olley 1990; Gokoglu 2002).

Diffusion is the most important mass transfer mechanism responsible for the transport of water and vapour in the muscle. Diffusion results from the concentration and pressure differences caused by protein denaturation, aggregation, and muscle shrinkage between inner and the outer layers of the muscle. Capillary forces are also thought to contribute to water migration within the fish muscle. The water evaporation rate on the muscle surface should be kept in balance with the speed of water passage from the inner layers to the surface layer. If the water on the surface of inner layer evaporates very quickly, this creates a layer that affects the drying process and extends the drying time. The heat transfer during the drying process can occur by convection, conduction, radiation, or a combination of these (Gavrila et al. 2008).

Water activity is defined as the ratio of the vapour pressure of the water in the caste and the vapour pressure of the pure water at the same temperature. At the beginning of the drying process (constant phase of drying rate), the water activity remains close to 1. However, the water activity decreases in the subsequent stages of the drying process. Water activity is of great importance for food storage. Water activity is the most important factor limiting chemical, biochemical and microbiological changes in foods. As water activity decreases, the durability of food increases. Bacteria that cause food spoilage cannot multiply in foods with water activity below 0.90. Some halophilic bacteria can operate up to 0.75 water activity. Moulds easily multiply until water activity drops to 0.80. In addition, the decrease in water activity also limits or prevents enzymatic changes. Enzymes such as peroxidases and phenoloxidases are inactive in water activity down to 0.85. However, lipase enzymes remain active until 0.25–0.30 (Gokoglu 2002; Rahman and Labuza 2007; Belessiotis and Delyannis 2011). To prevent microbial growth, it has been reported that the dried water activity should be kept below the critical value of 0.60 (Perera and Rahman 1997). The water activity of the dried shrimp at 27 °C was found to be below 0.80, which is considered the low limit for the growth of most food spoilage microorganisms. Moisture content in dried shrimps should be reduced to 20–25% (Tapaneyasin et al. 2005) (Fig. 2.9).

Drying Methods

The methods of drying seafood are found in a wide range from traditional drying methods in sun to high technology methods with computer control.

Sun Drying

Dried shrimp is one of the most important export products of Southeast Asia. Price depends largely on quality characteristics such as dryness, colour and size. Commonly traditional drying is carried out by boiling shrimps in salt water and then drying them in the sun. In the traditional drying method, the shrimp is dried under the sun for 3–5 h depending on the presence of sunlight. It is one of the most widely used drying methods in many parts of the world. It is one of the most widely used drying methods in many parts of the world. In this method, solar energy is used to reduce the amount of water in the product, that is, to ensure the evaporation of water. The pre-treatments before drying vary according to the region and the person who will perform the operation. During sun drying, the heat is transferred to the product by convection and radiation. The converted heat is partly transmitted to the interior, causing the internal temperature to increase, and is partly used for water and vapour migration from the inside to the surface. The remaining amount of energy is used to evaporate the surface water. The natural convection supported by wind force removes the evaporating water from the surrounding air. Due to these disadvantages of the conventional drying method, research has focused on the development of alternative drying methods. For this purpose, different methods such as spouted bed drying, heat pump drying, superheated steam drying, solar drying, drying with electromagnetic waves have been used (Nguyen et al. 2014).

Although it is an economically cheap method, it has the disadvantages such as the difficulty in controlling the drying process and parameters, the irregularity of the weather conditions, the need for a large drying area, insect infestation, mixing with dust and other foreign substances (Gokoglu 2002; Jain and Pathare 2007; Tirawanichakul et al. 2008). Also, there is a risk of unexpected rain or storm. In addition, open sun-dried products do not meet the international quality standards and lead to the product not being accepted in the international market (Sharma et al. 2009). The other disadvantages of open air drying are discontinuous drying and dependent on the weather conditions and the region. At night when the sun is off, the ambient temperature drops, relative humidity increases. In some cases, the product reabsorbs moisture and causes prolonged drying time and the formation of fungi in the product.

Solar Drying

Solar dryers are widely used in drying of seafood. Solar drying is a detailed process of drying in the sun and it is reported to be an energy efficient system. The term solar dryer is defined as a device that collects and concentrates solar energy and thus dries the product more quickly and improves its quality (Doe and Olley 1990). It has also been reported that solar drying not only saves energy but also saves a lot of time, does not require a large drying area, increases product quality, and protects the environment and makes the process more efficient (Atul et al. 2009; VijayaVenkataRaman et al. 2012). Compared to sun drying, solar dryers can provide higher air temperatures and lower relative humidity, which reduces the drying rate and the moisture content of the final product. This means less infestation and less spoilage. Therefore, it provides improved and more consistent product quality. As a result of the economic analysis of the tunnel type solar crushers compared to

Fig. 2.9 Dried shrimp in the market

the open sun drying method, the economic costs of dried peeled shrimp have been proven to be better than the sun dried method for solar tunnel dryer (Manjarekar and Mohode 2010).

In the tropical and subtropical regions, various types of solar dryers have been designed and developed for drying seafood. The solar energy dryers are distinguished in two main categories. These are natural convection, so-called passive dryers, and forced convection called active dryer. In active dryers, the air velocity is higher, and the drying time is shorter. In addition to all these, passive dryers require less investment costs and operate more simply (Nguyen et al. 2014).

In a study examining the properties of dried shrimps dried in a combined solar drying with low humidity air, the shrimps were dried in a daytime solar energy dryer (greenhouse dryer) and then dried in an air-conditioned controlled system overnight. At night, the low temperature and low relative humidity drying system kept the temperature between 28 and 30 °C for shrimp drying and relative humidity to 40–45% Rh. It has been determined that the products take 16 h to dry in this system (Wisaiproma et al. 2018).

Heat Pumping Drying

The heat pump is an effective device that removes thermal energy from a lower temperature source and leaves it to a higher temperature source. A heat pump drier is more costly than hot air drying, as it can extract and use the latent energy of air and water vapour to dry the product. However, it is a useful method for drying valuable and sensitive products such as shrimp due to its low temperature drying capability. Because the heat pump uses less energy, it offers an effective and environmentally friendly technology (Strømmen et al. 2000).

In a study in which peeled, headed or whole shrimp (*Pandalus borealis*) were dried in a heat pump drier at a temperature of −2 to 0 °C to 20 °C respectively, and the desorption isotherms of each shrimp sample group were characterized, drying rate of headed shrimp was found higher than that of whole shrimp (Guochen et al. 2009).

Freeze Drying

The freeze-drying method is also called lyophilisation. Freeze drying generally consists of two steps: (1) the product is frozen; (2) the product is dried by direct sublimation. Freeze drying is expensive, but in terms of quality it is the best drying method. Preserving the appearance, taste, colour, taste, structure of the raw material makes it the best drying method. The product also has good rehydration properties as it retains its initial shape and size. Very good preservation of aroma and flavour, high preservation of nutritional value, minimum shrinkage, and minimal change in shape, colour and appearance, negligible effect on structure and texture, and good rehydration properties due to the porosity of the final structure are the advantages of freezing drying (Aguilera et al. 2003). Also, as the water is about 98% removed, the weight of the product is reduced, which reduces the transport and storage costs.

However, freeze drying also has disadvantages such as high capital and operating costs and long processing time (Hammami et al. 1999). This makes the freeze-dried product a more expensive product for the consumer than the products processed by other methods.

In freeze drying, water in the products is removed by sublimation at low temperature and pressure below the triple point. First, the water in the products is frozen in the freezing step and then removed by sublimation of ice in the first stage of drying. Freeze drying requires very low pressures or high vacuum to achieve a satisfactory drying rate. If the water is pure, freeze-drying can be carried out at or near 0 °C under an absolute pressure of 4.58 mmHg. However, since water is usually present in combination or in a solution, it must be cooled below 0 °C to keep the water in the solid phase. The most critical stage of freeze drying is freezing. Freezing should be made fast at this stage. Because of the slow freezing occurs large ice crystals and this also deteriorates the quality of the final product. In the first stage of drying, the pressure is reduced by applying a high vacuum and heat is applied to provide the energy required to sublimate the ice. The water of 95% removes the first drying stage. This process can take from a few hours to 2 days. Moorjani and Dani (1968) dried the shrimps by freeze-drying method and compared to air drying method. Drying process took 12 h and the moisture content of dried shrimp was 3.0%. According to the sensory analysis results freeze dried shrimp were found better quality than air dried shrimps.

Freeze drying was also used in combination with other drying methods for shrimp drying. In a study combining freeze drying with electrohydrodynamic drying (EHD), a group of shrimps were dried at 18 °C by EHD, while one group was dried by freeze-drying (FD) and the other group was dried by combination of both methods (EFD). The samples were dried with EHD for 3 h and then the water content of the product was reduced to 12% by FD method. Less drying time was determined by the combined procedure compared to FD and EHD alone. The shrimp dried by the combined drying treatment exhibited lower shrinkage, higher rehydration rate and better sensory properties (Hu et al. 2013). In another study, freeze drying was combined with air drying. Similar results were reported with combined drying method (Bai et al. 2013).

Microwave Assisted Drying

Microwave radiation is a part of the electromagnetic spectrum with wavelengths ranging from 1 mm to 1 m. The frequency used for food applications is 2.450 MHz. Microwaves are not ionized in nature and are known as a fourth-generation technology. Microwave-assisted dehydration is increasingly noteworthy. Microwave-assisted drying is applied in conjunction with other techniques to solve problems encountered in conventional drying processes. It is reported that microwave assisted drying is applied in various types of food and good results are obtained. In the studies conducted with microwave-assisted drying, decrease in drying time and energy consumption, and increase in drying speed have been reported. Less space requirements and better overall process control are among the advantages of this method

(Moses et al. 2014). It is also reported that micro-wave assisted drying provides advantages such as rapid and homogeneous heating, high heat efficiency, sanitation, and quality finished product (Sharma and Prasad 2004; Reyes et al. 2007). Farhang et al. (2011) used microwave oven to dry shrimp. They used microwave power in the four levels of 200, 300, 400 and 500 W. As a result of the study, it was determined that the drying rate was in the period of fall and the increase in drying rate and decrease in moisture content were determined with the increase in microwave power. In addition, it has been reported that drying efficiency decreases by 22.54% and energy consumption by 28,394% with the increase of microwave power.

In another study, microwave drying was applied to the shrimp (*Penaeus indicus*). The vacuum microwave drying was conducted by a 4 kW, 2450 MHz vacuum microwave drier at microwave power of 4 or 2 kW, and vacuum level of 100- or 250-mmHg. Shrimps prepared with 4 kW, 100 mmHg vacuum microwave drying were found to have higher shrinkage potential and water retention ability compared to air dried samples. With the increase in microwave power, a reduction in the water-holding capacity and the rehydration potential was observed, resulting in a tougher texture (Lin et al. 1999).

Infrared Assisted Drying

The infrared radiation (IR) energy applied to the product for drying purposes penetrates the product surface and spreads to the material to form molecular vibrations. It provides thermal energy with relatively low energy loss compared to other types of heat sources. Infrared radiation (IR) is sent from the heaters to the product as an electromagnetic wave. The heat transfer between the product and the heater is directly proportional to the temperature difference of both materials. Like microwave energy, the FIR is also absorbed by foodstuffs and then converted to heat. Heating with FIR is effective on the cost and final product quality according to heating with traditional methods. The advantages of this method are effective heat transfer, less processing time and cost, operation at normal temperature, better control, reliable and small size equipment design. Infrared radiation is used successfully in seafood drying (Ratti and Mujumdar 1995).

It has been reported that infrared radiation must be combined with microwave and other general transmission or transfer methods so that it can be used efficiently in the food processing industry. In a study conducted with infrared-assisted vacuum freeze-dried method, tiger prawn (*Penaeus monodon*) was dried and the statistical model equation was developed by using the response surface method to predict the rehydration rate, final product temperature and final moisture content. 60-mm sample distance from IR heater, 65 °C IR temperature, 6.37 h freeze drying time and 10-mm sample thickness have been determined as the optimal drying conditions. In addition, IR-assisted freeze-dried shrimp was reported to be quite stable compared to fresh shrimp (Chakraborty et al. 2011).

Superheated Steam Drying

A steam having a temperature higher than the saturation temperature at a given pressure is defined as superheated steam. The superheated steam at atmospheric pressure is referred to as an alternative drying medium for materials that can withstand temperatures above 100 °C. The overheated steam drying method is distinguished from the conventional methods by having long constant rate periods and lower critical moisture content. High drying efficiency, no oxidation risk, no pollutants, smaller equipment use, better antimicrobial effect are the advantages of overheated steam drying (Devahastin and Mujumdar 2014).

This method is reported to be a suitable method for drying shrimps. This is because boiling and drying takes place in a single step, due to the use of overheated steam as the drying medium. Since it eliminates the boiling stage, it also prevents any nutritional losses during boiling (Prachayawarakorn et al. 2002).

Namsanguan et al. (2005) studied drying of shrimp using a superheated steam dryer and investigated a two-stage drying. In the first stage, the shrimps were dried by a superheated stem dryer and in the second stage by the hot air dryer. According to the results obtained, in the two-stage drying dryer compared with the single-stage superheated steam dryer, in the shrimps lower shrinkage degree, higher degree of rehydration, better colour, less hard and soft and more porous structure were observed.

Drying in Spouted Jet Bed

It was determined that required drying time and energy reduced for drying of the shrimp in a spouted jet bed dryer compared to conventional tray dryers. In addition, the drying temperature and shrimp size also affected the quality characteristics of the dried shrimp. For small and large shrimps, sensory sensitivity at temperatures of 100 and 120 °C was found to be at the highest level (Niamnuy et al. 2007).

Tapaneyasin et al. (2005) investigated the effects of inlet drying air temperature on the drying kinetics and various quality attributes of shrimp dried in a jet-spouted bed dryer. The results showed that low shrinkage, high rehydration, low maximum shear force, and high value of redness were observed when a constant inlet air temperature of 100 °C was used.

Changes in Shrimp Muscle During Drying

Protein denaturation that occurs during the drying process significantly affects the quality characteristics of the dried products. As a result of protein denaturation, protein chains are opened, and free carboxylic and amino groups appear. The exposed sulfhydryl groups are oxidized to form disulphide bonds. The solubility of the proteins decreases due to these changes in the protein structure. Drying methods and drying parameters have important effects on protein denaturation (Nguyen et al. 2014).

In addition, mailard reaction occurs during drying. Blacking reactions reduce the colour of the product and decrease its solubility and nutritional values; it causes undesirable structural changes.

During the drying process, lipolysis and oxidation in lipids occur. The presence of oxygen and water activity is effective on lipid oxidation during drying. Water activity has an accelerating effect on lipid oxidation. It is reported that reduction of water activity to about 0.3 significantly reduces oxidation reactions (Rahman and Labuza 2007). The effect of oxygen depends on the porosity of the product. As the porosity increases in freeze-dried products, they are more susceptible to oxidation.

During drying, aroma losses can also be seen in the products. Since the volatile compounds that provide the flavour have a lower boiling point than water, they can be removed from the product by evaporation during drying.

High quality dried shrimp should have low shrinkage and high rehydration capability. The shape and size of the products can vary considerably during drying. Shrinkage during drying plays an important role in determining the quality of the dried product. The shrinkage can be measured directly with a calliper or micrometre and can be determined by changes in relevant parameters such as porosity and density. Recently, many studies have been carried out to identify the shrinkage behaviour of products with prediction models. On the other hand, image processing techniques are used to measure the product shape (Yan et al. 2008). In a conducted research, shrimp was dried by a hot air thin-layer dryer and a near-atmospheric pressure superheated steam dryer based on computer vision. In that study, the effects of drying on appearance properties of products and relation between these properties and moisture content of products were investigated. It was determined that the temperature of the drying medium influenced the drying rate and drying time as well as on the field shrinkage and Feret's diameter. High temperatures caused more irregular edge deformation of the samples (Hosseinpour et al. 2011).

Quality Changes in Dried Shrimp During Storage

In dried products, changes in protein and lipids together with water loss cause textural changes. Drying methods and drying parameters affect the textural changes. The deterioration of the dried products occurs physical, autolytic, chemical, and microbial ways.

Physical Spoilage

Due to perishability of dried products should be taken care. They can be damaged very quickly by physical shocks. In addition, birds and other animals can damage dried products by eating. Losses that may occur in dried products because of physical damage can be reduced by taking care in packaging and other operations after drying.

Chemical Spoilage

The most important chemical degradation seen in dried products is fat oxidation. Undesirable rancid taste and odour caused by oxidation of fats can occur. Oxidation accelerates in low water activities. The heat, light and oxygen in the air accelerate the oxidation. The temperature has a significant impact on oxidation, especially during outdoor drying. Oil residues that have remained in the environment before drying also accelerate the oxidation. Contact with metals is another factor that accelerates the oxidation of the product. For this reason, the products should not be dried on the metal wires. To prevent oxidation in dried products, the drying temperature should be kept low, dried in the absence of light and some antioxidants should be used if possible.

Microbial Spoilage

As is known, the development of microorganisms in low water activities is prevented. Microbial deterioration can also be controlled by controlling the temperature and humidity conditions during the storage of dried products.

Preservation of dried products in suitable packages without moisture permeability is important for preventing microbiological growth.

To protect against oxidative degradation, dried products must be packed in a vacuum or inert gas with moisture and oxygen non-permeable packing material. Biede et al. (1982) investigated the quality changes of shrimp sun-dried, packed under modified atmosphere and vacuum and stored at 22 °C for 8 months. During the storage, it was reported that shrimp packaged with both methods decreased water activity and moisture contents, increased total volatile nitrogen and ammonia contents and the lowest oxidation determined in shrimps packed under vacuum. In a previous study, beheaded and dried fresh prawns, *Parapenaeopsis stylifera* were packed in polythene bags, and in cans under atmospheres of air, nitrogen, carbon dioxide and oxygen and then sealed hermetically. At the end of study, it was found that shelf life of shrimps packed in cans prolonged up to 5 months. However, some colour and flavour losses were observed in these samples. Lower quality was determined in case of packaging in polyethylene bags (Balachandran 1975).

Dried small marine shrimp was investigated in terms of quality parameter during the storage at 37 °C. It was determined the increase in trimethylamine (TMA) total volatile basic nitrogen (TVB-N), thiobarturic acid reactive substances (TBARS), total viable count (TVC) and sensory scores during the storage. In addition, the shelf life is reported to be less than 7 days when evaluated for sensory analysis results (Lu et al. 2011).

2.1.3.7 Smoking of Shrimps/Prawns

Smoking is a traditional process and preservation method used since the early ages. When the first people begin to consume the animals they hunt on an open wood fire shortly after discovering the fire, this was an important experience for them to realize that this method gives a taste to the food and prolongs the life time of foods. Smoking is a method which increases the shelf life of the product with the combined effect of drying process and chemical compounds which is formed by thermal degradation of the wood (Anderson and Pedersen 1951). Although it has an important protective feature, the main purpose of this method is to give a pleasant smell, taste, and appearance to the product. Mostly, the preservation remains in the second plan. After the smoking, the colour and the flavour of the fish changes and the odour, taste and colour of the smoked fish are formed (Swastawati et al. 2000; Visciano et al. 2008).

Salting, drying, and smoking are an each other complementary combination. Salting is one of the pre-treatments that are applied to the material before the smoking process and plays an important role in the quality of the final product. In salting, microorganism activity is prevented by removing water from the product and reducing the water activity of the product. Salt also gives the product the desired firm texture and flavor (Doe et al. 1998; Gokoglu 2002). Drying also reduces the water activity and prevents the passage of microorganisms by forming a light shell layer on the product surface. The substances in the smoke content have microbicide and microbiostatic effects, and antioxidative effects (Gokoglu 2002). The increase in the salt content of the fish muscle during the salting process, results in changes in the protein structure, which affects the water holding capacity, texture, and microstructure of the muscle (Thorarinsdottir et al. 2004). The smoking process causes weight loss due to dehydration. Depending on the raw material, the final product characteristics and the smoking method, a weight loss of about 10–25% is observed due to dehydration in the smoking process (Arason et al. 2014). During the smoking process, the pH of the fish muscle has been reported to decrease due to the absorption of acids from the smoke, moisture loss, and the reaction of phenolic compounds with proteins (Hassan 1988).

Smoke is obtained by the flameless burning of wood. The composition of the smoke depends on the conditions of smoke obtaining and the type of wood (Conde et al. 2006). The smoke contains both gas and dust particles. Substances present in the wood and can easily evaporate form vapour. The portion retained by the fish flesh during smoking is the vapour of smoke. They come into contact with the fish and transform to liquid state. As a result, the fish surface is moistened, and the smoke flow is accelerated. Phenolic compounds in the vapour phase of smoke, which provide biological stability in the smoked product, are acids, alcohols, hydrocarbons, aldehydes, ketones, esters, ethers, and carbonyls. These components first accumulate on the smoked product surfaces and then penetrate the meat (Guillen and Errecalde 2002; Gokoglu 2002; Goulas and Kontominas 2005). The volatile compounds in the smoke, especially phenols cause the changes in the taste and aroma of the fish muscle during the smoking process. Phenolic compounds, especially syringaldehyde and conifer aldehyde give smoked product aroma to the product. The intensity of odour and flavour of smoke are affected by the smoking temperature and method (Arason et al. 2014).

Smoking process is carried out by two different methods: hot and cold smoking. Hot smoking is a process in which the fish is smoked in a combination of temperature and time sufficient to cause coagulation of fish proteins. Hot smoking is usually sufficient to kill parasites, destroy non-spored bacterial pathogens, and eliminate the spores related to human health (FAO 2013). Hot smoking is the process of smoking the fish at 70–80 °C and thus cooking the fish. In order to stabilize the smoked product at room temperature, it is reported that the water activity (aw) of the hot smoked fish products should be less than 0.85 (Arason et al. 2014) Cold method is smoking of salted fish without exceeding the temperature of 33 °C by avoiding of cooking of meat and coagulation of proteins. The applied temperature is usually 12–15 °C (Gokoglu 2002). Cold smoking takes longer than hot smoking but provides higher efficiency and better protects the original textural properties than hot smoked products (Arason et al. 2014).

Almost any type of hard wood can be used for smoke production. The use of resinous wood is not suitable as it causes a bitter taste in the product. The choice of wood depends on the desired odour and taste in the final product. The most preferred wood type is oak. Oak, ash, beech, alder, acacia, poplar, apple, willow, orange, and lemon trees are stated to be the most suitable trees for smoking. Coniferous trees are not suitable for smoking. Coniferous trees cause the unpleasant terpentine taste and odour (Gokoglu 2002).

Liquid smoke solutions, which have been produced with the distillation of the wood and then concentrated to a certain extent, have been used for many years to give aroma to the meat and fish (Hattula et al. 2001; Arason et al. 2014). Concentrated smoke is dissolved in a solvent and can be used directly in products. The product is immersed in a liquid smoke solution. Liquid smoke solution transfers smoke to the product (Wheaton and Lawson 1985). Compared to conventional smoking methods, the use of liquid smoke has advantages such as lower cost, less environmental damage, easy availability, and variety of application methods (Hattula et al. 2001). Moreover, obtaining a uniform smoke aroma is faster and easier than conventional cold and hot smoking (Varlet et al. 2007). Polycyclic aromatic hydrocarbons (PAHs) are important carcinogenic compounds formed in smoke that can accumulate on the surface of smoked foods and then pass into the product. Purification of these compounds is possible by obtaining the natural smoke in either a generator or by condensation of its natural smoke. Liquid smoke is an aqueous solution of natural smoke condensate and is separated from the tar phase which is inconvenient for health. In the case of liquid smoke, low concentration of low molecular weight PAH compounds occur.

In a study in which shrimp (*M. reosenbergii*) was smoked using liquid smoke, shrimps were divided into two groups. A group of shrimps were removed from the head and shells, and the second group of shrimps was removed only from the head. After that, the shrimps were salted in 30% saline and then smoked using the traditional and liquid smoke methods. At the end of this study, it was reported that flavour and aroma specific for smoked shrimp were more detected in the shrimp smoked by the traditional method. Shrimp exposed to liquid smoke was characterized by an artificial and bitter taste. In beheaded shrimps treated with liquid smoke, the characteristic shrimp flavour was more visible (Portella et al. 2011).

2.2 Crabs

2.2.1 Introduction

Crabs are decapod crustaceans of the infraorder Brachyura. They have a very short projecting "tail" (abdomen) usually hidden under the thorax. They live in all oceans, in fresh water, and on land. They are generally covered with a thick exoskeleton and have a single pair of pincers. Crabs are decapod crustaceans of the infraorder Brachyura. They have a very short projecting "tail" (abdomen), usually hidden under the thorax. They live in all oceans, in fresh water, and on land. They are significantly evaluated economically in some countries of the world. It is consumed with pleasure due to its high nutritional value and attractive taste and flavour. They should be consumed quickly after harvest as they change the quality in a very short time after they die. They are processed by various methods for consumption as well as maintained by various methods.

Most of the crabs live in sea, freshwater and brackish waters. In addition to their basic duties in the ecosystem, crabs are consumed as nutrients in many countries. Generally marine species are consumed as nutrients. They are economically evaluated at a significant level in the countries such as China, France, Indonesia, Japan, Philippines, Spain, Thailand, and United States. It has also been reported that crab species in the world has been evaluated as feed additive and fertilizer purposes as well as direct consumption as food (Gulle et al. 2007).

Crabs are creatures that are cultured besides catching from their natural habitat. All cultured crabs are marine species. Although crab culture is not commonly done in Western countries, especially *Mithrax spinosissimus* (Lamarck, 1818), *Scylla serrata* (Forskal, 1775) and *Portunus pelagicus* (Linnaeus, 1758) cultivation is common in East Asian countries and Japan (Lee and Wickins 1992).

2.2.2 General Information About Crabs

2.2.2.1 Biology of Crabs

Crabs are covered with a hard shell called carapace made from chitin. This shell, which acts as an exoskeleton, is usually lead-coloured. Head and thorax parts are fused and covered with carapace. Their bodies are in enlarged form. The body of the crabs is enlarged on the cephalothorax in the Brachyura sub-order of the Decapoda order. Cephalothorax is seen almost from the top. Carapace is oval shaped. The abdomen, which consists of seven segments, is small and curled towards the bottom of the thorax and not seen from the top. This section is tapered towards the end in males, while in females it is wider and circular. In both male and female sex, the abdomen is folded over the sternum, starting from the starting area to the ventral side. In male individuals, narrow and pointed tips, female individuals are more oval, enlarged (Kaestner and Levi 1970). There are five pairs of walking legs from the thorax section. The first pair improved and has a strong pair of claws. They use their clamps as defence and nutritional purposes. The last walking legs of the marine species adapted to life in the pelagic areas of the sea have turned into swimming legs. Usually the right foot is larger. Crabs often walk sideway. They move in such a way that they can only bend their legs in one direction.

The basic crab design consists of an expanded carapace, and a strongly reduced abdomen that is tightly tucked underneath the thorax. In addition, the first pereiopods of brachyurans are fully chelate, and the walking legs are placed at the sides of the body. True brachyuran crabs are often confused with hermit and porcelain crabs belonging to the infraorder Anomura. In general, most anomuran crabs have only three pairs of walking legs clearly visible, with the last pair being very small and normally positioned under the abdomen and not visible externally (Jose 2015).

The true crabs (Decapoda: Eubrachyura) have a depressed carapace or cephalothorax and a much reduced, straight, and symmetrical abdomen which is closely bent under the cephalothorax; this abdomen is never used for swimming and lacks biramous uropods; in the female, during the spawning season, the eggs are attached to the abdominal appendages (berried crabs). The cephalothorax has five pairs of walking legs, the first of which is chelate (ending in pincers) and nearly always much stronger than the other legs (Tavares 1998).

When feeding of crabs is examined, it is understood that these creatures are both herbivorous and carnivorous. In general, crabs are fed with algae. Besides, many things such as molluscs, worms, other crustaceans, and sediments can also be a source of food for crabs according to their hunting abilities and species. Mixed nutrition (plant and animal) is a necessary condition for the development and metabolism of crabs.

Although most of them are benthic, only the larvae and some species of crabs that show general metamorphism in their developmental stages are pelagic. They are mostly found in rocky areas with plants and muddy and sandy grounds. Most of them are hunted alive and attacks on fish and amphibians. Cannibalism is also common among crabs (Carroll and Winn 1989).

Crabs are found in a distinctly shaped hard shell, and at regular intervals the shell is replaced by larger shells to allow the crabs to develop. This shell replacement process is called moulting or ecdysis. Shell replacement period is usually between July and October. Females begin a little later in males as they begin to change shells in July. In this period, crabs are protected between rocks. The shell cracks in such a way that it separates the lower and upper halves with a precise line. The soft crab slowly emerges from the slit. It then absorbs water and swells. When the shell replacement period is complete, the shell slowly hardens. Although the shell is not fully set within a few months, there is no change in size until the next shell change period. Shell replacement occurs frequently at the beginning of the life of the crab and occurs approximately every 2 years after reaching the specific size. It is impossible to determine the exact age of the crab. There is no guide in his body. But the average10–15 cm crab is about 4–5 years old (Edwards and Early 1978).

The sex of a crab can be easily determined. Females have a large abdomen, while males are narrower. Males are larger than females of the same size. The mating is carried out in the summer months and in the waters close to the shore while the females are still in the soft shell immediately after the shell change. For 2 weeks prior to shell replacement, male crabs are accompanied by hard-shell males. Spawning occurs immediately after the females discard the shell. Maturing of most of the females in July August spawn in November of the same year. However, in some cases, ovulation is delayed until next winter. Crabs usually choose soft sea bottoms to spawn. They often choose deep water. The eggs are kept in the abdomen of the crab for about 7 months. The number of eggs carried can vary from half a million to three million. In the spring and summer following the ovulation, the females move to the coastal waters where the eggs hatch (Edwards and Early 1978).

2.2.2.2 Trade and Marketing of Crabs

According to FAO (2016) report, the world fishery crabs' production is primarily comprised of *Scylla serrate*, *Scylla olivacea*, *Portunus pelagicus*, and *Callinectes sapidus*. Production has increased from 343 thousand t, in 1990, to more than 951,000 t, in 2015. China accounts for 71.9% of the world annual production (684,400 t), followed by Indonesia (77,700 t), USA (72,400 t), and Philippines (27,200). However, precise information of the global production is limited as most production systems are small-scale or family-owned, and when carried out on an industrial scale, there is usually no interest in the disclosure of data or, very often, trade secrets are involved. This difficulty limits the knowledge about real production data and often leads to the underestimation of the total amount produced (Hungria et al. 2017).

According to the FAO-Globefish (2019a, b) report, global imports of crab (all types) increased slightly in 2018, to 405,400 tonnes. Imports into the United States of America dropped by 3.3% to 104,400 tonnes, while Chinese imports grew slightly to 81,900 tonnes. Imports into the Republic of Korea also grew, from 45,400 tonnes to 50,100 tonnes. Shipments to the United States of America dropped from Canada and increased from the Russian Federation. Russian Federation crab exports increased by almost 10%, and the main markets were the Republic of Korea (43,000 tonnes or 61% of the total), the Netherlands (12,800 tonnes or 18% of the total) and China (11,500 tonnes or 16% of the total). China's crab exports fell slightly to 72,600 tonnes in 2018, from 75,000 in 2017. The main markets were the Republic of Korea, the United States of America, and Taiwan Province of China (FAO-Globefish 2019a).

Indonesia is the main supplier of blue swimming crab to the US market, but landings in Indonesia have been flat. In 2017, the United States of America imported a total of 26,300 tonnes. Indonesia shipped 46% of this volume, while China supplied 14% and the Philippines 12%. During the first 9 months of 2018, the United States of America imported 20,200 tonnes of blue and red swimming crab, around 6.8%

more than during the same period in 2017. Indonesia's share of this volume declined slightly from 50.1% in 2017 to 44.1% in 2018 (FAO-Globefish 2019b).

2.2.2.3 Some Important Crab Species

Crabs like shrimp and lobster also belong to the Decapoda order. Crabs can be divided into two main groups as brachyuran crabs (infraorder brachyura) and anomalous crabs (infraorder anomura). Most species of Brachyuran crabs, also called true crabs, can be easily distinguished by four pairs of well-developed walking legs from the crabs of the infraorder Anomura, called false crab (Ng 1998).

Crabs are creatures collected in Crustacea class, the Decapoda order and Majidae, Portunidae and Xanthidae families. The groups belonging to Brachyura infraorder and called "true crabs" are the most known and studied crabs. There are 4000 species in 26 families.

Swimming Crabs

Carapace hexagonal, transversely ovate to transversely hexagonal, sometimes circular; dorsal surface relatively flat to gently convex, usually ridged or granulose; front broad, margin usually multi dentate; usually 5–9 teeth on each antero lateral margin, postero lateral margins usually distinctly converging. Endopodite of second maxillipeds with strongly developed lobe on inner margin. Legs laterally flattened to varying degrees, last two segments of last pair paddle-like. Male abdominal segments 3–5 completely fused, immovable (Jose 2015). Ten species of interest to fisheries: *Arenaeus cribrarius* (Lamarck, 1818), *Callinectes bocourti* A. Milne Edwards, 1879; *Callinectes danae* Smith, 1869; *Callinectes exasperatus* (Gerstaecker, 1856); *Callinectes larvatus* Ordway, 1863; *Callinectes maracaboensis* Taissoun, 1969; *Callinectes ornatus* Ordway, 1863; *Callinectes rathbunae* Contreras, 1930; *Callinectes sapidus* Rathbun, 1896; *Callinectes similis* Williams, 1966 (Tavares 1998).

Callinectes sapidus **(Rathbun, 1896)**
Callinectes name means beautiful swimmer. The blue crab has five pairs of legs. Unlike other crabs, the last two legs are in the form of flat shovels to provide swimming. The first two legs end with sharp clamps. They use their clamps to catch and keep their food (Poppke 1977). They also hold them with their clamps to protect them from damage. If the crab is scared and wants to run away, it breaks or removes the clamp and improves the new one. The other three legs are used to walk.

Sex is easily understood by the colour of claws in live crabs. In the females, the colour of the claw is bright red orange. In males, blue and white are the source of the general blue crab concept. Gender is also understood from the shape of the abdomen. The lower part of the abdomen, called apron, is T-shaped. The females are large and circular, while the young females are triangles (Yerlikaya 2002).

If the enemy approaches, the blue crab gets a defensive position. Opens and removes clamps. The legs are ready for side movement. Blue crabs also prefer live food. Mussels and oyster beds are feeding areas of blue crabs. The other part of their diet is dead fish, algae, and seaweed. The life of the blue crab is about 2 years. The female can release a million larvae at a time. Shells do not grow. They regularly change the shell to grow. This event is called moulting. During moulting, the outer hard shell disappears. Underneath it forms a new soft shell and hardens. The moulting starts behind the crab eye, breaking the shell on both sides of the mouth and spreading over the entire shell. This break opens the shell. Each time, the lost limb grows a little larger until it is completely formed. Immediately after the shell is changed, the crab gets water. With the minerals taken from the water, the shell hardens within 12–48 h. Soft crabs are usually found in June, July, and August; however, they can be encouraged to change their shells in specially developed tanks. Thus, the soft-shell season can be extended. Crabs are sold at high prices right after the shell change (Dore 1991).

The male crab strolls over the female a few days before the female crab changed its shell. These cases are called doubler. It protects female crab which is defenceless and weakened during the moulting from external influences and harmful organisms. Blue crabs love the brackish waters in summer and the ocean depths in winter. Brackish water, mud and clayey lagoons and creeks are suitable for the growth of blue crabs (Gokoglu and Oray 1997).

Portunus pelagicus **(Linnaeus, 1758)**
Carapace rough to granulose, front with four acutely triangular teeth; nine teeth on each anterior lateral margin, the last tooth 2–4 times larger than preceding teeth. *Chelae elongate* in males, larger chela with conical tooth at base of fingers. Colour: males with blue markings, females dull green/greenish brown. It prefers sandy to sandy muddy substrates in shallow waters down to a depth of 50 m, including areas near reefs, mangroves and sea grass and algal beds. It is collected mainly by artisanal traps, trawls, beach seines, cylindrical wire traps, folding traps, pots, and drop nets. This species is most widely sold in markets in Southeast Asia. It is distributed throughout Indo-west Pacific (Ng 1998). It is a crab species that is included in the Mediterranean ecosystem through the Suez Canal. This species has delicious meat and high economic value (Fig. 2.10).

Scylla serrata **(Forskal, 1775)**
Carapace mouth, with strong transverse ridges; H-shaped gastric groove deep; relatively broad frontal lobes, all in line with each other; broad anterolateral teeth, projecting obliquely outwards. Well-developed spines present on outer surface of chelipedal carpus and anterior and posterior dorsal parts of palm.

S. serrata prefers more oceanic waters, usually found just offshore on soft muddy bottoms. All species of *Scylla* prefer soft substrates in shallow or intertidal water and mangroves. They are collected by trawls, traps, baited wire mesh pots, hooking and hand. Main markets are Taiwan, China, Hong Kong, and Singapore. They are always sold for high prices. All four species of *Scylla* apparently have a wide Indo-Pacific distribution.

Rock Crabs

Carapace broadly oval or hexagonal; front not produced in form of a rostrum but having a central tooth; anterior lateral margins toothed (nine quadrangular or pentagonal teeth in species listed herein); lateral spines not strongly developed; antennules folding length wise. Found only in northern part of area. This family comprises 1 genus, Cancer Linnaeus, 1758, and 4 living species in the Atlantic ocean, 2 of which are eastern Atlantic in distribution (*Cancer bellianus* Johnson, 1861, and *Cancer pagurus* Linnaeus, 1758) and 2 western Atlantic (*Cancer borealis* Stimpson, 1859, and *Cancer irroratus* Say, 1817). Nations' (1979) proposition of dividing the genus Cancer into subgenera is followed here. Two species of interest to fisheries marginally in the area: *Cancer irroratur* Say, 1817. *Cancer* (Metacarcinus) *borealis* Stimpson, 1859 (Tavares 1998).

Land Crabs

Land crabs carapace transversely oval, not strongly depressed, antero lateral margins strongly arched, not divided into teeth or lobes; front o-orbital margin (between outer orbital angles) very much shorter than greatest width of carapace; third millipedes gaping noticeably, exposing the mandibles; dactyls of walking legs ridged and spiny. Live on land, always at the reach of the water table. A single species of interest to fisheries in the area: *Cardisoma guanhumi* Latreille, 1828 (Tavares 1998).

Golden Crabs

Carapace hexagonal; dorsal surface relatively smooth to granular; frontal margin with 4 teeth; antero lateral margins distinctly convex, each with 3–5 low, sometimes indistinct teeth. Dactylus of walking legs is T-shaped incross-section. Male abdominal segments 3–5 fused, functionally immovable, but sutures still visible (Tavares 1998).

Fig. 2.10 *Portunus pelagicus*

Stone Crabs

Carapace transversely oval or transversely hexagonal, front broad and not ched centrally, never produced in form of a rostrum; antero lateral margin lobate (in the species listed herein) ortoothed; antennules folding transversely or obliquely. A single species of interest to fisheries in the area: *Menippemer cenaria* Say, 1818 (Tavares 1998).

Ghost Crabs

Carapace usually rectangular or nearly so, ortrapezoidal; front relatively narrow and somewhat bent downward; orbits occupying whole anterior border outside front, outer walls of orbits often open, eyes talks long (longer than width of front); third maxillipeds usually completely covering mouth cavity, concealing the mandibles; dactyls of walking legs smooth or ridged but not conspicuously spiny (Tavares 1998).

2.2.2.4 Production and Trade of Crabs

Crabs are found in all the oceans of the world. There are also many kinds of freshwater and land-type. Annual production in world seas is around 1.6 million tons. It constitutes 20% of all marine arthropods hunted all over the world. Crabs are found in all the oceans of the world; There are also many kinds of freshwater and land-type. Annual production in the world seas is around 1.7 million tons. It accounts for 27% of all crustaceans caught all over the world (FAO 2016). Total world imports of crab increased slightly to 294,000 tonnes during the first threequarters of 2017 compared to the same period in 2016. The main importers were the United States of America (91,700 tonnes), China (59,900 tonnes), Republic of Korea (29,800 tonnes) and Japan (29,200 tonnes). There were significant differences from country to country. Republic of Korean and Japanese imports fell by 8.1 and 16.7%, respectively, while Chinese imports increased by 23.2% to almost 60,000 tonnes, and Thai imports rose by 58.1% to 12,700 tonnes, during this period. Two of the most significant exporters of crab were the Russian Federation and China. In the Russian Federation, exports of crab rose by almost 29%, to 44,000 tonnes, during the first 9 months of 2017. The main market was the Republic of Korea, with as much as 70% of these exports. In contrast, Chinese exports fell by 9.4% during the same period. The Republic of Korea imported less Chinese crab, while Taiwan Province of China imported slightly more (FAO-Globefish 2018).

2.2.2.5 Fishing of Crabs

The crab fishing season is between March and September. The most quantity of crab is obtained between May and June. Fishing of crab is influenced by the movement of the crab into the deep waters in winter and into the shallow waters in the spring.

Crabs are caught with baited traps. These traps were designed to allow the crab to enter easily, but to trap it. The catching power of the traps depends on the attractiveness of the bait. Almost all kinds of fish and crustaceans can be used as bait. Fishermen make their own bait preferences. But the price and availability of bait affect the preference (Edwards and Early 1978).

Two types of traps are used in crab fishing: (1) Creel, (2) two-Inkwell pot.

Creel consists of rectangular timber base and frames. The frame is covered with natural or synthetic nets. There is a funnel-shaped opening which allows prey to enter easily. These opening narrows to prevent the prey from escaping into the end of the trap. Inkwell pot is round shape. It has a single eye on top. Most fishermen make their own traps and fix it. But they are made of different materials (Edwards and Early 1978).

2.2.2.6 Nutritive Composition of Crabs

Chemical composition of crab meat has been studied by several researchers. These studies have been conducted on different crab species. Table 2.4 shows proximate composition of crab meat reported by researchers. The table shows that crab meat with high moisture content is an important source of protein and on the other hand it is a dietary nutrient with low fat content. In a study, total lipid levels in *Paralithodes camtschaticus, Paralithodes platypus, Chionoecetes opilio, Chionoecetes angulatus* and *Chionoecetes japonicus* muscles were determined as 0.61, 0.60, 1.57, 0.96 and 0.53, respectively (Latyshev et al. 2009). Total lipid level was reported as 0.5% for *Carcinus maenus* (Skonberg and Perkins 2002), 0.75% for *Chionoecetes opilio* (Addison et al. 1972), 0.5% (Sidwell et al. 1974) and 2.6% (Otwell et al. 1984) for *P. camtschaticus*.

It was determined that the chemical composition of crab varies according to age, gender, and season. In *Carcinus aestuarii*, females were found to have high fat and protein content in gonad and hepatopancreas, and they contained lower moisture and ash compared to males. The highest fat content was found in winter and spring, while the lowest in the autumn. Due to the high protein and low-fat content it has been reported that the best season to consume the crab is autumn (Baklouti et al. 2013).

It is also reported by researchers that crab meat contains n-3 fatty acids at a significant level (Celik et al. 2004). The content of polyunsaturated fatty acid (PUFA) was found higher in *Portunus pelagicus* (43.8–45.3%) than in *Callinectes sapidus* (39.2–42.8%) (Ozogul et al. 2010). Baboli et al. (2016) found that total n-3 PUFA ranged between 11.31–20.26% in *P. pelagicus*. Ramamoorthy et al. (2016) found the levels of eicosapentaenoic acid (EPA) and docosahexaenoic acid (DHA) in *Portunus pelagicus* to be 0.89% and 0.63% respectively. On the other hand, EPA (5.47–7.22%) and DHA (7.27–11.96) contents of the same species (*P. pelagicus*) have been reported to be different according to breast meat, claw meat, male and female. The highest EPA and DHA were determined in the female claw (Baboli et al. 2016). The ratio of PUFA in total fatty acids in *Carcinus maenus* harvested from four different sites in Nova Scotia was found to be 47.1–50.5 and n-3 ratio was 37.4–40% (Naczk et al. 2004).

In addition, the crab meat is rich in mineral. It is determined that it is rich in sodium (Na), potassium (K), calcium (Ca) and phosphorus (P) (Gokoglu and Yerlikaya 2003; Musaiger and Al-Rumaidh 2005). In a previous research, it was reported that Na, P, Mg, K and Ca values in *Callinectes sapidus* were found to be 663.95, 176.24, 117.06, 69.15 and 398.2 mg/100 g, respectively (Kucukgulmez et al. 2006). In another study, Na, K, Mg, Ca, and P contents of *Eriphia verrucosa* were determined as 2300, 3035, 101.4, 896.8 and 3119 mg/100 g respectively (Erdem et al. 2015).

Many factors such as season, biological difference (species, height, age, sex, and sexual maturity), nutrition, environmental conditions (water chemistry, temperature, salinity, and contaminants) and processing method influence the concentration of minerals in aquatic products. Most aquatic products take and collect minerals from the environment. Raw crab meat contains higher mineral content than fish. Particularly, if the crust is mixed with the muscle meat of shellfish, the mineral content increases significantly (Gokoglu 2002). Mineral content may vary depending on the cooking process. In a study, it was reported that calcium, copper, iron, magnesium, phosphorous, potassium, sodium, and zinc contents changed significantly according to the cooking process (Lopez et al. 1981).

Table 2.4 Proximate composition of some crab species

Species	Moisture	Protein	Fat	Ash	Reference
			Body tissue		
Callinecres sapidus	81.58	14.71	0.79	1.89	Gokoglu and Yerlikaya (2003)
	79.05	18.81	0.90	2.15	Kucukgulmez et al. (2006)
	78.83	16.42	0.77	2.06	Anthony et al. (1983)
	77.80	17.50	0.37	2.40	Atar et al. (2001)
	77.4–86.7	11.9–19.2	0.4–1.5	1.3–1.8	Wheaton and Lawson (1985)
Sesarma brockii.	79.2	15.7	0.32	2.55	Marques et al. (2010)
Cancer magister	78.5–82.3	14.3–23.4	0.7–2.2	1.2–1.9	Wheaton and Lawson (1985)
Portunus pelagicus	75.28	22.64	1.21	2.24	Gokoglu and Yerlikaya (2003)
	86	10.3	0.67	1.96	Badawi (1971)
	77.57	18.83	1.45	2.34	Tureli et al. (2000)
Sesarma brockii.	79.2	15.7	0.32	2.55	Marques et al. (2010)
			Claw meat		
Callinectes sapidus	83.1	15	0.64	1.39	Gokoglu and Yerlikaya (2003)
	78.02	19.55	0.44	2.13	Kucukgulmez et al. (2006)
	80.12	15.13	–	1.63	Zotti et al. (2016)
Portunus pelagicus	77.09	21.54	0.81	2.52	Gokoglu and Yerlikaya (2003)
Carcinus maenus	79	16.8	0.5	2.2	Skonberg and Perkins (2002)
Cancer pagrus	77.96	17.02	–	2.25	Zotti et al. (2016)

2.2.3 Quality Changes and Spoilage of Crab Meat

Seafood spoilage is faster than other muscular foods. This can be attributed mainly to the fact that seafood contains more water than other muscular foods in the edible portion. High water activity and a loose connection between protein fibres provide a rich environment for microbial growth in seafood. This causes faster microbial spoilage in seafood than other muscle foods. Nitrogenous compounds found in shellfish cause spoilage metabolites.

The chemical degradation of crab meat is like that of other shellfish products, but it is reported to be different from that of fish meat. The free amino acid content of the crab meat is reported to be high and crab meat contains non-protein nitrogenous compounds. These nitrogenous compounds are converted to metabolites such as ammonia and biogenic amines (Sarnoski 2007). Enzymatic processes such as deamination of free amino acids, degradation of nucleotides, and oxidation of amines cause the increase in ammonia level in the tissue (Gill 1990).

Decomposition of macromolecules into small molecular compounds, constitutes good nutrient source for microorganisms and increases microbial growth and spoilage in crab. Growth of microorganisms is a main cause of shelf-life reduction. Sanitation during the processing of the crab will make a significant contribution to keeping the microbial load low. In a study in which 46 crab processing plants were examined, it was reported that crabs processed under good sanitation conditions had better bacteriological results than those processed under poor sanitation conditions and that microbial quality products could be obtained if the processors applied sanitation rules (Phillips and Peeler 1972). Manual processing of crabs can lead to cross-contamination of spoilage bacteria and pathogenic bacteria, as well as residues remained on the machine during machine operations may contaminate the next product to be processed.

Microbial flora of crabs is affected by environmental factors. The microbial flora reflects the environment in which it was harvested. In addition, the microbial flora of the crab has been affected by factors such as the region, season, water temperature and water quality. In a previous study, in all tissue of two crab species examined in terms of microbial population, gram negative bacteria, *Alicaligenes faecalis*, *Pseudomonas aeruginosa*, *Moraxella lacunata*, and *Aeromonas salmonicida* were determined to be predominant. Moreover, in the same study, *Pseudomonas aeruginosa* in shell tissue, *Bacillus cereus* in gill tissue, *Alcaligenes faecalis* in muscle tissue and *E. coli* in hepatopancreas were determined in more levels (Mahalaxmi et al. 2013). In another study, it was reported that the dominant aerobic microorganisms isolated from fresh crab meat were *Acinetobacter*, *Carnobacterium piscicola*, *Exiguo bacterium* and *Pseudomonas* which were both Gram positive and Gram-negative organisms. *Aeromonas, Arthrobacter, Bacillus, Brevibacillus, Brevibacterium, Chryseobacterium, Marcococcus, Providencia*, and *Staphylococcus* were reported to be less dominant (Suklim et al. 2008).

2.2.4 *Crab Processing and Preservation*

2.2.4.1 Alive Transportation of Crabs

Crab meat is very sensitive and can change its quality in a very short time. Therefore, it is very important to keep the crabs alive before consumption or before the process. Previously, the live consumption of crabs was limited to fishing zones. However, the fact that crabs must be transported live to other regions has resulted in the development of methods of transporting them alive.

Keeping alive of crabs starts first on the ship's deck at sea. Live crabs should be processed immediately after capture if possible. If they are exposed to sunlight and wind, they die quickly. If the time and weather conditions are appropriate, they should be stacked as soon as they are captured and, in any case, must be placed directly in the boxes or on the bottom of the ship. Crabs should be packed in top ventilated boxes as close as possible to each other. Thus, they are prevented from being damaged by fighting with each other during transportation. The full boxes must be protected from the sun and wind. Wet straw or sawdust on the bottom of the boxes will extend the shelf life. Crab transport should be carried out as soon as possible. Especially, it should be transported at night when the air temperature is low (Edwards and Early 1978; Borderias 2012). The crab, which is individually tied, is tightly packed in the carton in its vertical position with the claws at the top. In case of air transport, liner inserted to the boxes are used for sealing. A few layers of paper or newspapers are placed at the bottom of these boxes. Thus, the liquid to be leaked from the crab is impregnated. The mortality rate of crabs transported in this way is generally reported to be less than 10%. In some countries, such as the Philippines, the crab is placed in cartons covered with plastic sheets (after connecting separately). Cartons and plastic sheets have ventilation holes. The crab is also packaged in perforated polystyrene boxes (Mohamed and Devaraj 1997).

Under normal conditions, crabs can only live out of water for a short time. Longer life out of sea water depends on environmental factors such as temperature and humidity. If these factors are optimally adjusted, they can live longer. In a study, it was shown that crabs can live outside the water for 24–210 h depending on the humidity level in the atmosphere (Mohamed and Devaraj 1997). Relative humidity significantly affects the survival time of the crab. In the case of saturated relative humidity, the crab loses less weight. If the crab loses 10% of its weight, it dies (Gillespie and Burke 1991).

The activity of live crustaceans decreases as the temperature decreases. Handling of animals at low temperatures can be easier and any kind of transport should be less stressful. The decrease in the activity will reduce the amount of waste. The starvation period (24–48 h) will also reduce the production of waste. The gradual reduction in temperature and starvation time should be considered as pre-treatments for live transport or dry transport (Jacklin and Combes 2007a).

When taken out of the water, the crabs struggle for a few minutes and then become ineffective. The fishermen bind them or limit their claws. Unrestricted crabs are usually stored in plastic baskets. With this application, the crabs are taken alive to the shore. Undamaged crabs may be kept in a moistly, shady, or dark place and can be kept for hours without dying unless they are exposed to direct sunlight or heat. Crabs should not be unnecessarily disturbed or handled, and it is also important to avoid shocks such as dropping or hitting a crab holding container. Keeping the crabs cool from the moment they reach the boat will have a significant impact on the stress of keeping them the outside of the water. The deck is warmer than the water where crabs live. Keeping crabs relatively cool reduces oxygen requirements and allows a limited amount of oxygen to remain calm. Instead of keeping crabs in a cold room, cold sea water can also be sprayed on the crabs. To reduce the stress of crabs out of water, it is possible to use low temperature and water sprays only in large boats. Small speedboats running close to the shore can keep crabs away from water with minimal intervention to reduce stress (Paterson and Goodrick 1994). In a study, crabs were stored using three different methods (in air, in cold air (20 °C) and in an aerated flow-through seawater tank). Although it was expected that the crabs would be better kept in water than in the air, this was not the case. The best survival rate was found in crabs that were stored in cold air instead of crabs that were constantly stored in the renewed sea water tank (Paterson et al. 1994).

Algae are also used as packaging material in some parts of the world to keep the crabs alive. Algae allow the crabs to be kept moist. In India, crabs are reported to be brought to the market in baskets containing algae. However, it is also stated that this method carries the risk of algae deterioration in the case of long-term transport and the overlap of the baskets. On the other hand, some materials such as sea water-soaked cotton and sea-water-soaked wood were used for the live transport of crabs. These materials increase the lifetime by keeping the crabs moist. Under these conditions, the crabs have been reported to remain alive with wet cotton for 17–18 days and with wet wood for 7 days (Mohamed and Devaraj 1997).

2.2.4.2 Preliminary Processing

Sorting

Sorting process is carried out to separate the crabs which are to be kept and those are to be discarded. Diseased and weak ones should be extracted. Because they die in a short time and they start to deteriorate. If they are kept in separate places, it will be easier to separate the sick and dead ones. Those who recent moulted crabs should be removed from the storage medium. Because, they can be cannibalised by other hard-shelled ones. This will result in the loss of legs and antennas and will cause their values to fall even if they are alive (Jacklin and Combes 2007a).

Grading

Grading is a process of categorizing live crabs by size and gender to meet market demands. The buyer may have special preferences such as size, weight and gender. Meeting of such demands can be easier if storage is carried out in separate tanks (Jacklin and Combes 2007a).

Slaughtering

Stunning

Crabs that are kept alive should be stunned before they are killed. This is a sensitive, emotional, and controversial issue. Therefore, the welfare of the animal should be taken into consideration (Jacklin and Combes 2007a). In previous studies, it has been reported that crabs are killed or stunned by using gas or thermal shock or electric shock. In a study, it was reported that the electrical method was effective in stunning of crabs at the commercial level and that a minimum of 220 V was needed to stun the crabs within 1 s. It was reported that the combination of the electrical stunning method with the thermal shock was successful to stun edible crab (*Cancer pagurus*) (Roth and Grimsb 2013, 2016). In another study, it was reported that 10 s electroshock application paralyzed blue crab (*Callinectes sapidus*) (Weineck et al. 2018).

Spiking

Spiking is another method to kill crabs. They have two main nerve centres. Both nerve centres can be destructed by spiking using a pointed pithing tool or a sharp pointed knife.

Chilling in Air

As shellfish are cold-blooded animals, physiological functions are slowed down at temperatures as low as 4 °C. Even if they are alive, the metabolic rate decreases by decreasing the body temperature. However, the time required to slow down the activity of shellfish products with this method depends on the species, size, metabolic status, and cooling rate of the organism. The method of cooling in the air is the first stage of the two-stage killing method. This is followed by spiking that destroys the nervous system.

Boiling

The most common method used to kill crustaceans is to leave in boiling water in a live state. This causes them to die by causing physiological shock. During this procedure, autotomy can usually be seen. As a behavioural response, autotomy is the self-collapse of one or more limbs of animals. Generally, it is a self-defence mechanism applied to get rid of the hunter (Sherwin 2001; Gardner 2004). In a study, crabs were placed in boilers with cold water and heated slowly to 40 °C. As a result of the study, it was observed that the animals died silently without pain and nauseous symptoms (Gunter 1961).

Picking

The removal of crab meat is called picking. Crab meat can be removed manually or mechanically. It requires a serious experience in manual removal. Otherwise, the losses will be higher, and the efficiency will decrease. The yield of crabs removed by hand varies from Jumbo lump to small bits leg meat. First, the hinges are removed, and then back, gills and aprons are removed. This allows easy access to the carapace that can be taken with picks. To remove the claw's flesh, the shell is first broken with a wooden mallet. Roller is used to remove the remaining meat from the leg. An experienced person should be able to extract a crab every 30 s. In a good meat removal process, meat yield of Dungeness or blue crab can reach up to 25%. However, generally 10–20% yield can be obtained. This means about 80–85% waste (Galetti 2010).

Meats that cannot be completely removed by hand and remain as waste, particularly leg meats, may be removed with the help of compressed air jets. First, the body is pulled, and the brown meat is scraped off with a spoon. The claws are broken into segments and each segment is broken by a heavy rod. The white meat is collected with a short knife or a specially designed tool (Edwards and Early 1978).

Crabs are usually sold live. But it is also available in fresh, cooked, pasteurized, and frozen forms. Crabs are also canned.

Crabs should be kept alive at the beginning of the processing and must be killed before boiling. Crabs are killed either by drowning in fresh water or by inserting a sharp object into the centre of the brain. Crabs are kept in fresh water at 10 °C for 3–4 h. Small crabs are usually killed by this method. Crabs should be cooked immediately after killing. It should be boiled for 20–30 min in the water added with 2–3% salt. Boiling time may vary depending on the size of the crab. The cooking process will kill the existing bacteria and destroy the organisms that cause the disease (Edwards and Early 1978).

Prior to some operations, crabs are pre-cooked. In commercial crab cooking, two different cooking processes are applied, namely pressure steam and boiling water. In the pressurized steam cooking process, the cooking temperature is 121 °C and the duration is 3–20 min, while the temperature in cooking in boiling water is 100 °C and the time is 15–20 min (Thomas and Thomas 1983). Whilst the boiling process is preferred by the producers because of the higher crab yield, it is preferred by the consumers because the crab cooked with steam is tastier (Moody 1974).

2.2.4.3 Chilling

The post-harvest conditions of crabs can cause stress responses that can lead to reduced quality and survival. Chilled packaging method has been shown to minimize respiratory requirements and prolong transport time. Reduced oxygen consumption will reduce the sensitivity to stress (Ninlanon 2011).

Crabs are usually kept alive or in ice to prevent rapid quality deterioration. It is then stored in the refrigerator, which runs between 2 °C and 4 °C until melted ice or consumption (Fotedar and Evans 2011).

Raw or cooked storage of crabs also affects cold storage time. In a study, the shelf life of cooked crab (*Scylla serrata*) meat was determined as 8–9 days at 25–28 °C, 6 days at 6.5–7.5 °C and 11 days at 0 °C (George and Gopakumar 1988). The shelf life of the cooked and uncooked crab (*Portunus pelagicus*) stored in ice was determined as 14 and 10 days, respectively (Balasaraswathy et al. 2008).

In another study, live crabs were kept in tanks where sea water was provided at ambient temperature and then stored in four different conditions. The first group were kept on ice on 4 °C, the second group on 4 °C, the third group were kept on ice at 20 °C for 4 h and then at 4 °C for 16 h by simulating the market conditions and the fourth group at 20 °C was stored. According to the results of the study, the shelf lives of unprocessed crabs (*Carcinus maenas, Necora puber* and *Cancer pagurus*) were determined to be 9–11 days on ice at 4 °C; 13–29 days at 4 °C; under simulated supermarket conditions (4 °C and 20 °C) 5–7 days and 2–16 days respectively (Robson et al. 2007).

2.2.4.4 Freezing

Crabs should be frozen after cooking (not before). When crabs are frozen and subsequently cooked, it is extremely difficult to remove the meat and yield is low. When the crabs are frozen as whole, losses of the claws and legs are inevitable. The extracted meats are either frozen in cardboard boxes as consumer packs or in large blocks for catering manufacturers. They are usually glazed and packed in outer carton for storage. White and black meats can be frozen in separate packages. Or they can be packed in the same package with plastic film or parchment. Dressed crab can be frozen in plastic bags. Barbecue claws should be frozen in an airflow freezer due to their irregular shapes. However, homogeneous packages can be contact frozen. Freezing should be done quickly. First, the product must be cooled from 0 to −5 °C for no longer than 2 h. It should then be frozen. Unpackaged blocks should be glazed after being removed from the freezer. Immediately after freezing, the crabs should be taken to the cold store immediately. If possible, frozen crabs should be stored at −20 to −25 °C (Edwards and Early 1978).

In a study comparing freshly caught and then frozen crabs (*Scylla serrata*) after storage for 14 days on ice, the crabs were frozen in a plate freezer at −40 °C, coated with polyethylene paper and packaged in a cardboard box and stored at −23 °C. At the end of the study, fresh frozen crabs remained acceptable for up to 52 weeks. On the other hand, the samples stored on ice for 1 day could remain in good condition for 50 weeks, 7 days for 21 weeks, and 14 days for 8 weeks (George 1973).

The quality changes of leg meat and body meat of Jonah crabs, which were frozen in −23 °C in an air blast freezer after being cooked in steam for 8 min and then cooled in ice water, were examined and it was reported that earlier quality changes were observed in leg meat (Rebach et al. 1990). The maximum storage time of snow crab (*Chionoecetes japonicus*) at −20 °C was determined as 2 weeks (Jun et al. 2017). In a study conducted by Yerlikaya and Gokoglu (2004), it was determined that the whole blue crab (*Callinectes sapidus*) packaged in polyethylene bags and stored at −18 °C had a shelf life of 10 months.

While freezing increasing the shelf life of crabs one hand, on the other hand can accelerate the onset of melanosis. The freeze-thaw process is reported to increase melanosis by lysis of tissue cells. This leads to the rapid development of melanosis of PO enzymes that maintain functional integrity during the freeze-thaw process, together with appropriate substrate and oxygen. The fact that raw-frozen and cooked-frozen snow crabs (*Chionoecetes opilio*) exhibit significantly higher drip loss during storage compared to those that have not undergone freezing is indicated as a clear indication of the mechanical damage to cellular tissue membranes due to the freeze-thaw process (Lian et al. 2018).

Changes During Freezing and Frozen Storage

Changes in Proteins

Reactive sulphide group is generally reduced during frozen storage (Jiang et al. 1988). The reduction in the sulfhydryl group consists either of oxidation of sulfhydryl, disulphur changes, or the formation of hydrogen and hydrophobic bonds that mask the reactive sulfhydryl structure of actomyocin molecules (Benjakul and Bauer 2000). The sulfhydryl group content of hard- and soft-shell mud crab (*Scylla serrata*) was found to be reduced during 12 weeks of frozen storage (Benjakul and Sutthipan 2009).

An increase in the hydrophobicity of the proteins can be observed during frozen storage. This increase is due to the opening of proteins and the release of hydrophobic aliphatic and aromatic amino acids. As a matter of fact, Benjakul and Sutthipan (2009) found an increase in surface hydrophobicity of the samples during 6–8 weeks of frozen storage of mud crab (*Scylla serrata*).

Protein solubility is widely used as an indicator of structural changes in proteins. Protein solubility is indicative of protein denaturation. When the protein is denatured, the hydrophobic amino acids in the protein molecules will be exposed to the surface, and then the apolar groups will tend to interact with another to form protein aggregates that cause a decrease in solubility (Chan et al. 2011). A decrease in protein solubility was detected during the frozen storage of the mud crab (Benjakul and Sutthipan 2009).

During the freezing of the meats, cell membranes are damaged, which leads to a decrease in water holding capacity and a loss of cooking loss (Lagerstedt et al. 2008). It is reported that the aggregation and denaturation of proteins in crab muscle is heat induced and leads to loss of water retention capacity of proteins. In addition, denatured proteins formed during frozen storage are susceptible to heat denaturation.

Changes in Lipids

Lipid oxidation can be induced by frozen storage. Ice crystals formed during the freezing process can destroy the cells and release pro-oxidants for lipid oxidation, particularly free iron (Benjakul and Bauer 2001). The increase of thiobarbituric acid and values of the sea crab (*P. pelagicus*) (Subramanian 2007) and mud crab (*Scylla serrata*) (Benjakul and Sutthipan 2009) was determined during frozen storage.

2.2.4.5 Pasteurizing and Canning

Pasteurization is a heat treatment below 100 °C, which generally means light heating. Pasteurization increases the shelf life of crabs in refrigerated storage. The crab meat is commercially pasteurized at 85 °C and stored under refrigerated conditions (Innocen 2014). From this definition it is understood that the product is non-sterile and can therefore continue to contain microorganisms. As a result, pasteurized products must be kept in continuous cold so that the surviving microorganisms do not multiply very quickly. Crab meat is an excellent medium for bacterial growth due to its high water-activity, moderate pH, and low salinity. Most of the bacteria present on the crab are located on the shell, viscera, and gills. Pasteurization of foods other than crabmeat is often defined in terms of a target organism. There is no target organism for the pasteurization of crabmeat. The process evolved based on shelf life extension (Hackney et al. 1991). Crab meat is at risk for *Clostridium botulinum*, especially for type E. Contamination may occur due to post-cooking processes and environmental influences. The pasteurized crab meat is packaged with a hermetically strong seal, pasteurized, and stored in cold conditions. Pasteurized or canned crab products are marketed as jump, jumbo lump, back fin, claw meat and claw finger (Biji et al. 2013). Pasteurizing the crab can extend the shelf life at refrigerated temperatures from 70–10 days to 12 months (Matiella and Hsieh 1990).

In a study where microbial analyses were conducted at different control points in the pasteurized crab processing line, it was determined that the pasteurization process had a positive effect on the microbial quality of the final product (Olgunoglu 2010).

According to Codex Alimentarius international food standards canned crab is defined as "*Canned crab meat is prepared singly or in combination from the leg, claw, body and shoulder meat, from which the shell has been removed, of any of the edible species of the sub-order Brachyura of the order Decapoda and all species of the family Lithodidae*". Canned crab meat is placed in hermetically sealed containers and subjected to sufficient treatment to ensure commercial sterility (Anonymous 1981).

Crabs should be processed as quickly as possible for canning. Crabs should be processed under conditions that limit the enzymatic activity that causes the deterioration of fresh flavours. They should be kept wet and cool immediately after catching. Cooking may be in boiling water or in retorts using pressurized steam. After cooling, the edible meat is collected from the shells. The meat should be pressed before putting it in the can. The cans should be packed tightly because some shrinkage occurs during the process (Featherstone 2016b).

An important problem with canned crab meat is discoloration. In canned crabs, discoloration occurs with the formation of blue/black or grey/black pigments (Featherstone 2016b). During the canning process, while a blue colour forms due to the blood components of the crab meat, the brown associated with the mailard reaction, the black associated with the sulphite reactions of metals, the yellow associated with lipid oxidation and the red related to muscle pigments may be formed. In addition, metals such as copper iron, muscle pH and moisture content, heat treatment, gender, seasonal factors have been reported to be effective in colour change (Requena et al. 1999). It has been reported that blue discoloration can be reduced by bleeding the crab before processing (Boon 1975). In a study conducted by Vijayan and Balachandran (1981), it was found that there was a definite relationship between the copper content of the crabs (*Scylla serrata*) and the blue colour change. They also found that after thoroughly washing the crab carcass under running water, the copper level decreased, and the blue discoloration was prevented.

Crab meat contains a copper-based hemocyanin that can form grey to blue-black complexes when canned or pasteurized. Meats processed above about 87.7 °C often change colour. Contamination of crab meat with metals, especially iron, can greatly increase the blue discoloration. On the other hand, it is stated that ferric and ferrous salts added to canned product may cause blue discoloration after processing (Boon 1975).

Usually pasteurized crab meat is packed in aluminium cans. However, with the advent of thermally stable polyethylene packaging, the crab products were sold in laminated bags (Galetti 2010). Blue crab meat is pasteurized in different packages including steel boxes, aluminium boxes, plastic boxes, non-barrier bags and barrier bags. It was determined that crab meat pasteurized in plastic and aluminium cans had better quality and longer shelf life than those put in steel tin cans. The lowest quality and shortest shelf life are determined in the product packaged in oxygen barrier bags (Gates et al. 1993). Lacquered containers should be used for canned crab. Sometimes it is necessary to double-varnished boxes to completely protect the crabs from contact with the tin. Parchment is also used in some cases (Featherstone 2016b).

2.2.4.6 Crab Cakes

To produce crab cakes, the minced crab meat is mixed with spices or sauce, covered with dough or potatoes are added. Milk, eggs, mayonnaise can be used for the crab cake and green peppers and onions can be added. Crab cakes are either freshly marketed or packaged and frozen. Crab cakes can be produced by sautéing or frying in a small amount of oil. However, it is more preferred broiled crab cakes in terms of consumer health. However, consumers think that fried or sautéed crab cakes are even more delicious.

2.3 Lobsters

2.3.1 *Marine Lobsters*

2.3.1.1 General Information About Marine Lobsters

Biology of Marine Lobsters

Lobster is one of the arthropod marine species and is important seafood due to its interesting biology and high economic value. Since lobster is a very valuable flavour worldwide, there are ongoing technical advances to ensure that stocks in all major lobster producing countries remain sustainable for future generations. These include the development of hatchery techniques for the development of fishery, which is successful for the *Homarus* species and is in the experimental stage of rock lobster.

Lobsters are large crustaceans. Generally, they may vary in length from a few cm to more than 60 cm. Like shrimps, lobsters have a well-developed and elongated body and tail. In addition, the lobsters have a thicker shell. Lobsters are distinguished from shrimps by having a dorsoventrally depressed body, less developed pleopods and thoracic sternum wide and distinct (Ng 1998). The lobster body is divided into two parts. The first part is the body part consisting of the integrated head and thorax, the second part is the abdomen and the tail consisting of six parts. Head and thorax surrounding the shell is carapace.

The upper shell of the lobsters is flat. Behind the eyes are two pairs of spines. The outer pair is very small. The lobster has two moving eyes, two antennas, six mouth sections and 20 gills under the carapace. There are four or five teeth on the lateral edge of the mouth. Two grooves extend from the end of the mouth and the middle towards the middle edge of the shell. The abdomen is flat. Two pairs of arms and with large claws. One of them is a cutting claw and the other is a catch claw equipped with strong irregular teeth. The second and third pairs of arms are thinner. They also have clamps. The lobsters are bluish black in colour and the top is spotted and the bottom is greenish (Alpbaz 1993).

Most lobsters live in rocky and stony areas. They are especially hidden in intensive planktonic cavities. Some live on hard ground, while others are found among mud, sand, and moss. But they do not bury themselves in the mud. Lobsters are hidden in a large part of their time, looking for food for a very short time.

Lobsters reach maturity in the age of 4–6 years. Lobster mating occurs in the summer, during the female's moulting period. It is stated that mating can occur up to day 12 even if the mating is usually 48 h after the moulting. Success in mating depends on the balance between the size of the male and female. The size difference increases or eliminates the chance of mating. Male individuals may mate with one or two female individuals within a 2-day period. Females can also mate with more than one male. The sperm can be collected and stored in the seed section of the female for 9–13 months. When the female lays eggs at the time of ovulation, she also performs fertilization by releasing the sperm at the same time. Fertilized eggs adhere to the hairy section under the tail. The spawning efficiency of a female individual is approximately 300,000 eggs. The female carries her eggs in her body for 10–12 months as the incubation period (Alpbaz 1993). Hatching usually occurs between May and September with a peak in June and July depending on water temperature. The hatched larvae spend 4–6 weeks in the water column before moulting into a final stage. They then start to seek out a suitable rocky bottom habitat to settle into and develop into juvenile lobsters (Irish Sea Fisheries Board 2019).

Some Species of Marine Lobsters

The lobster category includes four main commercial species: European lobster (*Homarus gammarus*), American lobster (*Homarus americanus*), rock lobster (*Jasus* spp.), and tropical or spiny lobster (*Panulirus* spp. or *Palinurus* spp.) (Nguyen et al. 2017). Mostly harvested in America and Canada, *H. americanus*, known as American lobster, has been reported to be the most widely produced species in the world (Annie and McCarron 2006). The second most important species is the spiny lobster, which accounts for 38% of global production, and the third is Rock lobster, most of which is harvested in Australia and accounts for 6% of total production (Gary 2012).

Spiny lobster fishery is mainly from Australia, United Kingdom, United States of America, South Africa, Japan, France, New Zealand, and India. Spiny lobsters are found in warmer seas throughout the world. They are also called as rock lobster, spiny lobster, sea crawfish, langouste and langosta (Smith and Walton 1958). It is stated that there are about six species in the submerged rocky patches along the coast in India. The species in the order of commercial importance are *Panulirus homarus* (Linnaeus), *Panulirus polyphagus* (Herbst), *Panulirus ornatus* (Fabricus), *Panulirus versicolor* (Latrielle), *Panulirus penicillatus* (Olivier) and *Panulirus longipes* (Milne Edwards) (Mohan Rajan et al. 1981).

There are many lobster species belonging to many families. This book will provide information about some species of commercial importance. These are summarized as follows.

Family Nephropidae

They are moderate to large sized crustaceans. Carapace with well-developed rostrum, ornamented with spines or nodules. Eyes usually well developed and black or small and lacking pigmentation. First three pairs of legs form true pincers and the first pair are greatly enlarged and long. Both abdomen and tail fan well developed and powerful. All species of this family are deep sea forms and found from 150 to 1893 m. They prefer soft substrates (Chan 1998b).

Homarus americanus **(H. Milne-Edwards, 1837)**
American lobster is the main species of lobster and accounts for about 60% of total lobster landings. For almost 30 years, the amount of catching has increased from 37,000 tons to 140,000 tons (FAO 2017). It is one of the largest commercial species in terms of body size. It can be up to 64 cm in male and 61 cm in female. They are found in large amounts in the depths of 4–50 m. Most of the lobsters are caught on the shore at a depth of 40–100 m using traps (Squires 1990).

Homarus gammarus **(Linnaeus, 1758)**
It is known as "European or common lobster". It is a large crustacean. It has a maximum body length of 60 cm. Adult *H. gammarus* live on the continental shelf at depths of 0–150 m, although not normally deeper than 50 m. They prefer hard substrates, such as rocks or hard mud, and live in holes or crevices, emerging at night to feed. It is mostly fished using lobster pots, sometimes with baited lines. *Homarus gammarus* is traditionally "highly esteemed" as a foodstuff. It is sold very high prices as fresh, frozen, canned or powdered. The price of *H. gammarus* is up to three times higher than that of *H. americanus*, and the European species is more flavoured (Holthuis 1991). It has a broad geographical. In its northern range, it occurs from the Lofoten Islands in Northern Norway to south-eastern Sweden and Denmark, but is absent in the Baltic Sea probably due to lowered salinity and temperature extremes. Its distribution to southwards extends along the mainland European coast around Britain and Ireland (Prodohl et al. 2007).

Nephrops norvegicus **(Linnaeus, 1758)**
It is known as the "Norway lobster, Dublin Bay prawn, langoustine". It is the most important commercial crustacean in Europe. It lives in muddy seabed sediments, with more than 40% silt and clay (Bell et al. 2006). It is an important species for fisheries, being caught mostly by trawling. Around 60,000 tonnes are caught annually, half of it in the United Kingdom's waters (FAO 2004). It is distributed in the north-eastern Atlantic Ocean, and parts of the Mediterranean Sea, but is absent from the Baltic Sea and Black Sea (Bell et al. 2006). *Nephrops norvegicus* is found in the north-eastern Atlantic Ocean and North Sea as far north as Iceland and northern Norway, and south to Portugal. It is not common in the Mediterranean Sea except in the Adriatic Sea (Davidson 2002).

Family Palinuridae

Body of family members is tubular or dorsoventrally is flattened. Hairs are few and scattered. Rostrum is absent or reduced to a small spine. Carapace is sub-cylindrical or prismatic, laterally rounded, or straight. They have a pair of large frontal horns above eyes. Antennae are very long and rather thick. Legs are without true pincers and first pair is not or only slightly longer than the following legs (Chan 1998b).

Jasus edwardsii **(Hutton, 1875)**
It is a species of spiny lobster. Species is commonly called crayfish or crays in both Australia and New Zealand. The species lives in crevices of the rocky shores and among algae at depths between 5 and 200 m. Soft shelled specimen are occasionally caught in December and January. Maximum total body length is 58 cm (males), and 43 cm (females); maximum carapace lengths 23.5 cm (males), 18 cm (females); minimum legal carapace lengths 10 cm (males), and 9 cm (females). The species is usually caught with baited lobster pots, sometimes obtained by trawling and by diving. It is distributed in New Zealand, from Three Kings Islands (north west of-the northern tip of North Island) south to the Auckland Islands, also found at the Chatham Islands; most common off the south west part of South Island, and the east coast south of East Cape (Holthuis 1991).

Panulirus homarus **(Linnaeus, 1758)**
Carapace of this species is rounded and spiny and sometimes with branchiostegal areas slightly inflated. Rostrum is absent, anterior margin armed with four regularly spaced large spines other than frontal horns. This species lives in reef areas with sand in the surf zone and sometimes also in turbid waters at depths from 1 to 5 m. It is fished by hand, with traps, gill nets, cast nets and baited lines. It is widely distributed in the Indo-pacific from the eastern coast of Africa to Japan, Australia, and Marquesas Archipelago.

Panulirus polyphagus **(Herbst, 1793)**
It is known as "muddy spiny lobster". It has a maximum body length of 40 cm. Body length is commonly between 20 and 25 cm. It is mainly found on muddy bottoms in turbid waters near river mouths at depths from 3 to 90 m. Unlike other spiny lobster, this species is mainly taken by trawling, sometimes also by set nets, and seines, but rarely enters traps. Commercially and economically importance is mostly from the Gulf of Thailand. It is sold fresh or frozen in local markets and mounted dry specimens are sold as souvenirs to tourists. It is distributed in Indo-Pacific from Pakistan to India, Thailand, Vet Nam, Taiwan, Philippines, Indonesia, Papua New Guinea and northern Australia Indo-pacific from the eastern coast of Africa to the Red Sea, Japan, Australia, French Polynesia, Hawaii and Offshore islands near the western coasts of America (Chan 1998b).

Panulirus ornatus **(Fabricius, 1798)**
It is known as "ornate spiny lobster". It usually occurs at depths from 1 to 10 m, but can be found to a depth of 200 m. It is found in calm areas of coral and rocky reefs or reef slopes, sometimes also found on muddy substrate in river moths with turbid water. It lives solitary or in pairs. It is fished mostly by divers, sometimes by hand nets and trawls. It is sold mostly fresh or frozen in local markets. It is widely distributed in the Indo-Pacific from East Africa to Japan, Australia. It is also reported to enter the Mediterranean from Red Sea (Chan 1998b).

Panulirus versicolor **(Latreille, 1804)**
It is known as "painted spiny lobster". It has a maximum body length of 40 cm. Body length is commonly between 20 and 30 cm. It is found in reef areas at depths of usually less than 16 m in clear or sometimes turbid water with strong currents, often on seaward edges of reef plateau. Actively it is fished by divers, but apparently rarely enters traps or pots. The catches of this lobster are nowhere very large, and it is mainly locally consumed live, fresh, cooked whole, or tailed, but in some regions, such as from the Philippines, exported live or tailed. It is widely distributed in the Indo-Pacific from the eastern coast of Africa to the Red Sea, Japan, northern Australia, and French Polynesia (Chan 1998b).

Panulirus penicillatus **(Olivier, 1791)**
It is known as "pronghorn spiny lobster". It has a maximum body length of 40 cm. Body length is commonly between 20 and 30 cm. It is found in shallow waters, usually at depths from 1 to 4 m at seaward edges of reefs, in clean waters not influenced by rivers. Good catches are often possible during dark nights, particularly after the full moon. It is mostly taken during day or night diving by hand and spear, sometimes also by trammel nets and traps. However, the catches are generally not very abundant, and it is mostly sold fresh, live, cooked whole or tailed for local consumption, but also exported in some regions, such as Philippines and Indonesia. It is widely distributed in the Indo-pacific from the eastern coast of Africa to the Red Sea, Japan, Australia, French Polynesia, Hawaii, and Offshore islands near the western coasts of America (Chan 1998b).

Panulirus longipes **(A. Milne-Edwards, 1868)**
It is known as "long legged spiny lobster". It has a maximum body length of 60 cm. Body length is commonly between 20 and 35 cm. It is found in shallow coral or rocky reefs, usually in clear waters with moderate currents, sometimes in slightly turbid waters. It is fished mostly by divers during night diving, also with traps, tangle nets and lobster pots. The Fishery of this lobster is mostly of local interest. In some regions such as Philippines and Indonesia, occasionally is exported live, together with other species of this genus. It is sold mostly fresh or frozen in local markets. It is widely distributed in the Indo-Pacific from East Africa to Japan and Fiji (Chan 1998b).

Palinurus elephas **(Fabricius, 1787)**
Its maximum total body length is 50 cm, but usually not larger than 40 cm. The species lives on rocky bottoms, rarely on sand, in depths from 5 to 160 m, mostly between 10 and 70 m. Ovigerous females from September-October to February-March. is mostly caught with lobster pots, occasionally on hook-and-line and by spearing, rarely with trawls, tangles, or trammel nets. It is distributed in Eastern Atlantic, from southwestern Norway to Morocco, also in the Mediterranean, except the extreme eastern and south eastern parts (Holthuis 1991) (Fig. 2.11).

Palinurus delagoae **(Barnard 1926)**
Its maximum total body length is 35 cm, carapace length to 17 cm: average carapace length about 10 cm. It lives at 0 to 400 m depth, usually between 180 and 324 m. Off

South Africa it is found, on muddy or sandy substrates, sometimes with coral fragments; off Madagascar it has been reported from a rocky substrate. It is distributed in Indo-West Pacific region: East coast of Africa from (Mozambique) to (Natal, South Africa), south east Madagascar. Off south east Africa the species is taken by trawlers, while off Madagascar, lobster pots were used during experimental fishing (Holthuis 1991).

2.3.2 *Freshwater Lobsters (Crayfish)*

2.3.2.1 General Information About Crayfish

Crayfish are freshwater crustaceans resembling small lobsters (to which they are related). They are also known as crawfish, crawdads, freshwater lobsters, mountain lobsters, mudbugs, or yabbies. Taxonomically, they are members of the super families Astacoidea and Parastacoidea. They breathe through feather-like gills. Some species are found in brooks and streams where fresh water is running, while others thrive in swamps, ditches, and paddy fields. Most crayfish cannot tolerate polluted water, although some species such as *Procambarus clarkii* are hardier. Crayfish feed on animals and plants, either living or decomposing, and detritus. The name "crayfish" comes from the old French word escrevisse. Some kinds of crayfish are known locally as lobsters, crawdads, mudbugs, and yabbies. In the Eastern United States, "crayfish" is more common in the north, while "crawdad" is heard more in central and southwestern regions. In Australia (on the eastern seaboard), New Zealand and South Africa, the term crayfish or cray generally refers to a saltwater spiny lobster, of the genus Jasus that is indigenous to much of southern Oceania. In Singapore, the term crayfish typically refers to *Thenus orientalis*, a seawater crustacean from the slipper lobster family (Wikipedia 2019a).

Because of its important ecological role in freshwater ecosystems, crayfish have been called a keystone species. The impact of these crustaceans on ecosystems is more important than expected given their relative abundance or total biomass. Because these species are both prey and predators, they have a central position in the food chains of the ecosystem (Geiger et al. 2005).

Biology of Crayfish

Crayfish are found in freshwater environments such as lakes, dams, rivers, and irrigation canals ranging from sea level to sub-alpines. They can be found in most substrate types. Larger populations are found in areas where there are shelters, and low pollution, and no predators. Although crayfish are mainly a night animal, some crayfish are observed during the day, especially in dark waters that provide little protection from predators (Hollows 2016). Crayfish usually are seen in the darkness of the evening and morning or at night. During the day they are hiding in pits or

Fig. 2.11 *Palinurus elephas*

under stones. They do not like the sun. They prefer dark and dim places. In winter they are drawn to their own pits and nests. Crayfish hunt at night and in the evening. They usually go looking for their feeds after sunset and provide them at night at most (Alpbaz 2005). Crayfish must moult to grow and breed. Their growth is not continuous but depends on factors such as water temperature, calcium, feed availability and daylight (Hollows 2016). In crayfish, the exoskeleton is a support organ and acts as a protective armour against external factors. This hard layer of chitin and lime mixture completely covers the body. When the shell of the crayfish forming the skeleton hardens, growth stops. Growth is possible by changing of shell. When the old shell falls completely, a soft new shell that has already developed underneath it forms. This is called moulting. As this new shell grows so the crayfish grow. Moulting is the hardest period of crayfish life. In this period, the crayfish is very tired and rests in its nest until it becomes a new shell. During this time, it is vulnerable to its enemies. After maturity, females 1, males 2 times change their shell. Therefore, males are larger than females (Alpbaz 2005). They can grow all year round. Higher water temperature increases the growth. The frequency of moulting and thus the rate of growth decreases with age (Hollows 2016). The crayfish regularly gets too big for its skeleton, sheds it, and grows a new larger one. Such moulting occurs six to ten times during the first year of rapid growth, but less often during the second year. For a few days following each moult, crayfish have soft exoskeletons and are more vulnerable to predators (New World Encyclopedia 2019).

Females also require shell replacement for reproduction. A slightly bendable carapace is required to extrude the female's eggs. Once the female has laid eggs, the male stores the sperm package known as "spermatophore" near the breeding openings where the female's eggs are fertilized. Females produce between 20 and 400 eggs (Hollows 2016). Crayfish usually mature at age 5, depending on environmental conditions, and can reach 20–25 cm in length in 20 years. Since the beginning of autumn, there are improvements in the sex organs of adult lobsters. Eggs grown in females are easily attracted by their yellow-brown appearance. In males, sperm channels take the form of a white rope (Alpbaz 2005).

In the early stages of life, crayfish are not visible, because they are hidden under rocks or between woody vegetation to protect them from predators and larger crayfish. Crayfish eat almost everything and are therefore described as opportunistic omnivores. They eat everything, including filtering phytoplankton, grazing algae, invertebrates, and animal meats (Hollows 2016). Crayfish mostly prefer live foods. They do not want to eat stinking foods or dead animal wastes unless they are very hungry. But they eat everything that can be eaten in food difficulty (Alpbaz 2005).

The body of a decapod crustacean, such as a crab, lobster, or prawn, is made up of 19 body segments grouped into two main body parts, the cephalothorax, and the abdomen. Each segment may possess one pair of appendages, although in various groups these may be reduced or missing. The cephalothorax is covered by a carapace, which protects the internal organs and the gills; the section of the carapace that projects in front of the eyes is called the rostrum. Crayfish have two pairs of sensory antennae on the head, and the eyes are on movable stalks. On the thorax or pereon, there are four pairs of walking legs (pereiopods) and one pair of pereiopods (chelipes) armed with a claw (pinchers). On the abdomen, or pleon, there are pleopods (also called swimmerets), which are primarily swimming legs. At the end of the pleon is the tail fan, comprising a pair of biramous uropods and the telson, which bears the anus. Together, they are used for steering while swimming, and in the caridoid escape reaction. When they are in danger and need to flee, crayfish can swim backwards quickly by curling and uncurling their abdomen. They breathe through feather-like gills predators (New World Encyclopedia 2019).

Crayfish live in rivers, lakes, streams and even marshes. They are often housed on gravelly bottoms, under flat stones or in shallow pits. They love hollow, abundant vegetation, stony and slime-free waters. If the soil is not too hard, they will make a nest by carving the soil using their tails with digging movements. The crayfish lie in their nests leaving their clamps out and easily observing their enemies. Crayfish normally have a slow and shaky walk. They walk forward using their walking legs. But they swim backwards. Crayfish generally like abundant calcareous and discharge waters. Limestone helps the development of the shell (Alpbaz 2005).

Some Species of Crayfish

There are two centres of species diversity of freshwater crayfish. The first is in the South-eastern United States. The second centre is Victoria, Australia. Crayfish naturally occurs on all of the continent except Africa and Antarctica. The Astacidae are distributed in Europe, US, British Colombia, and Canada. The Cambaridae are found in the Eastern United States and South Mexico. Parastacidae are distributed in Australia, New Guinea, New Zealand, South America and Madagascar (Souty-Grosset and Fetsner 2016).

Species used for economic income are mainly *Procambarus clarkii*, *Pacifastacus leniusculus*, *Cherax destructor*, *C. quadricarinatus*, *Orconectes limosus*, *O. rusticus* and *Astacus leptodactylus*. Among these species, *Procambarus clarkii* can be found in Africa, Asia, America, California and Europe; *Procambarus leniusculus* to California in Japan, Europe and America; *C. destructor* in Africa and Australia; *C. quadricarinatus* was transported to South America, *O. rusticus* to North America, and *A. leptodactylus* was transported from Europe to the natural environment.

Procambarus clarkii ranks first among the crayfish species harvested from natural environments. The highest *P. clarkii* harvest is in China. *Procambarus clarkii* and *P. zonangulus* are the most important species in South America, mainly in the Louisiana region. The most important species cultivated in Australia is *Cherax destructor. C. quadricarinatus* is the second most important species produced in Australia. These species are larger than other crayfish species and therefore have higher commercial value.

There are many crayfish species belonging to many families. This book will provide information about some species of commercial importance. These are summarized as follows.

Astacus astacus **(Linnaeus, 1758)**
It is the most common species of crayfish in Europe, and a traditional food source. Like other true crayfish, A. *astacus* is restricted to fresh water, living only in unpolluted streams, rivers, and lakes. It is found from France throughout Central Europe, to the Balkan Peninsula, and north as far as parts of the British Isles, Scandinavia, and Eastern Europe. Males may grow up to 16 cm long, and females up to 12 cm. This species was once abundant in Europe, although it was expensive to buy, and is the finest edible crayfish (Wikipedia 2019b).

Astacus leptodactylus **(Eschscholtz, 1823)**
Astacus leptodactylus, also called "the Danube crayfish", "Galician crayfish", "Turkish crayfish" or narrow-clawed crayfish, is a species of brackish water crayfish imported and introduced to Central Europe in nineteenth century from the Caspian Sea region. It was originally distributed over an area corresponding to Turkey, the Ukraine, Turkmenistan and Southwest Russia, but is also found in Iran, Kazakhistan, Georgia, Belarus, Bulgaria, Romania, Hungary and Slovakia. However, this crayfish has been widely introduced into many countries as a replacement for noble crayfish populations lost due to crayfish plague (Souty-Grosset and Fetsner 2016).

Pacifastacus leniusculus **(Dana, 1852)**
It is known as "signal crayfish". It is a North American species of crayfish. It was introduced to Europe in the 1960s to supplement the North European *Astacus astacus* fisheries. They are bluish brown to reddish-brown in colour, with robust, large, smooth claws. *P. leniusculus* is distributed from British Columbia in Canada at the northern part of its range, South to central California and east to Utah and Montana in the USA. *P. leniusculus* can be found in a variety of habitats, from smell streams to large rivers and natural lakes (Cabi 2019a).

Cherax destructor **(Clark 1936)**
C. destructor is an aquatic crayfish, naturally distributed in a wide range of habitats in its native range of distribution throughout inland Australia. It has a relatively high commercial value, being a culinary delicacy ('baby lobster') and bait for sport fishing but it is also used as an aquarium species and in research. Destructor ranges over 2 million km^2 in its native range from South Australia and the southern parts of the Northern Territory in the west, to the Great Dividing Range in the east (Cabi 2019b).

Cherax quadricarinatus **(Von Martens, 1868)**
C. quadricarinatus is an Australian freshwater crayfish. *C. quadricarinatus* is an aquatic crayfish, naturally distributed in a wide variety of habitats in its native range of distribution (Queensland and Northern Territory in Australia and south-eastern Papua New Guinea). Due to the harsh physical conditions it has adapted to in its native range, *C. quadricarinatus* has a robust nature with broad tolerances to environmental extremes. Such environmental tolerance, combined with its rapid growth rate and relatively large dimensions, makes it an ideal species for aquaculture and aquarium trade. From Australia, the species has been exported to many other countries with subtropical to tropical climates where commercial production has been attempted (Cabi 2019c).

Orconectes limosus **(Rafinesque, 1817)**
O. limosus is a species of crayfish in the family Cambaridae. It is native to the east coast of North America, from Maine to the lower James River, Virginia, but has also been introduced to Europe. It is known commonly as the spiny cheek crayfish. It is unusual in that it lives in silty streams, rather than the clear water usually preferred by crayfish. Like *Pacifastacus leniusculus*, another invasive North American crayfish, *O. limosus* carries crayfish plague and is a threat to native European crayfish. They prefer flat, sandy, and rocky floors. They are also found outside the water on beaches or lawns near the pool of water. They use rocks to make burrows while in the water. This is a very common species of cray, especially on Northeast United States, and Southeast Canada (Wikipedia 2019c).

Procambarus clarkia **(Girard 1852)**
Procambarus clarkii is a species of cambarid freshwater crayfish, native to northern Mexico, and southern and south-eastern United States, but also introduced elsewhere (both in North America and other continents), where it is often an invasive pest. It is known variously as the red swamp crawfish, red swamp crayfish, Louisiana crawfish, Louisiana crayfish or mudbug *Procambarus clarkii* was originally distributed from northern Mexico to Florida, and north to southern Illinois and Ohio. It has been widely introduced in the USA, south and Central America, Europe, and other more dispersed areas such as Japan, China, and Taiwan. *P. clarkii* is most found in warm fresh water, such as slowly flowing rivers, marshes, reservoirs, irrigation systems and rice paddies. It is the most ecologically plastic species in the order Decapoda and is able to grow quickly even in only seasonally present water, being able to tolerate dry spells of up to 4 months (Wikipedia 2019d).

In the following sub-headings of this chapter, the term "lobster" will be used in both marine and freshwater lobsters. However, they will be expressed separately when it is necessary in the text.

Production and Trade of Lobsters

Lobster production in the world is 231,968 tons in 2013 and 304,806 tons in 2016. More than half of this production is provided by USA and Canada. The following countries are Indonesia, Australia, and Brazil (FAO 2015). Canada is the world's leading harvester of all lobster species. Canada is also the largest harvester of American lobster.

It is reported that American lobster (*Homarus americanus*) and Norwegian lobster (*Nephrops norvegicus*) account for more than 60% of the world's lobster presence, and *H. americanus* reached a record level of 16,000 tons in 2014 (FAO 2016). The average unit value corresponds to US $ 10 per kg for shrimp and less than US $ 5 for finfish, while it is expressed to be US $ 20 per kg for lobster (FAO 2017).

In 2014, the total amount of imports of lobster in the world was 170,156 tons. As of 2014, the top lobster importing countries are United States of America, Canada, and China. Among the European countries, the most important importers are France, Italy, and Spain.

In many countries, the increase of consumption and economic value of crayfish has accelerated the production of this product under aquaculture conditions. Large and detailed applications have been initiated for the creation of new crayfish sources with the fry produced and for the stocking of crayfish in the lakes deteriorated of population balance. There are many species of freshwater lobster found in all continents except Africa, especially in the USA. The most important genera are *Astacus, Pasifastacus, Procambarus, Lambarus, Orconectes, Cambarellus, Austropotamobius* and *Chrex*. Some of these genera, whose origins are reported to be specific to certain countries, are reported to have spread to various countries for many reasons (Alpbaz 2005).

In recent years, international trade in crayfish has increased dramatically because of farm production and export from China. Chinese crayfish are often exported as frozen peeled tails in different size categories. In addition to all cooked and spicy crawfish, special products for Scandinavian and USA are other product forms. Despite the growing importance of Chinese production, the three other important regions for fishing, aquaculture and consumption of crayfish are the USA, Europe, and Australia (Lee and Wickins 2002). China is the world's largest crayfish producer, according to a 2017 report by the then Ministry of Agriculture, now known as the Ministry of Agriculture and Rural Affairs. Its output skyrocketed to 852,300 tonnes in 2016 from 265,500 tonnes in 2007. Outside the domestic market, Chinese crayfish have found fans in the United States and Europe. In 2016, China exported 23,300 tonnes of crayfish worth 259 million US dollars. Nearly 40% went to the United States, while 90% of the crayfish consumed in Europe came from China (Huang 2018).

Fishery of Lobsters

A wide variety of catching gears are used to catch lobsters. These gears vary from design to fabrication and operation. Manual collection and trawl are used in various parts of the world. However, lobster traps are widely used in traditional catching. Traps and pots are interchangeable terms. Some primitive traps invented by ancient people are still in use in some parts of the world.

The catch ratio is a function of the abundance of lobster and catchability of the gear. The unit of effort for lobster fishing is bait traps, and its ability to catch is influenced by several factors related to lobster biology, the environment, mechanical design of traps, and fishing strategy (Drinkwater et al. 2006). Wind, which is one of the environmental factors, is reported to be an important factor affecting the catch rate. Winds from a certain direction provide good catch, while winds from the opposite direction can reduce the catch rate (Drinkwater et al. 2006). On the other hand, other researchers have reported the effect of temperature on catch rate. In a study, it was determined that the monthly catchability coefficient of western rock lobster (Morgan 1974) and American lobster (McLeese and Wilder 1958) had a positive correlation with temperature and salinity.

Crayfish are caught using various active and passive catching methods. These methods include hand-catching, bait-free traps, bait rods, nets, electro fishing and diving. There is no single effective method under all circumstances. For this reason, the characteristics of environmental conditions and sources of crayfish should be taken into consideration in the selection of the appropriate fishing method (Ulikowski et al. 2017).

Nutritive Composition of Lobsters

Lobsters, like other crustaceans, are important creatures with high protein and low fat content in terms of nutritional value (Table 2.5). The chemical composition of European lobster (*H. gammarus*) and American lobster (*H. americanus*) differed according to body parts, sex, and species. While muscle and gonads are rich in protein, hepatopancreas is rich in fat; cholesterol and energy content (Barrento et al. 2009).

Lobsters are rich in copper, selenium, zinc, phosphorus, vitamin B_{12}, magnesium, and vitamin E (Ayanda et al. 2018). The contents of vitamin A, Vitamin D_2, Vitamin D_3, and alpha tocopherol for *A. Leptodactylus* were reported as 0.43, 0.54, 0.26, and 1.19 μg/g, respectively (Harlioglu et al. 2012).

Crustaceans, shrimps, crabs, and lobsters are considered important sources of heavy metals for consumers. It has been discovered that they have high sensitivity to metals and can accumulate high concentrations of metal from their environment. These elements are ultimately transferred to higher levels of the food chain and finally to human beings.

The level of fatty acids varies between species. In one study, *H. americanus* was found to contain more MUFA, especially eicosenoic acid and less n-6 fatty acids than *H. Gammarus*. The main n-3 PUFA in all tissues were eicosapentaenoic (EPA, 20:5n-3) and docosahexaenoic (DHA, 22:6n-3) (Barrento et al. 2009). Lipids of Norway lobster (*N. norvegicus*) contain 42–48% PUFA, 26–35% MUFA and 23–27% SFA (Rosa and Nunes 2004). Samiee et al. (2017) found that the dominant omega-3 fatty acids in liver and muscle tissues of *Panulirus homarus* are Eicosapentaenoic acid (24.32–25.17%) and Docosahexaenoic acid (4.62–6.32%). Second dominant fatty acid was palmitic acid (19.18–22.14%). Baghlani et al.

(2018) found that palmitic, oleic, and docosahexaenoic acids were the predominant fatty acids in the meat of *Panulirus homarus*. In a study examining the fatty acids content of spiny-cheek crayfish, *Orconectes limosus*, which was caught from Goplo lake in Poland, it was reported that the highest recorded fatty acid was palmitic acid (C16: 0) from SFA group and oleic acid (C18: 1) from MUFA group. It has also been reported that the dominant group for spiny-cheek crayfish is PUFA, which accounts for approximately 49% of the total fatty acids and the highest rate of eicosapentaenoic acid (C20: 5 n3) (Stanek et al. 2011). Similar results have been reported for *A. leptodactylus* caught from different Turkish freshwater resources by Oksuz and Mazlum (2016).

Fatty acid content of lobster varies between edible tissues. The n-3 PUFA were found dominant in muscle (36.2%) and gonads (37.3%) of *H. gammarus* and *H. americanus*, compared to hepatopancreas (5.3%) (Barrento et al. 2009).

Other factors such as metabolism, moulting cycle, captivity duration, seawater temperature, and salinity as can also influence the distribution of lipids and fatty acids in crustaceans.

Essential amino acids (EAA), threonine, leucine, valine, lysine, and arginine are found in 40% of the raw meat of Norwegian lobster. In addition, non-essential lysine, alanine, and glutamic acid are present at levels of 57.9, 57.2, and 31.2 mg%, respectively (Rosa and Nunes 2004).

Crayfish, lobsters, and shrimps are known to have higher cholesterol content than other shellfish. The low-fat content of lobsters and crayfish has been reported not to cause an increase in blood cholesterol levels (Ladewig and Schaer 1993). Total cholesterol content of crustacean meat varies according to species, harvest period and tissues. Total cholesterol content in lobster meat is reported to range from 50 to 170 mg/100 g. In a study conducted with crayfish caught from Poland Goblo lake, total cholesterol content was determined as 64.84–72.11 mg/100 g and it was stated that the cholesterol content of crayfish caught in summer was higher than those caught in spring (Stanek et al. 2011).

Live Transport of Lobsters

Considering that live lobster is an invaluable export product, significant time and effort is being spent to develop storage methods and land and air cargo transport systems to ensure first-class access to the restaurant or consumer (FAO 2017).

The processed lobster market has grown significantly in recent years, with companies seeking to make their lobster products more accessible to consumers. Since live lobsters receive higher prices, more efforts are being made to ensure that species reach the market area alive (FAO 2017).

The transport of live lobster can cause stress and physical damage to animals, as well as significant differences in product quality at the destination. Factors that may cause stress include changes in temperature due to insufficient cooling in warm climates; insufficient temperature in cold climates; Low humidity; low oxygen; overcrowded; and rough handling. These stress factors can affect lobster health, reduce their quality, and even be lethal. There is still a view that the lobster that died during air transport can be eaten: While some sellers refused the sale of the dead, some of them use these for their cooked products. The risk from these products may increase related to the increase in the demand for raw and cooked lobsters in case poor hygienic conditions during production and trade (Tirloni et al. 2016). In addition, since lobsters are only animals and maintain their natural habitat, they can experience stress when placed in common environments. Protein levels and shell hardness in the blood of lobsters to be transported should be taken into consideration as they may change seasonally. Live lobsters should be kept cool during transport but avoid direct contact with coolants such as wet ice or gel packs as they may cause stress and death to the animal. Under the best shipping conditions, the moisture content in the packaging box will be about 70%. It has been reported that the most common transport method for live lobsters is expanded polystyrene coated liner boxes. An absorbent pad should be placed under the box. Lobsters can be separated by moist paper, chilled wet sponges, or gel packs. Seaweed is also used by some transporters, but it can be harmful to lobsters as some algae species may produce harmful gases in case of degradation. During transportation, the lobster loses weight and begins to accumulate nitrogenous waste, including ammonia. To prevent

Table 2.5 Proximate composition of different lobster species (g/100 g)

Species	Moisture	Protein	Fat	Ash	Reference
Body tissue					
Homarus gammarus	78.1–79.2	17.6–18.3	0.3–0.5	1.8–2.0	Barrento et al. (2009)
Homarus americanus	79.2–80.5	15.6–17.1	0.6–0.7	1.8–1.9	Barrento et al. (2009)
Procambarus clarkii	82.15	15.22	1.29	1.18	El-Sherif and El-Ghafour (2015)
	–	13.88	1.76	1.52	Zaglol and Eltadawy (2009)
	76.60	19.77	1.99	1.45	El-Kholie et al. (2012)
	82.7	15.6	0.59	1.51	El-Mossalami and Emara (1999)
	80.4	17.1	–	1.16	Alford et al. (2016)
Astacus leptodactylus	78.08–81.05	14.91–18.52	0.41–0.53	1.14–1.56	Gurel and Patır (2001)
	–	11.78–17.59	3.29–5.82	0.91–1.74	Berber et al. (2014)
	79.77	16.39	0.45	1.25	Gurel Inanlı and Coban (2007)
	81.76	13.96	0.53–1.31	1.72	Erkan et al. (2009)
	78.19	14.61	0.57	1.39	Harlioglu et al. (2012)
	80.36–82.06	15.41–17.46	1.09–1.62	0.64–0.97	Oksuz and Mazlum (2016)
Unspecified	71.63	24.10	1.03	2.28	Ayanda et al. (2018)
Unspecified		19.0	0.9	–	U.S. Dept. of Agriculture (USDA) (2019)

this from happening, live lobsters should not be fed since a few days prior to transport. Due to the danger of depletion of the animal's muscle tissue, care should be taken not to starve for a long time (FAO 2017).

If the lobster gills remain wet or moistly, they can live for 4–5 days out of the water. The best environment in which they breathe most easily is where the humidity approaches 100%. Since oxygen consumption increases as the temperature rises, cooling is essential during transportation. A temperature of 0–5 °C provides the optimum environment. Jussila et al. (2013) reported that cooling crayfish immediately after harvest will reduce the likelihood of stress-related death during transport stages until later delivery to the market. The same researchers reported that crayfish deaths were high during subsequent transport and holding practices due to prolonged catching hours and hot summer days. From this point of view, these researchers examined the factors affecting the survival of crayfish (*Pacifastacus leniusculus*) during transport on board and developed an on-board transport method accordingly. In the summer months, they conducted a study on Saimaa Lake in Finland and found that the critical factors for the survival of the crayfish are water temperature during catching and air temperature during on board transport. They have developed a new method of rapid cooling to reduce the deaths caused by high temperatures and compared this with the conventional method. With this cooling method, they lowered the crayfish temperature from ambient temperature to 5–7 °C and achieved 100% survival rate.

Lobster colour affects the price of live export of lobsters. Dark red lobsters are sold at a higher price than pale ones. Red-coloured lobsters have been reported to be in high demand in the Chinese market, which covers 90% of the Australian market (Wood et al. 2007). Lobsters are also kept alive in restaurants and therefore appearance is critical. Pale-coloured lobsters are either sold at discounted prices or evaluated for processing into cooked products (Harrison 2004). Chandrapavan et al. (2009) reported that the colour of lobsters is associated with the depth of hunting. They reported that they observed dark red colour in shallow depths and pale colour dominance in deeper waters.

Lobsters are usually shipped in wooden cages in Canada and America. Their trees are selected from loosely structured species that allow some air circulation. Wet sacks can also be a good alternative. Crushed ice is filled into and around the cage and wrapped. Appropriate holes must be drilled to release the melting ice. Refrigerated trucks are ideal for this type of shipment. The main problem arises when transporting lobster in open vehicles and in hot weather. Long distances need to be shipped by plane. For all shipments up to 24 h or in air cargo, is generally cardboard packaging insulated with Styrofoam of 20–25 kg is generally used (Estrella 2002).

Killing of Lobsters

Crustaceans show consistent reactions with signs of pain and sadness. They also have the cognitive capacity to remember and avoid unpleasant stimuli. For this reason, they should be caught, handled, stored, transported, and killed humanely.

Killing involves loss of sensitivity, followed by death. The animal should be desensitized without discomfort and pain. Lobsters should be done insensible using the following methods before killing (RSPCA Australia 2017).

Electrical Stunning

Sufficient electrical current to be applied to the lobster can render insensitive within 1 s. In the UK, Simon Buchaven developed "crustastun" (is a device designed to administer a lethal electric shock to shellfish before cooking). It is reported that the application of electric current of 110 V and 2–5 A by this device prevents pain and distress by destroying the nervous system of the animal (Yue 2014).

Chilling

Crustaceans enter a torpor state at temperatures of 4 °C and below. They become insensitive when body temperatures drop sufficiently with cooling. Although the relationship between the absence of stress and pain and cooling has not been fully established, it has been reported that the cooled lobster does not exhibit the stressful behaviour of those killed by the other method. One major benefit of chilling is that it reduces mobility. This makes crustaceans easier to handle and humanely kill, and prevents individuals from injuring each other (RSPCA Australia 2017).

Splitting

All nerve centres of crustaceans with multiple nerve centres should be destroyed. Lobster has a chain of nerve centres running through longitudinal midline of animal. Splitting is the fast cutting of the centre line of the lobster head, abdomen, and thorax with a large and sharp knife (Yue 2014).

Spiking

Spiking is not suitable for lobsters because of its long chain of nerve centres. Crabs are suitable for this method.

Boiling

The most common method is to leave live lobsters in boiling water. Although this process kills the animal, it causes physiological shock and autotomy. In addition, the lobsters struggle for about 2 min after being placed in boiling water. Therefore, this method is not considered as humanely method.

Drowning

Lobsters can be killed by changing salinity. When they are immersed in fresh water, they slowly die as salt may be lost from their blood. Hyposaline conditions prevent loss of extremity in lobsters.

In a study, crayfish *Astacus* (*Astacus astacus*, and *Astacus leptodactilus*) and American lobster (*Homarus americanus*) were anesthetized using four different methods (addition of $MgCl_2$ to the holding water, cooling, bubbling holding water with CO_2 and electric stunning). Neuronal responses of the animals were measured via implanting electrodes in central nervous system (CNS) According to the results of the study, it was determined that cooling with $MgCl_2$ (10%) did not anaesthetized lobster and crayfish, and CO_2 was partially effective due to low pH in water. On the other hand, cooling did not show any anaesthetic effect. Anaesthetic effect of electric stunning was determined (Fregin and Bickmeyer 2016).

2.3.3 *Quality Changes in Lobsters*

Shellfish have a short shelf life due to their small size, lack of intestine removal, high levels of non-protein nitrogen compounds, polyunsaturated fatty acids and melanin pigments. Thus, post-mortem autolytic changes, microbial spoilage, melanosis development and lipid oxidation are affected quality of crustaceans and limited their shelf life (Rosa and Nunes 2004; Ashie et al. 1996b).

Lobsters are marketed as whole or as tails. Lobster tails are sometimes removed after catch and the heads discarded. It is thought that tailing process can affect the quality and the shelf-life of the product. It is reported that if the lobsters are damaged during catching in Icelandic fishery, muscle disruption called "skyrhumar" can be seen in Norway lobsters. This has been reported to be caused by the release of proteases from the hepatopancreas of the nephrops into the abdominal muscles (tail flesh). It is stated that this process is related to the crushing of animals that may cause crushing of cephalothorax during trawling and releasing hepatopancreas enzymes to abdominal muscle (Neil 2012). Accordingly, removal of the head (containing the hepatopancreas) from the tail should reduce the onset of "skyrhumar". Especially after catching, tailing is very important in terms of onset of "skyrhumar" at sea. In one study, lobster tails which were tailed immediately after being caught were first cooled at 0 °C and then frozen at −24 °C. For comparison purposes, the lobsters were taken from the same catch and tailed in a processing plant on land and frozen. As a result of the research, the "skyrhumar" index of the lobsters whose was tailed immediately after catching was found to be very low (Gunnarsson 2010). Removal of the internal organs is effective in reducing the bacterial load of some seafood. Tailing can also be thought to be effective in reducing the internal bacteria in the stomach of lobster. In this study, it was determined that tailing had no effect on the total plate count of Norwegian lobster. Although the stomach was removed by tailing, it was stated that the intestinal tract was not removed, and the tailing was not effective in reducing bacteria in lobster (Bekaert et al. 2015).

Stress during capture has a significant impact on the animal's muscle quality. When different nucleotide profiles and other related metabolites were examined, it was determined that the capture method had a significant effect on the energy state of the abdominal muscles of Norwegian lobsters. In this study where the effect of trawling and creel catching on the physiological state of the lobster was examined, a better physiological state was observed in the lobster catching with creel. Creeled lobsters managed to maintain muscle ATP levels; although a tendency for lower arginine phosphate was detected when compared with rested undisturbed lobsters. However, larger changes were observed in trawled animals. Norway lobsters, in order to escape from the trawl, were immediately using anaerobic metabolism to obtain energy, a situation that led rapidly to their exhaustion. It has been reported that after 15 min in the trawl, the animals are already exhausted and no further physiological changes are observed as the trawl time increases. Stress-related and anaerobic metabolites together with muscle pH indicated that trawled animals, even at the shortest time tested, were using anaerobic metabolism (Albalat et al. 2009).

To maintain the quality of lobsters, the following points should be considered after catching. The deck where the lobsters are brought in should be kept clean. Lobsters should be carefully removed from the pots as they may lose their claws and legs due to stress. Any bleeding or damage to lobsters will reduce the chance of survival during transport and storage. To prevent damage, lobsters should be carefully placed in tanks or boxes. Damaged lobsters should be returned to the sea immediately. Lobsters should be placed in boxes covered with water-absorbing and porous materials such as hemp sacks. These boxes should be stored in a place protected from wind and rain, not exposed to direct sunlight. Lobsters should never be placed in a tank with still water. As the oxygen in the water will be exhausted rapidly and wastes will be formed in the water, it will have toxic effect. The lobsters should be banded by hand to prevent claw losses. Recently moulted lobsters should be kept separate until they harden. Otherwise they may be damaged (Irish Sea Fisheries Board 2019).

Rapid microbial degradation during post-mortem storage such as in fish is an important problem in the processing of cooked lobster. Components such as non-protein nitrogenous compounds and free amino acids in crayfish are a source of nutrients for microbial growth. As a result of microbial, enzymatic and oxidative processes, lobster undergoes quality changes and loses its freshness. This significantly reduces its market value (Cremades et al. 2011). *Micrococcus*, *Staphylococcus*, *Alcaligenes* and *Achromobacter* have been reported to be dominant strains in peeled crayfish (*Procanbarus clarkii*) tails stored at 0 and 5 °C, and *Pseudomonas* and *Achromabacter* have been described as spoilers (Cox and Lovell 1973).

Melanosis Formation in Lobsters

Melanosis, also known as blackspot, is an important cause of loss of lobster quality during iced storage. Blackspot occurs in shellfish during storage as a result of the action of polyphenoloxidase (PPO). PPO oxidizes diphenols to quinones, which undergo autoxidation and polymerization to form dark pigments. PPO is responsible for the sclerotization (hardening) of the exoskeleton after moulting. Cuticle hardening occurs when the quinones produced from diphenol oxidation form cross-links between adjacent protein chains (Stevenson 1985). Factors affecting blackspot in shrimps and lobsters include method of capture, catch handling, age, and sex. polyphenoloxidase (PPO) activity was measured in cuticles of Norwegian lobsters caught over a 7-month period. It was observed that PPO activity increased in early August and early October, but the high PPO activity detected in May showed the spring moulting period. Exposure to damage and stress of Norwegian lobsters was associated with an increase in PPO activity. No correlation was observed between initial PPO activity and black spot development during the storage of Norwegian lobsters. This showed that biochemical events during storage were more than the initial PPO level for black spot development. The times during the year during which the Norway lobsters are more likely to moult are dependent on sex and maturity, but also differ from region to region. On reaching sexual maturity, male Norway lobsters moult twice a year until they are 6–7 years old, whereas females moult once, or not at all, in a given year. Longer trawl hauls and rough handling on board would lead to the loss of claws, whereas, conversely, short hauls and gentle handling of the catch on board would preserve more Norway lobsters intact. Thus, missing claws reflect rough treatment of the Norway lobsters both during and after the haul. A linear correlation was found between the number of remaining claws on the Norway lobsters and PPO activity. This suggests that rough handling of the Norway lobster catch is associated with an increase in PPO activity in the lobsters (Bartolo and Birk 2002).

Melanosis has been reported to occur due to the presence of high levels of polyphenoloxidase (PPO) in cephalotorax of lobsters (Martínez-Alvarez et al. 2008b). Rough handling of lobsters and other traumatic events trigger the defence mechanism of these organisms having PPO activity, resulting in increased blackening. Even when they are alive, they can be encouraged to develop melanosis because of any injury (Ogawa et al. 1984). The blackening of the broken clamped legs, parapods and carapax is caused by hemocyanin from PPO (Gimenez et al. 2010). Inhibition of melanosis catalyzed by polyphenoloxidases is inhibited by bisulphites by reaction with quinones forming sulphoquinones (Ferrer et al. 1989).

In various studies, different inhibitors were used for inhibition of melanosis in lobsters and positive results were obtained. Brack et al. (2008) found that Mimosine inhibits monophenol and diphenoloxidase activity of European spiny lobster (*Palinurus elephas*). Opoku-Gyamfua et al. (1992) reported inhibition effect of EDTA (ethylenediamine tetra acetic acid) on PPO from lobster (*Homarus americanus*). In another study, it was reported that dusting of Norwegian lobster (*Nephrops norvegicus*) with sulphides delayed the formation of melanosis in the chilled storage for at least 7 days (Martínez-Alvarez et al. 2007, 2008b). The combination of organic acids and chelating agents and 4-hexyresorcinol was found to inhibit the PPO activity of Norwegian lobster (*Nephrops norvegicus*) (Lopez-Caballero et al. 2006).

In addition to being valuable and delicious, lobsters are products that can easily undergo quality changes and therefore easily perish. Biogenic amines can be formed by decarboxylation of free amino acids that are formed by decarboxylation of low molecular weight amino acids and are naturally present in tissues or post-mortem. Lobsters can be rejected if high levels of biogenic amines are detected (Arulkumar 2018).

Live Storage of Lobsters

The main advantage of live storage is that lobsters can be kept under optimal conditions. Another advantage is that it can be purchased when it is abundant and relatively inexpensive and then stored for sale when prices increase. Live storage also provides marketing and sales flexibility. Various systems for storing lobsters are available. These changes from the simple ponds to the bays on the shores, areas created at sea, and areas with recirculation system using artificial sea water at land. They are suitable for holding a small number or many lobsters for periods ranging from several days to several months. They are even kept in decorative units in hotels and restaurants. The ponds and bays are generally geographically isolated areas and the lobsters are temporarily stored there before being shipped to the market. The survival of lobsters during storage depends on the quality of the water in which they are found. If the water quality is poor, then the survival rate will decrease and there will be an increased risk of mass deaths. For successful lobster storage, water quality factors should be kept in balance at the same time. Negative changes in any of these conditions will undoubtedly affect the survival of the lobster. The main water quality factors for storing lobster are oxygen, temperature, and salinity. On the other hand, waste materials such as ammonia excreted by the lobsters are also considered as water quality factors (Beard and McGregor 2004).

Boiling Live Lobsters

The most popular and traditional way to cook live lobsters is boiling and steaming. If you want to bring lobster to the table as whole, cooking is the most appropriate method. Boiling is a bit quicker and easier. Separating of meat from the shell is easier with boiling compared to the steaming. In contrast, steaming is gentler, yielding slightly more tender meat. It preserves a little more flavour and it is more forgiving on the timing front. It is better to cook lobsters before chilling or freezing than to store them raw. Cooking is carried out by immersing lobsters placed in wire baskets in water containing about 2–3% salt. Since the lobsters added to the water reduce the water temperature, the cooking time should be measured from the moment the water temperature returns to the boiling point. The cooking time varies depending on the size of the lobsters and usually takes 15–30 min. Weight loss of 5–25% occurs during cooking. Once the lobster is cooked, it must be removed from the water. In fact, the lobster continues to cook in its shell even when it is removed from the cooking water. So, it is best to take it out just before cooking. After cooking, the lobsters should be cooled down quickly therefore they place in ice water bath. If lobster is not chilled fast enough, it loses as much as 12% of its weight. Rapid chilling after it is cooked prevents this weight loss. If lobsters are to be served cold, they should be kept in the refrigerator until use. The lobster should be stored on its back to prevent fluid loss (Anonymous 2019). Cooked crayfish tails generally have a limited shelf life because of endogenous proteolytic processes, microbial spoilage and oxidative processes such as the lipid and protein oxidation associated with physicochemical changes and off-flavours All of this causes a loss in freshness and lowers the quality of the product, causing a market loss of crayfish. Therefore, the development of storage procedures that reduce or delay the loss of freshness and the quality of cooked crayfish tails is important to the crayfish industry (Cremades et al. 2011).

Refrigeration of Live Lobsters

The establishment and maintenance of lobster holding systems is vital for the lobster industry. The simplest type of lobster holding tank system is an open or flow-through system. Lobsters may also be held in a closed or recirculating system. The tank should be kept as cold as possible, as long as it's above freezing. Because water retains more oxygen at low temperatures, lobsters need less oxygen and become less active. Thus, cannibalism and the threat of disease are reduced. Low temperatures allow lobsters to be processed more easily. Waste quantities will also be reduced due to their reduced activity. The gradual reduction in temperature and fasting time should be considered as pre-treatments for live transport, as animals can protect themselves from waste products under these transport conditions (Jacklin and Combes 2007b).

2.3.4 *Lobster Processing and Preservation*

Lobster is consumed all over the world, but the producer country is limited to a few. The export market of lobster is increasing rapidly in the world. Lobsters are generally preferred alive by the consumers. However, there are some problems in the export of live lobster, such as high cost and the high mortality rates and losses during transport. In contrast, processed lobsters have many advantages such as ease of handling and storage, long shelf life, easy supply, and ease of preparation of products and added value. Therefore, lobster producing countries process more than half of the lobsters they harvested. Lobsters are commercially processed into various products such as fresh lobster meat, picked lobster meat, canned lobster, lobster medallion, whole cooked lobsters, and frozen lobsters (Nguyen et al. 2017).

The most important part of the crayfish in the processing industry is the tail. On the other hand, arms meats are also used. It is reported that tail meat constitutes 9–13% of the body weight of the crayfish. In crayfish plants, the tail meats are usually cooked in boiling water for a few minutes, then the shell is removed, as well as the intestines are removed from the tail meat and then frozen in packs (Fard et al. 2015).

Pre-treatments in Processing of Lobsters

Selection

Selection is the sorting process that separates lobsters to be kept alive and discarded. The selection process is essential for maintaining quality. Sick or dead lobsters should be removed. Because they will die in a short time and then deteriorate. The easiest way to separate dead and sick animals is to keep lobsters individually in separate tanks or tank-like sections. Recently moulted lobsters should be removed from the storage tank due to cannibalism risk. Even if these lobsters survive, their value will be reduced as their legs and antennae will be eaten (Jacklin and Combes 2007b).

Grading

Grading is a selection process that separates lobsters into sizes and genders to meet market demands. In the sale of live lobsters, the buyer may have specifications for size and gender. In this case, such requests can easily be met by the storage facilities organizing separate tanks for certain grades (Jacklin and Combes 2007b).

Chilling of Lobsters

Crustaceans must be chilled to the point of insensitivity (unconsciousness) prior to being killed. They should be stored in a refrigerator at approximately 5 °C, both for food safety reasons and for the welfare of the animal. If lobsters are kept at −10 to −18 °C for 45 min they become unconscious but do not freeze.

It is reported that in *N. norvegicus*, the storage temperature is the most crucial factor affecting microbial growth, microbial activity, and spoilage potential. Therefore, N. *norvegicus* tails should be stored in a cool environment immediately after catch (even prior to 'tailing') and should be stored in ice (at 0 °C) as soon as possible after on-board processing/handling. Once the tails are stored this way, great care should be taken not to warm up the product during later transport and processing. In this way, bacterial growth may be slowed, the spoilage activity and spoilage potential of the intrinsic spoilage-associated microflora may be reduced, and thus, shelf life might be prolonged (Gornik et al. 2011).

Nephrops norvegicus was stored at 0 °C, 5 °C and 20 °C for 10, 10 and 7 days respectively. During the storage quality of lobster was investigated. Lobsters stored at 0, 5 and 20 °C were rejected sensorially after 48, 72 and 96 h respectively. *Pseudomonas* sp., H_2S producing bacteria and Enterobacteriaceae caused microbial spoilage. TVB-N was stated as a reliable quality indicator for Norway lobster (Boziaris et al. 2011).

Juvenile Australian red claw crayfish (*Cherax quadricarinatus*) packed in sealed freezer bag were stored on crushed ice (0 °C) in a cold room of 5 °C for 14 days. According to the results of the study, lipid oxidation and protein denaturation were determined in cold storage in red claw muscle. Cooking yield and tenderness decreased with storage during storage. It is reported that crayfish should be kept in less than 7 days on ice as in retail conditions (Tseng et al. 2002). Lobster (*N. norvegicus*) tails (within their shells) were placed in plastic resealable bags and either kept continuously on ice or held at 16 °C for 4, 8 or 24 h prior to storage on ice. The effects of icing delays on the quality of tail meat were evaluated. Changes in *K*-values, microbial load, muscle pH and TMA indicated that the delay to icing should be no more than 4 h (at 16 °C) to ensure that quality is not compromised during subsequent post-harvest storage (Albalat et al. 2011). Delays before icing and inadequate icing increase the commercial rate of rejection. Trawled Nephrops are highly perishable and need to be promptly iced at sea and landed soon after catching to ensure high yields of best quality product (Anonymous 1994).

Studies on the preservation of lobsters in the cold have shown that not only raw lobsters but also cooked lobsters can be stored in cold conditions. In one study, the crayfish (*P. clarkii*) were washed with tap water, then cooked in boiling water at 95–100 °C for 180–190 s and then cooled rapidly with cold water. After the peeling, the crayfish were packed under modified atmospheres and then stored at 2 °C. According to the results of the study, it was determined that the shelf life of crayfish packaged in modified atmosphere can be increased from 6 to 12 days at 2 °C compared to the samples packed in atmospheric air used as control (Cremades et al. 2011). In a similar study conducted on precooked and peeled red claw crayfish (*Cherax quadricarinatus*), modified atmosphere packaging almost totally inhibited the growth of aerobic bacteria and coliforms in crayfish stored at 2 °C compared with aerobic and vacuum packaging (Chen et al. 2007; Chen and Xiong 2008). Furthermore, it has been reported that the shelf life of *N. norvegicus* can be extended from 6.5 to 13 days when packaged in MAP (10% O_2, 80% CO_2 and 10% N_2) and storage temperatures do not exceed 1 °C (Gornik et al. 2013).

Rapid chilling of seafood after harvest is important for both quality and safety. Icing is the most common cooling method for seafood. Slurry ice has been used in recent years in addition to traditional icing methods for fast and effective cooling. Slurry ice also has the advantages of not causing physical damage, being able to cool to lower temperatures, increasing the shelf life of the product and reducing dehydration. It has been found that Norwegian lobster (*Nephrops norvegicus*) stored on slurry ice maintains its sensory quality better and has a longer shelf life compared to those stored with flake ice (Aubourg et al. 2007; Losada et al. 2006).

Freezing of Lobsters

Freezing is an important method for preservation of seafood since frozen product can be stored for long periods by preventing microbial spoilage. Freezing inhibits the activity of microorganisms that cause food spoilage and food spoilage. Lowering the temperature below freezing inhibits the growth and activity of most microorganisms. Decreasing the storage temperature greatly slows down enzymatic and biochemical reactions that normally occur in unfrozen foods. A good freezing process followed by cold storage ensures that shellfish are kept for months or even a year or longer.

Freezing lobster is attractive for restaurant operators and supermarkets because it makes it easier to buy and stock lobster. It is better to freeze them because it is difficult to keep the lobsters alive until they are consumed. Freezing also makes it easier to ship lobsters to more places: sellers do not have to worry about lobsters dying if the temperature changes or a truck or plane is delayed. Because processors can stockpile frozen lobster more readily, the price also tends to be more stable (Porter 2019).

An important problem in the freezing of live lobsters is the adhesion of meat to the shell and the difficulty in removing the shell by consumers. Therefore, it is recommended to heat the lobsters before freezing. Thus, the destruction of the tissues that connect the flesh to the shell occurs (Work et al. 1997). In order to activate the protection of the freezing process, suitable pre-treatments must be carried out according to the type of crustacean. The pre-treatment for lobsters is usually cooked in brine solution. The lobsters are cooked in 3% saline for 10–20 min. In some lobster species, the tail is separated from the body to remove the intestines. Frozen lobsters are reported to have a shelf life of 3 months at a temperature of −23 °C or less (Noomhorm and Vongsawasdi 2004).

The effects of freezing methods on the quality of lobster are important. Rapid freezing has been recommended to prevent the formation of large ice crystals. The large crystals can damage cells and allow the loss of moisture during the frozen storage and upon thawing. Slow freezing result in an inferior quality product due mainly to protein denaturation. The design of the freezer depends largely on the type of product to be frozen, the duration of freezing and the amount of raw material. Various freezing methods are currently available. These are sharp freezing, plate freezing, blast freezing, cryogenic freezing, liquid nitrogen freezing, carbon dioxide freezing, spray or dipping freezing and pressure shift freezing. These methods have advantages and disadvantages and therefore the appropriate method for the product should be determined (Noomhorm and Vongsawasdi 2004). It has been reported that cooked crayfish (*Procambarus clarkii*) can be stored at −20 °C for 6 months after freezing at different freezing rates (Zhao et al. 2014).

In a study (Work et al. 1997), soft- and hard-shell lobsters (whole, claw and tail) were cryogenically frozen as using liquid CO_2 and compared their quality during frozen storage for 9 months. In hard shell claw meat, higher lobster flavour was determined with soft shell claw. The soft-shell claws contained more moisture. During storage, soft- and hard-shell tail meats had a good lobster flavour. Higher tenderness scores were obtained for soft shell tails.

In the other study, a part of the lobster krill (*Munida* spp.) was packaged in polyethylene bags and immediately frozen at −30 °C. Other part was divided into two groups. The first group was immersed into water and the second group was immersed into sodium metabisulphite solution. All samples were packaged in polyethylene bags and stored at −30 °C. After 2 days, at all bags were placed at −18 °C. According to the results of the study, it was determined that pre-immersion in SMB on one hand increased sensory acceptance and prevented lipid hydrolysis, on the other hand increased trimethylamine formation (Garcia-Soto et al. 2015). Similar good results were obtained with soft-shell lobster (*Homarus americanus*) treated with sodium tripolyphosphate (STP) prior to freezing. Lobsters previously were injected with STP then cooked in a steam cooker followed by frozen with liquid nitrogen. As a result, it was determined that STP treatment increases the shelf life of lobsters, preserves texture colour and flavour properties, and reduces drip loss (Calder et al. 2005).

In another study, the crayfish (*Astacus leptodactylus*) caught from the reservoir lake of Aras dam were immediately transported to the laboratory with ice, washed with water, their tail peeled, and their tail fillets removed. Fillets placed in polyamide bags were packaged by exposure to air and stored at −18 °C for 6 months in a freezer. The results showed that freshwater crayfish tail fillets packaged in air can be stored maximum 5 months at −18 °C (Fard et al. 2015).

Unfavourable conditions may occur in freezing of lobsters in the shell, such as toughening in meat, off-flavour in storage, and difficulty in separating meat after thawing. In lobster, blue or black discoloration is one of the most troublesome problems. After freezing or during frozen storage, blueing may occur. This phenomenon may even occur after it has been thawed and subsequently exposed to air or shortly after cooking. This bluish black change can be caused by biuret type reactions between copper pigments and heat denatured protein. Other discoloration problems found in lobster meats are yellowing and fading of the red or orange-red carotenoid. Both indicate oxidation during processing or prolonged storage. Air exposure time, temperature, freezing and storage conditions are effective on this change. Frozen lobsters gradually lose their juice during freezing and frozen storage. This leads to textural changes. In addition, protein denaturation during freezing and storage leads to textural changes (Noomhorm and Vongsawasdi 2004).

Canning of Lobsters

Boiling is the first stage of processing lobsters for canning. Live lobsters are placed in a basket and then placed in boiling water. Boiling water usually contains 3% salt. After 10–30 min of boiling, the lobsters are cooled in 2–6% chilled brine. Boiling time varies according to lobster size. Once the lobsters have cooled sufficiently, claws and tails are severed. After the body shell is opened, the stomach, liver and coral are removed, and the body is removed from the shell. The claws are broken, and the meat is taken as whole as possible. Tails are split and intestines are pulled. In the canning process, lacquered tin cans are used to package lobsters. Usually, the arm meat and tails are placed at the bottom of the can, while the claws meat is placed on top. Then a small amount of salt is added. If the lobsters are to be packaged in cold boxes, the cans must be heated before closing. Heating is carried out by passing the cans through an exhaust tunnel, thereby achieving a temperature of 60 °C in the centre of the cans. Vacuum sealing of the cans is recommended using a mechanical vacuum container sealing machine (Anonymous 2016).

References

Abu Bakar F, Salleh AB, Razak CNA, Basri M, Ching MK, Son R (2008) Biochemical changes of fresh and preserved freshwater prawns (*Macrobrachium rosenbergii*) during storage. Int Food Res J 15(2):181–191

Addison RF, Ackman RG, Hingley J (1972) Lipid composition of the queen crab (*Chionoecetes opilio*). J Fish Res Board Can 29:407–411

Aguilera LM, Chiralt A, Fito P (2003) Food dehydration and structure. Trends Food Sci Technol 14:432–437

Akintola S, Bakare S (2011) Microbiological changes in freshwater prawn (*Macrobrachium vollenhovenii*, Herklots 1857) stored in ice. Am J Food Technol 68:500–506

Akonor PT, Ofori H, Dziedzoave NT, Kortei NK (2016) Drying characteristics and physical and nutritional properties of shrimp meat as affected by different traditional drying techniques. Int J Food Sci Article ID 7879097:5 p

Aladaileh S, Rodney P, Nair SV, Raftos DA (2007) Characterization of polyphenoloxidase activity in Sydney rock oysters (*Saccostrea glomerata*). Comp Biochem Physiol B 148:470–480

Albalat A, Gornik SG, Atkinson RJA, Coombs GH, Nel DM (2009) Effect of capture method on the physiology and nucleotide breakdown products in the Norway lobster (*Nephrops norvegicus*). Mar Biol Res 5:441450

Albalat A, Gornik SG, Mullen W, Crozier A, Atkinson RJA, Coombs GH, Neil DM (2011) Quality changes in chilled Norway lobster (*Nephrops norvegicus*) tail meat and the effects of delayed icing. Int J Food Sci Technol 46:1413–1421

Alford AB, Schilling MW, Kaminski RM (2016) Consumer acceptability and composition of crayfish harvested from commercial production fields and moist-soil wetlands. J Aquat Food Prod Technol 25(6):875–884

Al-Maliki GM, Al-Khafaji KK (2018) Comparativestudy of chemical composition of three species of commercial shrimp *Penaeus semisulcatus* (De Haan, 1844), *Metapenaeus affinis* and *Parapenaeopsis stylifera* (H. Milne-Edward) in North West Arabian Gulf. Oceanogr Fish Open Access J. MS ID 555725

Alpbaz A (1993) Crustaceans and Arthropods culture. Ege University Fisheries Faculty Publications No:26. İzmir, pp 300–302

Alpbaz A (2005) Aquaculture. Alp Publications, Izmir, pp 331–343

Amegovu AK, Serunjogi ML, Ogwok P, Makokha V (2012) Nucleotited degradation products, total volatile basic nitrogen, sensory and microbiological quality of nile pearch (*Lates niloticus*) fillets under chilled storage. J Microbiol Biotechnol Food Sci 2:653–666

Anderson CL, Pedersen RK (1951) The preservation of fish by smoking and drying. In: Tressler DK, Lemon JM (eds) Marine products of commerce. Reinhold Publishing, New York, pp 394–417

Ando M, Nakamura H, Harada R, Yamane A (2004) Effect of super chilling storage on mof freshness of kuruma prawn. Food Sci Technol Res 10(1):25–31

Annie T, McCarron P (2006) Lobster market overview. Maine Lobstermen's Association

Anonymous (1972) Handling and processing shrimp. Torry Advisory Note No: 54. Department of Trade and Industry Torry Research Station. 18 p

Anonymous (1981) Standard for canned crab meat CXS 90-1981. Codex Alimentarius International Food Standards

Anonymous (1993) Canned—canned shrimp standard. TSE 6015 (in Turkish)

Anonymous (1994) The effect of icing on the quality of trawled whole Nephrops. SeaFish house, 1994 / 29 / 1st edition

Anonymous (2016) Processing procedures for canned food products. A Complete Course in Canning and Related Processes (Fourteenth Edition). Woodhead Publishing Series in Food Science, Technology and Nutrition. pp 231–265

Anonymous (2019) Atlantic lobster. Food Service Guide. http://lobstercouncilcanada.ca/wp-cvontent/uploads/2017/03/Atlantic_Lobster_Guide_P12_compressed.pdf. Accessed 6 Sept 2019

Anthony JE, Hadgis PN, Milam RS, Herzfeld GA, Taper LJ, Ritchey SJ (1983) Yields, proximate composition and mineral content of finfish and shellfish. J Food Sci 48:313

Arancibia MY, Lopez-Caballero ME, Gomez-Guillen MC, Montero P (2015) Chitosan coatings enriched with active shrimp waste for shrimp preservation. Food Control 54:259–266

Arason S, Nguyen MV, Thorarinsdottir KA, Thorkelsson G (2014) Preservation of fish by curing. In: Boziaris IS (ed) Seafood processing, technology, quality and safety. Wiley, Blackwell, West Sussex, pp 129–160

Artuz ML (2005) Türkiye Denizlerinde Bulunan Karides Türleri Üzerine Etüt. Zoo-Natantia, Publications Scientifiques, 22 p

Arulkumar A (2018) Evaluation of psychrophilic, mesophilic, histamine forming bacteria and biogenic amine content in the muscle of mud spiny lobster, *Panulirus polyphagus* (HERBST, 1793) during ice storage. J Food Saf 39(1):e12582

Ashie INA, Simpson BK, Smith JP (1996a) Mechanisms for controlling enzymatic reactions in foods. Crit Rev Food Sci Nutr 36:1–30

Ashie I, Smith J, Simpson B (1996b) Spoilage and shelf-life extension of fresh fish and shellfish. Crit Rev Food Sci Nutr 36:87–121

Atar HH, Olmez M, Bekcan S, Seçer S (2001) A research on efficiency and nutrient content of blue crab (*Callinectes sapidus* Rathbun 1896). Food 26(3):195–201. (In Turkish)

Atay D (1997) Shellfish and production technique. University of Ankara Publications. No: 1478. Textbook: 441. Ankara

Atkinson DA, Sim TC, Grant JA (1993) Sodium metabisulphite and SO_2 release: an under-recognized hazard among shrimp fishermen. Ann Allergy 71(6):563–566

Attala OA (2012) Impact of using organic acids and sodium sulfite on the quality of unpeeled shrimp. World J Fish Mar Sci 4(3):284–289

Atul S, Chen CR, Nguyen VL (2009) Solar-energy drying systems: a review. Renew Sust Energ Rev 13:1185–1210

Aubourg SP, Losada V, Prado M, Miranda JM, Barros-Velázquez J (2007) Improvement of the commercial quality of chilled Norway lobster (*Nephrops norvegicus*) stored in slurry ice: effects of a preliminary treatment with an antimelanosic agent on enzymatic browning. Food Chem 103:741–748

Ayanda IO, Dedeke GA, Ekhator UI, Etiebet MK (2018) Proximate composition and heavy metal analysis of three aquatic foods in Makoko River, Lagos, Nigeria. J Food Qual Article ID 2362843:6 p

Aydın I, Gokoglu N (2014) Effects of temperature and time of freezing on lipid oxidation in anchovy (*Engraulis encrasicholus*) during frozen storage. Eur J Lipid Sci Technol 116:996–1001

Azam KMS, Mansur MA, Asadujjaman M, Rahman M, Sarwer MG (2013) Quality and safety aspects of fresh and frozen prawn (*Macrobrachium rosenbergii*), Bangladesh. Am J Food Sci Technol 1(4):77–81

Baboli J, Velayatzadeh M, Roomiani L, Khoramabadi A (2016) Effect of sex and tissue on fatty acid composition in the meat of blue swimming crab (*Portunus pelagicus*) from the Persian Gulf, Iran. Iran J Fish Sci 15(2):818–826

Badawi HK (1971) On the chemical composition of the red crab *Portunus pelagicus*. Mar Biol 11:198–200

Baghlani F, Raissy M, Roomiani J (2018) Fatty acid profile and chemical composition of lobster (*Panulirus homarus*) caught from the Persian Gulf. J Food Technol Nutr 15(3):43–50

Bai Y, Hu Y, Huang Q (2013) Study on a combination drying technique of shrimp. Appl Mech Mater Online 260–261:804–807

Bak LS, Andersen AB, Andersen EM, Bertelsen G (1999) Effect of modified atmosphere packaging on oxidative changes in frozen stored cold water shrimp (*Pandalus borealis*). Food Chem 64(2):169–175

Baklouti S, Derbali A, Dhieb K, Kammoun W, Jarboui O (2013) Proximate composition and its seasonality of the Mediterranean green crab: *Carcinus aestuarii* Nardo, 1847 (Brachyura, Portunidae), in Southern Tunisian Waters (Central Mediterranean). J Mar Biol 2013:989467

Balachandran KK (1975) Storage studies on prawns dried in rotary drum dryer. Fish Technol 12(2):116–119

Balasaraswathy N, Sugumar G, Selvan A, Ramesh U, Velayutham P (2008) Changes in quality characteristics of cooked and uncooked crab meat (*Portunus pelagicus*) under ice storage. Asian Fish Sci 21:101–112

Banerjee S (2006) Inhibition of mackerel (*Scomber scombrus*) muscle lipoxygenase by green tea polyphenols. Food Res Int 39(4):486–491

Barrento S, Marques A, Teixeira B, Vaz-Pres P, Nunes ML (2009) Nutritional quality of the edible tissues of European lobster *Homarus gammarus* and American Lobster *Homarus americanus*. J Agric Food Chem 57:3645–3652

Bartolo I, Birk EO (2002) Some factors affecting Norway lobster (*Nephrops norvegicus*) cuticle polyphenol oxidase activity and black spot development. Int J Food Sci Technol 33(3):329–336

Bascinar NS (2004) Shrimp. SÜMAE YUNUS Araştırma Bülteni 4:1–3

Beard TV, McGregor D (2004) Storage and care of live lobsters. Laboratory Leaflet Number 66 (Revised) CEFAS, Lowestoft, p 26

Bekaert K, Devriese L, Maes S, Robbens J (2015) Characterization of the dominant bacterial communities during storage of Norway lobster and Norway lobster tails (*Nephrops norvegicus*) based on 16S rDNA analysis by PCR-DGGE. Food Microbiol 46:132–138

Belessiotis V, Delyannis E (2011) Solar drying. Sol Energy 85:1665–1691

Bell MC, Redant F, Tuck I (2006) Nephrops species. In: Phillips BF (ed) Lobsters: biology, management, aquaculture and fisheries. Wiley-Blackwell, Oxford, pp 412–461

Benjakul S, Bauer F (2000) Physicochemical and enzymatic changes of cod muscle proteins subjected to different freeze–thaw cycles. J Sci Food Agric 80:1143–1150

Benjakul S, Bauer F (2001) Biochemical and physicochemical changes in catfish (*Silurus glanis* Linne) muscle as influenced by different freeze–thaw cycles. Food Chem 72:207–217

Benjakul S, Sutthipan N (2009) Muscle changes in hard and soft shell crabs during frozen storage. LWT-Food Sci Technol 42:723–729

Benjakul S, Visessanguan W, Tanaka M (2005) Properties of phenoloxidase isolated from the cephalothorax of kuruma prawn (*Penaeus japonicus*). J Food Biochem 29:470–485

Benjakul S, Visessanguan W, Tanaka M (2006) Inhibitory effect of cysteine and glutathione on phenoloxidase from kruma prawn (*Penaeus japonicus*). Food Chem 98:158–163

Benner RA, Miget R, Fınne G, Acuff GR (1994) Lactic acid/melanosis inhibitors to improve shelf life of brown shrimp (*Penaeus aztecus*). J Food Sci 59(2):242–245

Berber S, Turel S, Yılmaz S (2014) The chemical composition of the crayfish (*Astacus leptodactylus*) in pond Yenice. In: Proceedings of the 5th international symposium on sustainable development, Sarajevo, pp 75–86

Biede SL, Himelbloom BH, Rutledge JE (1982) Influence of storage atmosphere on several chemical parameters of sun-dried shrimp. J Food Sci 47(3):1030–1031

Biji KB, Chako ST, Yathavomoorthi R, Ravishankar CN, Bindu J, Mathew S (2013) Optimization of process parameters for ready to serve bread spread from blue swimmer crab *Portunus pelagicus* in tin-free steel cans. Fish Technol 50:237–244

Biji KB, Ravishankar CN, Venkateswarlu R, Mohan CO, Srinivasa Gopal TK (2016) Biogenic amines in seafood: a review. J Food Sci Technol 53(5):2210–2218

Bingol EB, Bostan K, Varlık C, Uran H, Üçok-Alakavuk D, Sivri N (2015) Effects of chitosan treatment on the quality parameters of shrimp (*Parapenaeus longirostris*) during chilled storage. Turk J Fish Aquat Sci 15:821–831

Bono G, Gai F, Peiretti PG, Badalucco C, Brugiapaglia A, Siragusa G, Palmegiano GB (2012) Chemical and nutritional characterisation of the Central Mediterranean giant red shrimp (*Aristaeomorpha foliacea*): influence of trophic and geographical factors on flesh quality. Food Chem 130:104–110

Boon DD (1975) Discoloration in processed crab meat: a review. J Food Sci 40:756–761

Boonsumrej S, Chaiwanichsiri S, Tantratian S, Suzuki T, Takai R (2007) Effect of freezing and thawing on the quality changes of tiger shrimp (*Penaeus monodon*) frozen by air-blast and cryogenic freezing. J Food Eng 80(1):292–299

Borderias J (2012) Pre-treatments for fish and seafood in general. In: Mascharoni RH (ed) Operation in food refrigeration. CRC Press, Taylor and Franchis Group, Boca Raton, pp 173–181

Botta JR (1995) Chemical methods of evaluating freshness quality. In: Evaluation of seafood freshness quality. VCH Publishers, Inc., New York

Bottino NR, Lilly ML, Finne G (1979) Fatty acid stability of Gulf of Mexico brown shrimp (*Penaeus aztecus*) held on ice and in frozen storage. J Food Sci 44:1778–1779

Boziaris IS, Kordila A, Neofitou C (2011) Microbial spoilage analysis and its effect on chemical changes and shelf-life of Norway lobster (*Nephrops norvegicus*) stored in air at various temperatures. Int J Food Sci Technol 46:887–895

Brack A, Hellman N, Decker H (2008) Kinetix properties of hexameric tyrosinase from crustaceans *Palinurus elephas*. Photochem Photobiol 84:682–699

Bragagnolo N, Rodriguez-Amaya DB (2001) Total lipid, cholesterol, and fatty acids of farmed freshwater prawn (*Macrobrachium rosenbergii*) and wild marine shrimp (*Penaeus brasiliensis*, *Penaeus schimitti*, *Xiphopenaeus kroyer*). J Food Compos Anal 14:359–369

Buricova L, Reblova Z (2008) Czech medical plants as possible sources of antioxidants. Czech J Food Sci 26:132–138

Bykowski PJ (1990) The preparation of the catch for preservation of marketing. In: Sikorski ZE (ed) Seafood resources, nutritional composition, and preservation. CRC Press Inc., Boca Raton, pp 77–93

Cabi (2019a) Invasive Species Compendium. https://www.cabi.org/isc/datasheet/70581. Accessed 13 Sept 2019

Cabi (2019b) Invasive Species Compendium. https://www.cabi.org/isc/datasheet/89134. Accessed 13 Sept 2019

Cabi (2019c) Cherax quadricarinatus (redclaw crayfish). https://www.cabi.org/isc/datasheet/89135. Accessed 16 Sept 2019

Calder BL, Bushway AA, Bayer RC, Davis-Dentici K, Camire ME (2005) Quality of whole lobster (*Homarus americanus*) treated with sodium tripolyphosphate before cooking and frozen storage. J Food Sci 70(9):C523–C528

Carroll JC, Winn RN (1989) Species profiles: life histories and environmental requirements of coastal fishes and invertebrates (Pacific Southwest)—brown rock crab, red rock crab, and yellow crab. U.S. Fish Wildl. Serv. Biol. Rep. 82 (11.117). U.S. Army Corps of Engineers, TR EL-82-4, 16 p

Celik M, Tureli C, Celik M, Yanar Y, Erdem U, Kucukgulmez A (2004) Fatty acid composition of the blue crab (*Callinectes sapidus* Rathbun, 1896) in the north eastern Mediterranean. Food Chem 88:271–273

Cemeroglu B, Karadeniz F, Özkan M (2003) In: Cemeroglu B (ed) Fruit and vegetable processing technology. Başkent Klişe Printing, Ankara, pp 37–134

Cemeroglu B, Yemenicioglu A, Ozkan M (2004) In: Cemeroglu B (ed) Fruit and vegetable processing technology, vol 1. Başkent Klişe Printing, Ankara, pp 116–144

Chakraborty R, Mukhopadhyay P, Bera M, Suman S (2011) Infrared-assisted freeze drying of tiger prawn: parameter optimization and quality assessment. J Dry Technol 29(5):508–519

Chan TY (1998a) Shrimps and prawns. The living marine resources of the western central pacific. In: Carpenter KE, Niem H (eds) Cephalopods, crustaceans, holothurians, and sharks. FAO fisheries identification guide for fishery purposes, vol 2. FAO, Rome, pp 851–971

Chan TY (1998b) Lobsters. In: Carpenter KE, Niem VH (eds) The living marine resources of the Western Central Pacific, vol 2. Cephalopods, crustaceans, holothurians, and sharks. FAO species identification guide for fishery purposes. FAO, Rome, pp 973–1028

Chan ST, Yao MWY, Wong YC, Wong T, Mok CS, Sin DWM (2006) Evaluation of chemical indicators for monitoring freshness of food and determination of volatile amines in fish by headspace solid-phase microextraction and gas chromatography-mass spectrometry. Eur Food Res Technol 224:67–74

Chan JTY, Dileep AO, Betti M (2011) Effect of ultimate pH and freezing on the biochemical properties of proteins in turkey breast meat. Food Chem 120:109–117

Chandrapavan A, Gardner C, Linnane A, Hobday D (2009) Colour variation in the southern rock lobster *Jasus edwardsii* and its economic impact on the commercial industry. N Z J Mar Freshw Res 43:537–545

Cheftel JC, Culioli J (1997) Effects of high pressure on meat: a review. Meat Sci 46(3):211–236

Chen G, Xiong YL (2008) Shelf-stability enhancement of precooked red claw crayfish (*Cherax quadricarinatus*) by modified CO2/O2/N2 gas packaging. LWT-Food Sci Technol 41:1431–1436

Chen JS, Wei CI, Rolle RS, Otwell WS, Balaban MO, Marshall MR (1991) Inhibitory effect of kojic acid on some plant and crustacean polyphenoloxidases. J Agric Food Chem 39:1396–1401

Chen G, Xiong YL, Kong B, Newman MC, Thompson KR, Metts LS, Webster CD (2007) Microbiological and physicochemical properties of red claw crayfish (*Cherax quadricarinatus*) stored in different package systems at 2°C. J Food Sci 72(8):442–449

Cheng L, Sun D-W, Zhu Z, Zhang Z (2017a) Emerging techniques for assisting and accelerating food freezing processes: a review of recent research progresses. Crit Rev Food Sci Nutr 57(4):769–781

Cheng L, Sun D-W, Zhu Z, Zhang Z (2017b) Effects of high pressure freezing (HPF) on denaturation of natural actomyosin extracted from prawn (*Metapenaeus ensis*). Food Chem 229:252–259

Chevalier D, Sequeira-Munoz A, Le Bail A, Simpson B, Ghoul M (2000) Effect of pressure shift freezing, airblast freezing and storage on some biochemical and physical properties of turbot (*Scophthalmus maximus*). Lebenmitt Wissenschaft Technol 33:570–577

Chiar SS, Bakerj C, Hotchkiss H (2006) Quality comparison of thermo processed fishery products in cans and retortable pouches. J Food Sci 48(5):1521–1525

Clark AC, Vestner J, Barril C, Maury C, Prenzler PD, Scollary GR (2009) The influence of stereochemistry of antioxidants and flavanols on oxidation processes in a model wine system: ascorbic acid, erythorbic acid, (+)-catechin and ()-epicatechin. J Agric Food Chem 58(2):1004–1011

Cobb BF, Vanderzant C, Thompson CA, Custer CS (1973) Chemical characteristics, bacterial counts and potential shelf-life of shrimp from various locations on the northwestern Gulf of Mexico. J Milk Food Technol 36:463–468

Cobb BF, Vanderzant C, Hanna MO, Yeh CPS (1976) Effect of ice storage on microbiological and chemical changes in shrimp and melting ice in a model system. J Food Sci 41(1):29–34

Çolakoğlu FA, Ormancı HB, Altın A (2006) Determination of the shelf life of fresh shrimp (*Parapaneus longirostris*) treated with Frische-Star. *E.U.* J Fish Aquat Sci 23(Suppl (1/3)):383–386

Conde FJ, Afonso AM, González V, Ayala JH (2006) Optimization of an analytical methodology for the determination of alkyl- and methoxy-phenolic compounds by HS-SPME in biomass smoke. Anal Bioanal Chem 385:1162–1171

Condurso C, Tripodi G, Cincotta F, Lanza CM, Mazzaglia A, Verzera A (2016) Quality assessment of Mediterranean shrimps during frozen storage. Ital J Food Sci 28:497–509

Connell JJ (1995) Quality deterioration and extrinsic quality defects in raw material. In: Connell JJ (ed) Control of fish quality. Fishing News Books Ltd, Surrey, pp 31–35

Corzo-Martinez M, Corzo N, Villamiel V, del Castillo MD (2012) Browning reactions. In: Simpson BK, Nollet LML, Toldra F, Benjakul S, Paliyath P, Hui YH (eds) Food biochemistry and food processing. Wiley, Oxford, pp 56–83

Cox NA, Lovell RT (1973) Identification and characterization of the microflora and spoilage bacteria in freshwater crayfish *Procambarus clarkii* (Girard). J Food Sci 38(4):679–681

Cremades O, Alvarez-Ossorio C, Gutierrez-Gil JF, Parrado J, Bautista J (2011) Quality changes of cooked crayfish (*Procambarus clarkii*) tails without additives during storage under protective atmospheres. J Food Process Preserv 543:898–906

Dabade DS (2015) Shrimp quality and safety management along the supply chain in Benin. PhD thesis, Wageningen University, Wageningen, 158 p

Dabade DS, den Besten HN, Azokpota P, Nout MR, Hounhouigan DJ, Zwietering MH (2015) Spoilage evaluation, shelf life prediction, and potential spoilage organisms of tropical brackish water shrimp (*Penaeus notialis*) at different storage temperatures. Food Microbiol 48:8–16

Dalgaard P (2000) Fresh and lightly preserved seafood. In: Man CMD, Jones AA (eds) Shelf-life evaluation of foods. Aspen Publishers, London, pp 110–139

Dalgaard P, Gram L, Huss HH (1993) Spoilage and shelf life of cod fillets packed in vacuum and modified atmosphere. Int J Food Microbiol 19(4):283–294
Dall W, Hill B, Rothlisberg P, Sharples D (1990) The biology of the Penaeidae. In: Blaxter JHS, Southward AJ (eds) Advances in marine biology, vol 27. Academic, London, 488 p
Dang TT, Gringer N, Jessen F, Olsen K, Bøknæs N, Nielsen PL, Orlien V (2018a) Emerging and potential technologies for facilitating shrimp peeling: a review. Innov Food Sci Emerg Technol 45:228–240
Dang TT, Gringer N, Jessen F, Olsen K, Bøknæs N, Nielsen PL, Orlien V (2018b) Facilitating shrimp (*Pandalus borealis*) peeling by power ultrasound and proteolytic enzyme. Innov Food Sci Emerg Technol 47:525–534
Dang TT, Gringer N, Jessen F, Olsen K, Bøknæs N, Nielsen PL, Orlien V (2018c) Enzyme-assisted peeling of cold water shrimps (*Pandalus borealis*). Innov Food Sci Emerg Technol 47:127–135
Davidson A (2002) Mediterranean seafood. A comprehensive guide with recipes. Ten Speed Press, Berkeley, 431 p
Dayal JS, Ponniah AG, Khan HI, Babu EPM, Ambasankar K, Vasagam KPK (2013) Shrimps—a nutritional perspective. Curr Sci 104(11):1487–1491
Devahastin S, Mujumdar AS (2014) Superheated steam drying of foods and biomaterials. In: Tsotsas E, Mujumdar AS (eds) Modern drying technology. Wiley-VCH, Verlag, pp 57–84
Diaz-Tenorio LM, Garcia-Carreno FL, Pacheco-Aguilar R (2007) Comparison of freezing and thawing treatments on muscle properties of whiteleg shrimp (*Litopenaeus vannamei*). J Food Biochem 31:563–576
Dincer MT, Aydin I (2014) Proximate composition and mineral and fatty acid profiles of male and female jinga shrimps (*Metapenaeus affinis*, H. Milne Edwards, 1837). Turk J Vet Anim Sci 38:445–451
Doe P, Olley J (1990) Drying and dried fish products. In: Sikorski Z (ed) Seafood resources, nutritional composition, and preservation. CRC Press Inc., Boca Raton, pp 126–140
Doe P, Sikorski Z, Haard NF, Olley J, Pan BS (1998) Basic principles. In: Doe P (ed) Fish drying and smoking: production and quality. Technomic Publishing Company Inc., Lancaster, pp 13–45
Don S, Xavier KAM, Devi ST, Nayak BB, Kannuchamy N (2018) Identification of potential spoilage bacteria in farmed shrimp (*Litopenaeus vannamei*): application of relative rate of spoilage models in shelf life prediction. LWT-Food Sci Technol 97:295–301
Dore I (1991) The new fresh seafood buyer's guide a manual for distributors, restaurants and retailers. New York, 280 p
Drinkwater KF, Tremblay MJ, Comeau M (2006) The influence of wind and temperature on the catch rate of the American lobster (*Homarus americanus*) during spring fisheries off eastern Canada. Fish Oceanogr 15(2):150–165
Du L, Fan X, Liu F, Zhou Q, Yuan J, Ju X (2017) Changes of dominant spoilage bacteria and biogenic amines of taihu white prawn (*Exopalaemon modestus*) during ice storage. J Food Prot 80(12):2099–2104
Edwards E, Early JC (1978) Catching, handling and processing crabs. Torry Advisory Note No.26. Ministry of Technology Torry Research Station, Edinburg, HMSO Press
El-Kholie EM, Khader SA, Abdelreheem MAT (2012) Chemical, physical, microbiological and quality attributes studies on River Nile crayfish. Afr J Biotechnol 11(51):11262–11270
El-Mossalami MK, Emara MT (1999) Safety and quality of freshwater crayfish *Procambarus clarkii* in the river Nile (Article). Nahrung/Food 43(2):126–128
El-Sherif SA-H, El-Ghafour SA (2015) Nutritive value of canned River Nile Crayfish (*Procambarus clarkii*) products. Egypt J Aquat Res 41:265–272
Encarnacion AB, Fagutao F, Hirono I, Ushio H, Ohshima T (2010) Effects of ergothioneine from mushrooms (*Flammulina velutipes*) on melanosis and lipid oxidation of kuruma shrimp (*Marsupenaeus japonicus*). J Agric Food Chem 58(4):2577–2585
Erdem ME, Bilgin S (2004) Researches on quality changes of shrimp (*Palaemon adspersus* Rathke, 1837) stored at refrigerated temperatures as boiling and raw *Fırat University*. J Sci Eng Sci 16(4):687–694

Erdem ME, Turan H, Kaya Y (2015) Mineral and trace element contents of warty crab (*Eriphia verrucosa*) and brown shrimp (*Crangon crangon*). J Fish Aquat Sci 30(2):26–31

Erkan N, Özden Ö, Alakavuk ÜD, Tosun ŞY, Varlık C, Baygar T (2007) Determination of sodium metabisulphide level of shrimps sold in Istanbul. J Fish Sci 1(1):26–33

Erkan N, Mol S, Tosun SY, Alakavuk DU (2009) Seasonol variations of the yield and proximate composition of freshwater crayfish, *Astacus leptodactylus* (Eschscholtz, 1823). Adv Food Sci 31(1):41–44

Eskin NAM, Henderson HM, Toznsend RJ (1971) Browning reactions in foods. In: Eskin NAM, Henderson HM, Toznsend RJ (eds) Biochemistry of foods. Academic, New York, pp 69–83

Essien EU (1995) Lipid content and fatty acid profiles of some lesser known Nigerian foods. J Food Biochem 19:153–159

Estrella BT (2002) Techniques for live storage & shipping of American lobster. Department of Fisheries, Wildlife & Environmental Law Enforcement Division of Marine Fisheries, Massachusetts Division of Marine Fisheries Technical Report TR-8, 17 p

Fang XB, Sun HY, Huang BY, Gao-Feng Y (2013) Effect of pomegranate peel extract on the melanosis of Pacific white shrimp (*Litopenaeus vannamei*) during iced storage. J Food Agric Environ 11(1):105–109

FAO (2004) *Nephrops norvegicus*. FAO, Fisheries Global Information System (FIGIS)

FAO (2013) Standard for smoked fish, smoke-flavoured fish and smoke-dried fish CXS 311-2013. Codex Alimentarius Food Standards

FAO (2015) Fishstatj. FAO Rome. http://www.fao.org/fishery/statistics/software/fishstat/en

FAO (2016) Yearbook of fishery statistics summary tables fisheries and aquaculture department. Rome, 200 p. http://www.fao.org/3/i9942t/I9942T.pdf

FAO (2017) The world lobster market, by Graciela Pereira and Helga Josupeit, FAO Consultants. Globefish Research Programme, vol 123. Rome, Italy

FAO (2018) The State of World Fisheries and Aquaculture 2018-Meeting the sustainable development goals. Rome

FAO-Globefish (2018) Globefish highlights. A quarterly update on world seafood markets. FAO, Rome, p 6

FAO-Globefish (2019a) Globefish highlights: a quarterly update on world seafood markets. April 2019 Issue, with Jan–Dec 2018 Statistics, No: 2/2019

FAO-Globefish (2019b) Globefish highlights: a quarterly update on world seafood markets. January 2019 Issue, with Jan–Sep 2018 Statistics, No: 1/2019

Fard AN, Seidgar M, Azadikhah D (2015) The study of frozen *Astacus leptodactylus* tail fillet quality changes. AACL Bioflux 8(6):988–998

Farfante IP (1988) Illustrated key to penaeoid shrimps of commerce in the America. NOAA Technical Report NMFS 64. U.S. Department of Commerce National Oceanic and Atmospheric Administration National Marine Fisheries Service, pp 6–32

Farhang A, Adel H, Darvishi H, Nargesi F (2011) Shrimp drying characterizes undergoing microwave treatment. J Agric Sci 3(2):157–164

Fatima R, Khan MA, Qadri RB (1988) Shelf life of shrimp (*Penaeus merguiensis*) stored in ice (0°C) and partially frozen (-3°C). J Sci Food Agric 42(3):235–247

Featherstone S (2016a) Canning of fish and seafood. In: A complete course in canning and related process. Cambridge, Woodhead Publishing Series in Food Science, Technology and Nutrition, Number 282, pp 231–265

Featherstone S (2016b) A complete course in canning and related processes, 14th edn, vol 3. Processing procedures for canned food products woodhead publishing series in food science, technology and nutrition. pp 231–265

Fennema OR, Powrie WD, Marth EH (1973) Low temperature preservation of foods and living matter. Dekker, New York, 598 p

Ferrer AJ, Koburger JA, Otwell WS, Gleeson RA, Simpson BK, Marshall MR (1989) Phenoloxidase from the cuticle of Florida spiny lobster (*Panulirus argus*): mode of activation and characterization. J Food Sci 54:63–67

Food and Drug Administration (FDA) (1982) The food defect action levels: current levels for natural or unavoidable defects in food for human use that present no health hazard. 200 C Street, S.W., Washington, DC, 20204

Food and Drug Administration (FDA) (2000). Sulfites: An important food safety issue, an update on regulatory status and methodologies. http://www.cfsan.fda.gov/~dms/fssulfit.html.

Fotedar S, Evans L (2011) Health management during handling and live transport of crustaceans: a review. J Invertebr Pathol 106:143–152

Fregin T, Bickmeyer U (2016) Electrophysiological investigation of different methods of anesthesia in lobster and crayfish. PLoS One 11(9):e0162894

Galetti JA (2010) Mechanical processing of European green crab (*Carcinus maenas*), the development of a value-added product and the use of restructuring additives to increase the functional properties of green crab patties. Electronic thesis and dissertations. p 78. http://digitalcommons.library.umaine.edu/etd/78

Garcia-Soto B, Miranda JM, Barros-Velazquez J, Aubourg SP (2015) Quality changes during the frozen storage of the crustacean lobster krill (*Munida spp.*). Eur J Lipid Sci Technol 117:431–439

Gardner C (2004) Treating the prawn well on its way to the barbie: welfare of aquatic crustaceans. In: Jones B (ed) Welfare underwater: issues with aquatic animals. Proceedings of the 2004 RSPCA Australia scientific seminar, Canberra, February 26, RSPCA Australia Inc., Deakin West, Australia, pp 21–24

Gary M (2012) Southern rock lobster strategic plan 2011–2016. Southern Rock Lobster Ltd

Gates KW, Parker AH, Bauer DL, Huang YW (1993) Storage changes of fresh and pasteurized blue crab meat in different types of packaging. J Food Sci 58(2):314–317

Gavrila C, Ghiaus AG, Gruia I (2008) Heat and mass transfer in convective drying processes. In: Proceedings of the 2nd COMSOL Conference, 4–6 November, Hannover

Geiger W, Alcorlo P, Baltana A, Montes C (2005) Impact of an introduced Crustacean on the trophic webs of Mediterranean wetlands. Biol Invasions 7:49–73

George C (1973) Technological aspects of preservation and processing of edible shellfishes. III: factors influencing the keeping quality of crab (*Scylla serrata*) during freezing and frozen storage. Fish Technol 10(1):15–20

George C, Gopakumar K (1988) Spoilage changes in the muscle of crab, Scylla serrata stored at three different temperatures. In: The first Indian fisheries forum, proceedings. Karnataka Press, Mangalore, pp 347–349

Ghaly AE, Dave D, Budge S, Brooks MS (2010) Fish spoilage mechanisms and preservation techniques: review. Am J Appl Sci 7(7):859–877

Gil MI, Gomy JR, Kader AA (1998) Responses of 'fuji' apple slices to ascorbic acid treatments and low-oxygen atmospheres. Hortic Sci 33:305–309

Gill T (1990) Objective analysis of seafood quality. Food Rev Int 6(4):687–697

Gillespie NC, Burke JB (1991) Report on the seminar on the mud crab culture and trade. BOBP/REP/51, pp 207–209

Gillett R (2008) Global study of shrimp fisheries. FAO Fisheries Technical Paper 475. Food and Agriculture Organization of the United Nations, Rome, p 313

Gimenez B, Martinez-Alvarez O, Montero P, Gomez-Guillen MC (2010) Characterization of phenoloxidase activity of carapace and viscera from cephalothorax of Norway lobster (*Nephrops norvegicus*). LWT-Food Sci Technol 43:1240–1245

Gokoglu N (2002) Seafood processing technology. Su Vakfı Publications, Book, 157 p. (In Turkish)

Gokoglu N (2004) The effect of organic acid treatments on the melanosis inhibition and shelf-life in shrimp (*Penaeus japonicus*). Acta Aliment 33(2):191–199

Gokoglu M, Oray IK (1997) A study on the fishing of blue crab (*Callinectes sapidus* Rathbun) in the Gulf of Antalya. Workshop on fishing and processing technology organised by University of Istanbul Faculty of Fisheries, Istanbul (In Turkish)

Gokoglu N, Toktas B (2018) Investigation of the inhibitory effect of erythorbic acid on melanosis in red shrimp (*Aristomorpho foliac*ea). In: Hydromedit 2018, 3rd international congress on applies ichtyology and aquatic environment, 8–11 Nov 2018. Proceedings Book, Volos, pp 300-302

Gokoglu N, Yerlikaya P (2003) Determinaton of proximate composition and mineral contents of blue crab (*Callinectes sapidus*) and swim crab (*Portunus pelagicus*) caught off the Gulf of Antalya. Food Chem 80:495–498

Gokoglu N, Yerlikaya P (2008) Inhibition effects of grape seed extracts on melanosis formation in shrimp (*Parapenaeus longirostris*). Int J Food Sci Technol 43(6):1004–1008

Gokoglu N, Yerlikaya P (2015) Quality changes and spoilage of fish. In: Seafood chilling, refrigeration and freezing. Wiley, Chichester, pp 38–54

Golan-Goldhirsh A, Whitaker JR (1984) Effect of ascorbic acid, sodium bisulfite and thiol compounds on mushroom polyphenol oxidase. J Agric Food Chem 32:1003–1009

Gomez-Guillen MC, Martinez-Alvarez O, Llamas A, Montero P (2005) Melanosis inhibition and SO_2 residual levels in shrimps (*Parapenaeus longirostris*) after different sulfite-based treatments. J Sci Food Agric 85:1143–1148

Gonçalves AA, De Oliveira ARM (2016) Melanosis in crustaceans: a review. LWT-Food Sci Technol 65:791–799

Gonçalves AC, López-Caballero ME, Nunes ML (2003) Quality changes of deepwater pink shrimp (*Parapenaeus longirostris*) packed in modified atmosphere. J Food Sci 68:2586–2590

Goncalves AA, Santiago C, Gindri Jr G (2009) The effect of glaze uptake on storage quality of frozen shrimp. *J Food Eng, 90* :285–290.

Goodrick GB, Paterson BD, Grauf S (1995) Alive and kipping. Storage and transport of live prawns. Aust Fish 54:18–21

Gornik SG, Albalat A, Macpherson H, Birkbeck H, Neil DM (2011) The effect of temperature on the bacterial load and microbial composition in Norway lobster (*Nephrops norvegicus*) tail meat during storage. J Appl Microbiol 111:582–592

Gornik SG, Albalat A, Theethakaew C, Neil DM (2013) Shelf life extension of whole Norway lobster *Nephrops norvegicus* using modified atmosphere packaging. Int J Food Microbiol 167:369–377

Goulas AE, Kontominas MG (2005) Effect of salting and smoking-method on the keeping quality of chub mackerel (*Scomber japonicus*): biochemical and sensory attributes. Food Chem 93:511–520

Gram L, Huss HH (1996) Microbiological spoilage of fish and fish products. Int J Food Microbiol 33:121–137

Guandalini E, Ioppolo A, Mantovani A, Stacchini P, Giovannini C (1998) 4-Hexylresorcinol as inhibitor of shrimp melanosis: efficacy and residues studies; Evaluation of possible toxic effect in a human intestinal in vitro model (Caco-2); Preliminary safety assessment. Food Addit Contam 15(2):171–180

Guillen MD, Errecalde C (2002) Volatile components of raw and smoked black bream (*Brama rail*) and rainbow trout (*Oncorhynchus mykiss*) studied by means of solid phase microextraction and gas chromatography/mass spectrometry. J Sci Food Agric 82:945–952

Gulle P, Turna İİ, Gulle I (2007) Some reproductive and population characteristics of freshwater crab (*Potamon potamios* Olivier. 1804) in Egirdir Lake. Süleyman Demirel University, J Sci 11(2):134–139

Gunnarsson GH (2010) Muscle spoilage in Nephrops. Skýrsla Matís 32-10, Matís ohf/Matis—Food Research, Innovation & Safety, Iceland, 37 p

Gunter G (1961) Painless killing of crabs and other large crustaceans. Science 133(3449):327

Guochen Z, Arason S, Árnason SV (2009) Dehydration property of Shrimp (*Pandalus borealis*) undergoing heat-pump drying process. Int J Agric Biol Eng 2(4):92–94

Gurel Inanlı A, Coban OE (2007) Çemişgezek Keban dam lake region freshwater crayfish (*Astacus leptodactylus*, ESCH., 1823) Meat Yield and Chemical Quality. Eastern Anatolia Reg Stud 5(2):79–82

Gurel A, Patır B (2001) Meat yield and chemical quality of crayfish (*Astacus leptodactyills* Esch., 1823) in Keban dam lake. Selçuk Univ J Vet Sci 17(2):23–30. (in Turkish)

Haard NF, Simpson BK (2000) Seafood enzymes, utilization and influence on postharvest seafood quality. Marcel Dekker Inc, New York, 696 p

Hackney CR, Rippen TE, Ward DR (1991) Principles of pasteurization and minimally processed seafoods. In: Ward DR, Hackney CR (eds) Microbiology of marine food products. Van Nostrand Reinhold, New York, pp 335–371

Hammami C, Floris R, Marin M (1999) Process-quality optimization of the vacuum freeze drying of apple slices by the response surface method. Int J Food Sci Technol 34:145–160

Hansen OD, Nielsen BK (1988) U.S Patent No. 4,639,976. U.S. Patent and Trademark Office, Washington, DC

Hardisson A, Rubio C, Frias I, Rodrigez I, Reguera IJ (2002) Content of sulphite in frozen prawns and shrimps. Food Control 13:275–279

Harlioglu AG, Aydin S, Yilmaz O (2012) Fatty acid, cholesterol and fat-soluble vitamin composition of wild and captive freshwater crayfish (*Astacus leptodactylus*). Food Sci Technol Int 18(1):93–100

Harrison AJ (2004) A report on the abalone and rock lobster post harvesting industry in Tasmania. Hobart, Tasmania Abalone Council, 121 p

Hassan IM (1988) Processing of smoked common carp fish and its relation to some chemical, physical and organoleptic properties. Food Chem 27:95–106

Hattula T, Elfving K, Mroueh U, Luoma T (2001) Use of liquid smoke flavouring as an alternative to traditional flue gas smoking of rainbow trout fillets (*Oncorhynchus mykiss*). Lebensmitt Wissenschaft Technol 34:521–525

Heinsz LJ, Harrison MA, Leiting VA (1988) Microflora of brown shrimp (*Penaeus aztecus*) from Georgia coastal waters. Food Microbiol 5(3):141–145

Hernandez-Padilla JC, Zetina-Rejon MJ, Arreguín-Sanchez F, Nieto-Navarro JT, Hernandez-Lope A (2018) Spatio-temporal distribution of four commercial shrimp species in the southeastern Gulf of California, Mexico. Lat Am J Aquat Res 46(4):727–734

Hernandez-Romero D, Sanchez-Amat A, Solano F (2006) A tyrosinase with an abnormally high tyrosine hydroxylase/dopa oxidase ratio/role of the seventh histidine and accessibility to the active site. FEBS J 273:257–270

Hollows J (2016) Freshwater crayfish farming—a guide to getting started. Unpublished, Ernslaw One Ltd

Holthuis LB (1991) "*Homarus gammarus*". Marine lobsters of the world. FAO Species Catalogue, vol 13. FAO Fisheries Synopsis No. 125. Food and Agriculture Organization, p 60

Horner WFA (1992) Canning of fish and fish products. In: Hall GM (ed) Fish processing technology. Blackie Academic & Professional, New York, pp 114–153

Hosseinpour S, Rafiee S, Mohtasebi SS (2011) Application of image processing to analyze shrinkage and shape changes of shrimp batch during drying. Dry Technol 29:1416–1438

Hsu AF, Shieh JJ, Bill DD, White K (1988) Inhibition of mushroom polyphenoloxidase by ascorbic acid derivatives. J Food Sci 53:765

Hu Y, Huang Q, Bai Y (2013) Combined electrohydrodynamic (EHD) and vacuum freeze drying of shrimp. *J Phys Conf Ser* 418(2013):012143

Huang Z (2018) The role of crayfish in the economic and social spheres of Chinese food industry. https://scholarblogs.emory.edu/noodles/2018/06/29/the-role-of-crayfish-in-the-economic-and-social-spheres-of-chinese-food-industry/. Accessed 12 Sept 2019

Huang YW, He LP, Gates KW (1996) Qualities of fresh and previously marinated shrimp. In: Proceedings of the tropical and subtropical seafood and technology society of the Americas, Florida Sea Grant Program, University of Florida, Gainesville, pp 218–232

Huang J, Chen Q, Qiu M, Li S (2012) Chitosan-based Edible Coatings for Quality Preservation of postharvest whiteleg shrimp (*Litopenaeus vannamei*). J Food Sci 77(4):491–496

Huidobro A, Cabellarao ML, Nunes ML (2002) Onboard processing of deepwater pink shrimp (*Parapenaeus longirostris*) with liquid ice: Effect on quality. Eur Food Res Technol 214(6):469–475

Hultmann L (2003) Endogenous proteolytic enzymes—studies of their impact on fish muscle proteins and texture. Dr. ing.thesis 2003:110. Faculty of Natural Sciences and Technology Department of Biotechnology. Norwegian University Science and Technology

Hungria DB, Tavares CPS, Pereira LA, Silva UAT, Ostrensky A (2017) Global status of production and commercialization of soft-shell crabs. Aquac Int 25(6):2213–2226

Huss HH (1988) Fresh fish quality and quality changes. FAO/DANIDA.A training manual, FAO Fisheries Series No. 29

Huss HH (1995) Quality and quality changes in fresh fish. Fisheries Technical Paper. United Nations, Food and Agricultural Organization. FAO, Rome

Hwang C, Lee Y, Huang Y, Lin C, Shiau C, Hwang D, Tsa Y (2010) Biogenic amines content, histamine-forming bacteria and adulteration of bonito in tuna candy products. Food Control 21:845–850

Innocen A (2014) Changes in amino acid composition in pasteurised blue swimming crab (*Portunus pelagicus*) during chilled storage. Master thesis, Department of Fish Processing Technology, Fisheries College and Research Institute Tamil Nadu Fisheries University Thoothukudi, 628008

Iranto HE, Giyatmi (1997) Post harvest technology of shrimp: review of Indonesian experience. J Aquat Food Prod Technol 6(1):5–20

Irish Sea Fisheries Board (2019) European Lobster (*Homarus gammarus*). Handling and quality guide. http://www.bim.ie/media/bim/content/downloads/BIM,Lobster,Handling,and,Quality,Guide.pdf. Accessed 10 Sept 2019

Jabbour T, Hognason G (2011) U.S Patent No. 7,871,314 B2. U.S. Patent and Trademark Office, Washington, DC

Jacklin M, Combes J (2007a) The good practice guide to handling and storing live crustacea. A European Union project, Research and Development, Seafish, Scotland

Jacklin M, Combes J (2007b) the good practice guide to handling and storing live Crustacea. SEAFISH report. http://www.seafish.org/media/Publications/CrustaceaGPG_0505.pdf

Jaffres E, Lalanne V, Mace S, Cornet J, Cardinal M, Serot T, Dousset X, Joffraud J (2011) Sensory characteristics of spoilage and volatile compounds associated with bacteria isolated from cooked and peeled tropical shrimps using SPME–GC–MS analysis. Int J Food Microbiol 147(3):195–202

Jain D, Pathare PB (2007) Study the drying kinetics of open sun drying of fish. J Food Eng 78:1315–1319

Jang MS, Sanada A, Ushio H, Tanaka M, Ohshima T (2003) Inhibitory effect of enokitake extract on melanosis of shrimp. Fish Sci 69(2):379–384

Jayathilakan K, Sharma KG, Radhakrishna K, Bawa AS (2007) Antioxidant potential of synthetic and natural antioxidants and its effect on warmed-over-flavour in different species of meat. Food Chem 105:908–916

Jeyasekaran G, Ganesan P, Anandaraj R, Jeya Shakila R, Sukumar D (2006) Quantitive and qualitative studies on the bacteriological quality of Indian white shrimp (*Penaeus indicus*) stored in dry ice. Food Microbiol 23:526–533

Jeyaweera V, Subasinghe S (1988) Some chemical and microbiological changes during chilled storage of prawns (*Penaeus indicus*). FAO Fisheries Report No. 401, pp 19–22

Jiang S, Hwang D, Chen C (1988) Effect of storage temperatures on the formation of disulfides and denaturation of milkfish actomyosin (*Chanos chanos*). J Food Sci 53:1333–1335

Jose J (2015) Classification, biodiversity, and conservation of marine crabs. In: Gopalakrishnan A (ed) Summer school on recent advances in marine biodiversity conservation & management. CMFRI Lecture Note Series No, 1/2015 Central Marine Fisheries Research Institute, DARE, Govt. of India, Kerala, India

Jun JY, Jung MJ, Kim DS, Jeong IH, Kim BM (2017) Post mortem changes in physiochemical and sensory properties of red snow crab (*Chionoecetes japonicus*) leg muscle during freeze storage. Fish Aquat Sci 20:13

Jussila J, Tiitinen V, Fotedar R, Kokko H (2013) A Simple and efficient cooling method for post-harvest transport of the commercial crayfish catch. Freshw Crayfish 19(1):15–19

Kaale LD, Eikevik TM, Rustad T, Kolsaker K (2011) Super chilling of food: a review. J Food Eng 107:141–146

Kaestner A, Levi H (1970) Invertebrate zoology. Interscience Publishers, New York, 523 p

Kalleda RK, Han IY, Toler JE, Chen F, Kim HJ, Dawson PL (2013) Shelf life extension of shrimp (white) using modified atmosphere packaging. Pol J Food Nutr Sci 63(2):87–94

Kasai T, Sakai H (2004) Seasonal changes in eicosapentaenoic acid and docosahexaenoic acid contents in muscle lipids of Hokkai prawn *Pandalus kessleri*. Fish Sci 70:527–529

Kauffeld M, Wang MJ, Goldstein V, Kasza KE (2010) Ice slurry applications. Int J Refrig 33(8):1491–1505

Kiani H, Zhang Z, Sun DW (2013) Effect of ultrasound irradiation on ice crystal size distribution in frozen agar gel samples. Innov Food Sci Emerg Technol 18:126–131

Kim J, Marshall MR, Wei C (2000) Polyphenoloxidase. In: Haard NF, Simpson BK (eds) Seafood enzymes: utilization and influence on postharvest seafood quality. Marcel Dekker, New York, pp 271–315

King I, Childs MT, Dorsett C, Ostrander JG, Monsen ER (1990) Shellfish: proximate composition, minerals, fatty acids, and sterols. J Am Diet Assoc 90:677–685

Knorr D, Zenker M, Heinz V, Lee DU (2004) Applications and potential of ultrasonics in food processing. Trends Food Sci Technol 15:261–266

Kocataş A, Katağan T, Uçal O, Benli HA (1991) Turkey prawns and shrimp breeding, vol 4. T. C. Ministry of Agriculture and Rural Affairs, Directorate of Aquaculture Research Institute, pp 1–144

Krzeczkowski RA (1970) Fatty acids in raw and processed Alaska pink shrimp. J Am Oil Chem Soc 47:451–452

Krzynowek J, Panunzio LJ (1989) Cholesterol and fatty acids in several species of shrimp. J Food Sci 54:237–239

Kucukgulmez A, Celik M, Yanar Y, Ersoy B, Cikrikci M (2006) Proximate composition and mineral contents of the blue crab (*Callinectes sapidus*) breast meat, claw meat and hepatopancreas. Int J Food Sci Technol 41:1023–1026

Kumlu M (2001) Shrimp, lobster, and mussels. Textbook. Çukurova University Faculty of Fisheries Publications No: 6, pp 25–42

Kung H, Wang T, Huang Y, Lin C, Wu W, Lin C, Tsai Y (2009) Isolation and identification of histamine-forming bacteria in tuna sandwiches. Food Control 20:1013–1017

Ladewig KF, Schaer SL (1993) Crwfish. A healthy choice. Southern Regional Aquaculture Center. Publication N: 243, 4 p

Lagerstedt A, Enfa L, Johansson L, Lundstro K (2008) Effect of freezing on sensory quality, shear force and water loss in beef *M. Longissimus dorsi*. Meat Sci 80:457–461

Lakshmanan R, Shakila RJ, Jeyasekaran G (2002) Survival of amine-forming bacteria during the ice storage of fish and shrimp. Food Microbiol 19:617–625

Lalitha KV, Surendran PK (2006) Microbiological changes in farm reared freshwater prawns (*Macrobrachium rosenbergii* de Man) in ice. Food Control 17:802–807

Lannelongue M, Finne G, Hanna MO, Nickelson R, Vanderzant C (1982) Storage characteristics of brown shrimp (*Penaeus aztecus*) stored in retail packages containing CO_2-enriched atmosphere. J Food Sci 47:911–913, 923

Lapeyre J, Lapeyre R, Profumo F, Rauls S (2004) The Lapeyre automatic shrimp peeling machine, model 'A', no. 572, 1979. American Society of Mechanical Engineers, Biloxi

Latyshev NA, Kasyanov SP, Kharlamenko VI, Svetashev VI (2009) Lipids and of fatty acids of edible crabs of the North-Western Pacific. Food Chem 116:657–661

Lee DOC, Wickins JF (1992) Techniques Species/groups. Crustacean farming. Blackwell Scient, Oxford, pp 216–222

Leitao MFF, Rios DPA (2000) Microbiological and chemical changes in freshwater prawn (*Macrobrachium rosenbergii)* stored under refrigeration. Braz J Microbiol 31:178–183

Li G, Sinclair AJ, Li D (2011) Comparison of lipid content and fatty acid composition in the edible meat of wild and cultured freshwater and marine fish and shrimps from China. J Agric Food Chem 59(5):1871–1881

Lian F, Magdeb I, Lorentzena G, Siikavuopioa SI, Øverbøa K, Vanga B, Lindberg D (2018) Exploring the effect of inhibitors, cooking, and freezing on melanosis in snow crab (*Chionoecetes opilio*) clusters. Food Control 92:255–266

Lin TM, Durance TD, Scaman CH (1999) Physical and sensory properties of vacuum microwave dehydrated shrimp. J Aquat Food Prod Technol 8:41–53

Lin T, Wang JJ, Li JB, Liao C, Pan YJ, Zhao Y (2013) Use of acidic electrolyzed water ice for preserving the quality of shrimp. J Agric Food Chem 61(36):8695–8702

Linton JB, Gordon JM (1992) U.S. Patent No. 5,087,221 A. U.S.Patent and Trademark Office, Washington, DC

Londahl G (1997) Technological aspects of freezing and glazing shrimp. Infofish Int 3:49–56

Lopez A, Williams HL, Ward DR (1981) Essential elements in raw, boiled, steamed, and pasteurized crabmeat. J Food Sci 46:1128–1131

Lopez-Caballero M, Gonçalves A, Nunes M (2002) Effect of CO2/O2-containing modified atmospheres on packed deepwater pink shrimp (*Parapenaeus longirostris*). Eur Food Res Technol 214(3):192–197

Lopez-Caballero ME, Martinez-Alvarez O, Gomez-Guillen MC, Montero P (2006) Quality of Norway lobster (*Nephrops norwegicus*) treated with a 4-hexylresorcinol-based formulation. Eur Food Res Technol 222:425–431

Losada V, Rodriguez O, Miranda JM, Barros-Veazquez J, Aubourg SP (2006) Development of different damage pathways in Norway lobster (*Nephrops norvegicus*) stored under different chilling systems. J Sci Food Agric 86(10):1552–1558

Lu F, Zhang JY, Liu SL, Wang Y, Ding YT (2011) Chemical, microbiological, and sensory changes of dried Acetes chinensis during accelerated storage. Food Chem 127(1):159–168

Ma LY, Deng JC, Ahmed EM, Adams JP (1983) Canned shrimp texture as a function of its heat history. J Food Sci 42(2):360–363

Mahalaxmi B, Revathy K, Raghunathan C, Anjalai K, Subashini A (2013) Distribution of microbial population associated with crabs from Ennore Sea coast Bay of Bengal North East coast of India. Int J Curr Microbiol App Sci 2(5):290–305

Majumdar RK, Roy D, Saha A (2017) Textural and sensory characteristics of retort-processed freshwater prawn (*Macrobrachium rosenbergii*) in curry medium. Int J Food Prop 20(11):2487–2498

Maldhavi DL, Deshpande SS, Salunkhe DK (1995) In: Maldhavi DL, Deshpande SS, Salunkhe DK (eds) Food antioxidants. Technological, toxicological and health perspective. Marcel Dekker, Inc., New York

Manheem K, Benjakul S, Kijroongrojana K, Visessanguan W (2012) The effect of heating conditions on polyphenol oxidase, proteases and melanosis in precooked Pacific white shrimp during refrigerated storage. Food Chem 131(4):1370–1375

Manjarekar RG, Mohode AG (2010) Economic evaluation of solar tunnel dryer for drying peeled prawns. Int J Agric Eng 3(1):68–72

Manso T, Texeira L, Correia PMR (2013) Frozen fish: control of glazing operation. Int J Biol Food Vet Agric Eng 7:310–314

Marques A, Teixeira B, Barrento S, Anacleto P, Carvalho ML, Nunes ML (2010) Chemical composition of Atlantic spider crab *Maja brachydactyla*: Human health implications. J Food Compos Anal 23:230–237

Marra A, Mona S, Sa RM, D'Onghia G, Maiorano P (2015) Population genetic history of *Aristeus antennatus* (Crustacea: Decapoda) in the Western and Central Mediterranean Sea. PLoS One 10(5):e0128609

Marshall MR, Kim J, Wei CI (2000) Enzymatic browning in fruits, vegetables and seafoods. Food Agric Organ 41:259–312

Martínez-Alvarez O, Lopez-Caballero ME, Montero P, Gomez-Guillen MC (2005a) A 4-hexylresorcinol based formulation to prevent melanosis and microbial growth in chilled tiger prawns (*Marsupenaeus japonicus*) from aquaculture. J Food Sci 70:M415–M422

Martínez-Alvarez O, Montero P, Gomez-Guillen M d C (2005b) Controlled atmosphere as coadjuvant to chilled storage for prevention of melanosis in shrimps (*Parapenaeus longirostris*). Eur Food Res Technol 220(2):125–130

Martínez-Alvarez O, López-Caballero ME, Montero P, Gómez-Guillén MC (2007) Spraying of 4-hexylresorcinol based formulations to prevent enzymatic browning in Norway lobsters (*Nephrops norvegicus*) during chilled storage. Food Chem 100:147–155

Martínez-Alvarez O, Gomez-Guillen C, Montero P (2008a) Effect of different chemical compounds as coadjutants of 4-hexylresorcinol on the appearance of deepwater pink shrimp (*Parapenaeus longirostris*) during chilled storage. Int J Food Sci Technol 43:2010–2018

Martínez-Alvarez O, Gomez-Guillen MC, Montero P (2008b) Chemical and microbial quality indexed of Norwegian lobster (*Nephrops norvegicus*) dusted with sulfites. Int J Food Sci Technol 43:1099–1010

Martínez-Alvarez O, Lopez-Caballero ME, Gomez-Guillen M d C, Montero P (2009) The effect of several cooking treatments on subsequent chilled storage of thawed deepwater pink shrimp (*Parapenaeus longirostris*) treated with different melanosis-inhibiting formulas. LWT-Food Sci Technol 42(8):1335–1344

Matiella J, Hsieh CY (1990) Analysis of crab meat volatile compounds. J Food Sci 55(44):962–966

Matsumoto JJ (1979) Denaturation of fish muscle protein during frozen storage. In: Fennema OR (ed) Protein at low-temperatures. American Chemical Society, Washington, DC, pp 205–224

McEvily AJ, Iyengar R, Otwell WS (1991) 4-Hexylresorcinol controls enzymatic browning in shrimp and has potential for application in a variety of other foods and beverages. Food Technol 45:80–86

McLeese DW, Wilder DG (1958) The activity and catchability of the lobster (*Homarus americanus*) in relation to temperature. J Fish Res Board Can 15:1345–1354

Mendes R (2006) Guidebook on melanosis inhibitors and processing technology of crustaceans. INIAP/IPIMAR: Project QLK1-CT-2002e71517 (CRUSTAMEL new approaches to the crustaceans prevention of melanosis and quality indices), pp 41

Mendes R, Silva HA, Nunes ML, Empis JMA (2000) Deteriorative changes during ice storage of irradiated blue jack mackerel (*Trachurus picturatus*). J Food Biochem 24:89–105

Mendes R, Gonçalves A, Pestana J, Pestana C (2005) Indole production and deepwater pink shrimp (*Parapenaeus longirostris*) decomposition. Eur Food Res Technol 221:320–328

Mietz JL, Karmas E (1978) Polyamines and histamine content in rockfish, salmon, lobster and shrimp as an indicator of decomposition. J Assoc Off Anal Chem 61:139–145

Mika A, Skorkowski E, Stepnowski P, Golebiowski M (2013) The lipid composition of the abdominal muscle of shrimp *Crangon crangon* from the Gulf of Gdansk in spring and winter periods. J Mar Biol Assoc U K 93(7):1825–1183

Mohamed MP, Devaraj M (1997) Transportation of live finfishes and shellfishes. Central Marine Fisheries Research Institute Indian Council of Agricultural Research (CMFRI) Special Publication Number 66, pp 29–32

Mohan Rajan KV, Meenakumari B, Balasubramanyan R (1981) Spiny lobster and their fishing techniques. Fish Technol 18:1–11

Mohan CO, Ravishankar CN, Bindu J, Geethalakshmi V, Srinivasa Gopal TK (2006) Effect of thermal process time on quality of "shrimp kuruma" in retortable pouches and aluminium cans. J Food Sci 71(6):S496–S500

Montero P, Avalos A, Perez-Mateos M (2001a) Characterization of polyphenoloxidase of prawns (*Penaeus japonicus*). Alternatives to inhibition: additives and high-pressure treatment. Food Chem 75:317–324

Montero P, Lopez-Caballero ME, Perez-Mateos M (2001b) The effect of inhibitors and high pressure treatment to prevent melanosis and microbial growth on chilled prawns (*Penaeus japonicus*). J Food Sci 66(8):1201–1206

Montero P, Martínez-Alvarez O, Gomez-Guillen M d C (2004) Effectiveness of onboard application of 4-hexylresorcinol in inhibiting melanosis in shrimp (*Parapenaeus longirostris*). J Food Sci 69(8):C643–C647

Montero P, Martínez-Alvarez O, Zamorano J, Alique R, Gomez-Guillen M d C (2006) Melanosis inhibition and 4-hexylresorcinol residual levels in deepwater pink shrimp (*Parapenaeus longirostris*) following various treatments. Eur Food Res Technol 223(1):16–21

Moody WM (1974) Louisiana seafood delight, the blue crab. Louisiana State University Agricultural Centre, vol 7, p 86

Moorjani MN, Dani NP (1968) Textural and reconstitutional properties of freeze-dried shrimp. Food Technol (Champaign) 22(7):886–888

Morgan GR (1974) Aspects of the population dynamics of the Western rock lobster, *Panulirus cygnus* George. II Seasonal changes in the catchability coefficient. Aust J Mar Freshw Res 25:249–259

Moses JA, Norton T, Alagusundaram K, Tiwar BK (2014) Novel drying techniques for the food industry. Food Eng Rev 6:43–55

Musaiger AO, Al-Rumaidh RJ (2005) Proximate and mineral composition of crab meat consumed in Bahrain. Int J Food Sci Nutr 56(4):231–235

Naczk M, William J, Brennan K, Liyanapathirana C, Shahidi F (2004) Compositional characteristics of green crab (*Carcinus maenas*). Food Chem 88(3):429–434

Namsanguan Y, Tichangton S, Tia W, Soponronnarit S, Devahastin S (2005) Simulation of a mixed air and superheated steam drying system. Dry Technol 23:25–248

Nayar NR, Iyer KM, Appukuttan NM, Jacob S (1962) Storage characteristics of prawns held in crushed ice and chilled sea water. Proc-Indo-Pac Fish Counc 10(2):294–297

Neil DM (2012) Ensuring crustacean product quality in the post-harvest phase. J Invertebr Pathol 110:267–275

New World Encyclopedia (2019) Crayfish. https://www.newworldencyclopedia.org/entry/Crayfish. Accessed 11 Sept 2019

Ng PKL (1998) Crabs. In: Carpenter KE, Niem VH (eds) The living marine resources of the western central Pacific, vol 2. Cephalopods, crustaceans, holothurians and sharks. FAO species identification guide for fishery purposes. FAO, Rome, pp 1045–1154

Nguyen MV, Arason S, Eikevik TM (2014) Drying of fish. In: Boziaris IS (ed) Seafood processing technology quality and safety. Wiley, Chichester, pp 161–207

Nguyen TT, Barber AR, Corbin K, Zhang W (2017) Lobster processing by-products as valuable bioresource of marine functional ingredients, nutraceuticals, and pharmaceuticals. Bioresour Bioprocess 4(1):27

Niamnuy C, Devahastin S, Soponronnarit S (2007) Effects of process parameters on quality changes of shrimp during drying in a jet-spouted bed dryer. J Food Sci 72(9):553–563

Nicolas J, Billaud C, Rouet-Mayer M-A, Philippon J (2003) Enzymatic browning-technical aspects and assays. In: Caballero B, Trugo L, Finglas PM (eds) Encyclopaedia of food sciences and nutrition. Academic, London, pp 686–692

Ninlanon W (2011) Effect of pre-cooling on muscle yield of mud crab, Scylla serrata, during emersion storage. In: 2nd international conference on biotechnology and food science, IPCBEE vol 7, IACSIT Press, Singapore

Nirmal NP, Benjakul S (2009a) Melanosis and quality changes of Pacific White shrimp (*Litopenaeus vannamei*) treated with catechin during iced storage. J Agric Food Chem 57(9):3578–3586

Nirmal NP, Benjakul S (2009b) Effect of ferulic acid on inhibition of polyphenoloxidase and quality changes of Pacific white shrimp (*Litopenaeus vannamei*) during iced storage. Food Chem 116:323–331

Nirmal N, Benjakul S (2010) Effect of green tea extract in combination with ascorbic acid on the retardation of melanosis and quality changes of pacific white shrimp during ıced storage. Food Bioprocess Technol 5(8):1–11

Nirmal N, Benjakul S (2011a) Retardation of quality changes of Pacific white shrimp by green tea extract treatment and modified atmosphere packaging during refrigerated storage. Int J Food Microbiol 149(3):247–253

Nirmal N, Benjakul S (2011b) Inhibition of melanosis formation in Pacific white shrimp by the extract of lead (*Leucaena leucocephala*) seed. Food Chem 128(2):427–432

Nirmal NP, Benjakul S (2012) Biochemical properties of polyphenoloxidase from the cephalothorax of Pacific white shrimp (*Litopenaeus vannamei*). Int Aquat Res 4(1):6–13

Noomhorm A, Vongsawasdi P (1998) Effect of preservation methods on qualities of giant freshwater prawns (*Macrobrachium rosenbergii*). J Food Qual 21:145–154

Noomhorm A, Vongsawasdi P (2004) Freezing shellfish. In: Hui YH, Cornillon P, Legarretta IG, Lim MH, Murrell KD, Nip WK (eds) Food science and technology—handbook of frozen foods. Marcel Dekker, New York, p 133

Noordin WNM, Shunmugam N, Huda N, Adzıtey F (2018) The Effects of essential oils and organic acids on microbiological and physicochemical properties of whole shrimps at refrigerated storage. Curr Res Nutr Food Sci 6(2):273–283

O'grady MN, Maher M, Troy DJ, Moloney AP, Kerry JP (2006) An assessment of dietary supplementation with tea catechins and rosemary extract on the quality of fresh beef. Meat Sci 73:132–143

Odilichukwu C, Okpala R, Choo WS, Dykes AG (2014) Quality and shelf life assessment of Pacific white shrimp (*Litopenaeus vannamei*) freshly harvested and stored on ice. LWT-Food Sci Technol 55:110–116

Oehlenschläger J (1997) Volatile amines as freshness/spoilage indicators. A literature review. In: Luten J, Borresen T, Oehlenschläger J (eds) Seafood from producer to consumer, integrated approach to quality. Elsevier, Amsterdam, 571 p

Ogawa M, Perdigao NB, de Santiago ME, Kozima TT (1984) On physiological aspects of black spot appearance in shrimp. Bull Jpn Soc Sci Fish 50(10):1763–1769

Okpala COR (2014a) Investigation of quality attributes of ice-stored Pacific white shrimp (*Litopenaeus vannamei*) as affected by sequential minimal ozone treatment. LWT-Food Sci Technol 57:538–547

Okpala COR (2014b) Changes of Physico-chemical properties of sequential minimal ozone treated ice-stored Pacific white shrimp (*Litopenaeus vannamei*). Curr Nutr Food Sci 10:218–227

Okpala COR (2015) Quality evaluation and shelf life of minimal ozone-treated Pacific white shrimp (*Litopenaeus vannamei*) stored on ice. J Verbr Lebensm 10(1):49–57

Okpala COR (2017) Changes in some proximate, colour and textural characteristics of ozone-processed shrimp: combined effects of increasing ozone discharge and iced storage. Iran J Fish Sci 16(2):625–638

Okpala COR, Choo WS, Dykes GA (2014) Quality and shelf life assessment of Pacific white shrimp (*Litopenaeus vannamei*) freshly harvested and stored on ice. LWT-Food Sci Technol 55:110–116

Oksuz A, Mazlum Y (2016) Determination of proximate composition and fatty acid profiles of *Astacus leprodactylus* Eschscholtz, 1823 in Turkish freshwater resources. Crustaceana 89(10):1135–1147

Oksuz A, Ozyilmaz A, Aktas M, Gerçek G, Motte J (2009) A comparative syudy on proximate, mineral and fatty acid compositions of deep seawater rose shrimp (*Parapenaeus longirostris*, Lucas 1846) and golden shrimp (*Plesionica martia*, Milne-Edwards, 1883). J Anim Vet Adv 8(1):183–189

Olafsdottir G, Nesvadba P, Di Natale C, Careche M, Oehlenschläger J, Tryggvadóttir SV, Schubring R, Kroeger M, Heia K, Esaissen M, Macagnano A, Jørgensen BM (2004) Multisensor for fish quality determination. Trends Food Sci Technol 15(2):86–93

Olgunoglu IA (2010) Determination of microbiological contamination sources of blue crab meat (*Callinectes sapidus* Rathbun, 1896) during pasteurization process. Pak J Zool 42(5):545–550

Olgunoglu IA, Artar E, Gocer M (2015) Comparison of fatty acid profiles of male and female giant red shrimps (*Aristaeomorpha foliacea* RISSO, 1827) obtained from Mediterranean Sea. Ital J Food Sci 27:432–436

Opoku-Gyamfua A, Simpson BK, Squires EJ (1992) Comparative studies on the polyphelooxidase fraction of lobster and tyrosinase. J Agric Food Chem 40:772–775

Osuga D, Van der Schaaf A, Whitaker JR (1994) Control of polyphenol oxidase activity using a catalytic mechanism. In: Yada RY, Jaclunan RL, Smith JJ (eds) Protein structure-function relationships in foods. Blackie Academy and Professional, Glasgow, pp 62–88

Otwell WS, Bellairs JJ, Sweat D (1984) Initial development of a deep-sea crab fishery in the Gulf of Mexico. Florida Sea Grant College. Report Number 61, 29 p

Otwell WS, Iyengar R, McEvily AJ (1992) Inhibition of shrimp melanosis by 4-hexylresorcinol. J Aquat Food Prod Technol 1(1):53–65

Ozogul Y, Ayas D, Ozogul F, Ozyurt G, Kuley EB, Yazgan H (2010) Comparison of fatty acid, trace element and proximate compositions of male and female of blue crabs and swim crabs from Mersin Bay, Turkey. Rapp Comm Int Mer Médit 39:618

Pacquit A, Frisby J, Diamond D, Lau KT, Farrell A, Quilty B, Diamond D (2007) Development of a smart packaging for the monitoring of fish spoilage. Food Chem 102:466–470

Pardio VT, Waliszewski KN, Zuniga P (2011) Biochemical, microbiological and sensory changes in shrimp (*Panaeus aztecus*) dipped in different solutions using face-centred central composite design. Int J Food Sci Technol 46:305–314

Paterson BD, Goodrick GB (1994) Out of the frying pan into the flyer: harvesting, storage and transport of live spanner crabs. In: Paterson B, Exley P, Smith R (eds) Live transporr of crustceans in air prolonging the survival of crabs, Report to the Fisheries Research and Development Corporation (Project 92/71), Department of Primary Industries, Queensland, December 1994, pp 1–5

Paterson BD, Goodrich B, Exley P, Smith R (1994) Storing spanner crabs (Ranina ranina) in air at low temperatures on boats reduces physiological stress and improves survival in storage tanks on shore. In: Paterson B, Exley P, Smith R (eds) Live transporr of crustceans in air prolonging the survival of crabs, Report to the Fisheries Research and Development Corporation (Project 92/71), Department of Primary Industries, Queensland, pp 7–21

Paul DK (2015) Effect of different ratios of ice on quality and shelf life of cold water shrimp (*Pandalus borealis*). United Nations University Fisheries Training Programme, Iceland [final project]

Perera CO, Rahman MS (1997) Heat pump dehumidifier drying of food. Trends Food Sci Technol 8:75–79

Phillips FA, Peeler JT (1972) Bacteriological survey of the blue crab industry. Appl Microbiol 24(6):958–966

Piratheepa S, Chitravadivelu K, Edrisinghe U (2012) Updates on the species of shrimps in Kakkaithivu coastal waters, Jaffna Estuary, Sri Lanka. Trop Agric Res 24(1):82–90

Pires DR, de Morais ACN, Coelho CCS, Marinho AF, Góes LCDSA, Augusta IM, Ferreira FS, Saldanha T (2018) Nutritional composition, fatty acids and cholesterol levels in Atlantic white shrimp (*Litopenaeus schimitti*). Int Food Res J 25(1):151–157

Ponting JD, Jackson R, Watters G (1972) Refrigerated apple slices: preservative effects of ascorbic acid, calcium and sulfites. J Food Sci 37:434–436

Poppke RW (1977) How to catch a crab. Stein and Day Publishers Scarborough House, Briarcliff Manor, New York, 1277 p

Portella CG, Sant'Ana LS, Valenti WC (2011) Sensory aspects of liquid smoking of giant river prawn: comparison with traditional smoking. Int J Food Sci Technol 46:834–839

Porter A (2019) New process patented for freezing lobster. http://ellsworthamerican.com/archive/news2000/03-23-00/ea_story4_03-23-00.html. Accessed 9 Sept 2019

Prabhu NM, Rajasekar P, Nisha RG, Yasminebegam AA, Kannapiran E, Manikandan R (2016) Effect of turmeric on shrimp (*Penaeus semisulcatus*) shelf life extension in chilled storage. J Fish Sci 15(1):39–52

Prachayawarakorn S, Soponronnarit S, Wetchacama S, Jaisut D (2002) Desorption isotherms and drying characteristics of shrimp in superheated steam and hot air. Dry Technol 20(3):669–684

Prodohl PA, Jorstad KE, Triantafyllidis A, Katsares V, Triantaphyllidis C (2007) European lobster–*Homarus gammarus*. In: Svåsand T, la Crosetti D, García-Vázquez E, Verspoor E (eds) Genetic impact of aquaculture activities on native populations, Gen impact final scientific report (EU contraction. RICA-CT-2005-022802), pp 91–98

Qingzhu Z (2003) Quality Indicators of Northern Shrimp (*Pandalus borealis*) stored under different cooling conditions. UNU-Fisheries Training Programme Final Project, 45 p

Quang NY (2005) Guidelines for handling and preservation of fresh fish for further processing in Vietnam. Final project. The United Nations University, Fisheries Training programme. Reykjavik, Iceland

Rahman MS, Labuza TP (2007) Water activity and food preservation. In: Rahman MS (ed) Handbook of food preservation. CRC Press, Boca Raton, pp 447–476

Rahman MS, Velez-Ruiz JF (2007) Food preservation by freezing. In: Rahman MS (ed) Handbook of food preservation. CRC Press, Taylor and Francis Group, Boca Raton, pp 636–657

Rahman MM, Yasmin L, Kamal M, Islam MN, Ahmed SU, Mazid MA (2000) Studies on the post mortem changes in shrimp and prawn during ice storage: I. Organoleptic and physical changes. Bangladesh J Fish Res 4(1):83–89

Ramamoorthy N, Karuppasamy PK, Sri Sakthi Priyadarshini R (2016) Proximate, amino acid and fatty acid composition the marine crabs from the southeast coast of India. J Mar Biosci 2(1):91–98

Ramesh Babu K, GovindaRao V, Krishna NM, Geetha S, Kakara RR (2017) Assessment of bacteriological quality in selected commercially ımportant shrimps of visakhapatnam, east coast of India. Int J Microbiol Biotechnol 2(2):102–105

Ratti C, Mujumdar AS (1995) Infrared drying. In: Mujumdar AS (ed) Handbook of industrial drying. Marcel Dekker, New York, pp 567–588

Rebach JS, Stribling J, Wilber M (1990) Frozen storage quality changes in whole jonah crabs. J Food Qual 13:203–208

Requena DD, Hale SA, Green DP, McClure WF, Farkas BE (1999) Detection of discoloration in thermally processed blue crab meat. J Sci Food Agric 79:786–791

Reyes A, Ceron S, Zuniga R, Moyano P (2007) A comparative study of microwave-assisted air drying of potato slices. Biosyst Eng 98:310–318

Riaz M, Qadri RB (1990) Time temperature tolerance of frozen shrimp. 2. Biochemical and microbiological changes during storage of frozen glazed shrimps. Trop Sci 30:343–349

Richards-Rajadurai PN (1989) Live storage and distribution of fish and shellfish. Infofish Int 3:23–27

Robson AA, Kelly MS, Latchford CW (2007) Effect of temperature on the spoilage rate of whole, unprocessed crabs: *Carcinus maenas*, *Necora puber* and *Cancer pagurus*. Food Microbiol 24:419–424

Rosa R, Nunes ML (2004) Nutritional quality of red shrimp *Aristeus antennatus* Risso, pink shrimp *Parapenaeus longirostris* Lucas and Norway lobster *Nephrops norvegicus* Linnaeus. J Sci Food Agric 8:89–94

Roth B, Grimsb E (2013) Electical stunning of edible crabs. NOFIMA report No: 18/2013. https://www.nofima.no/filearchive/Rapport%2018-2013.pdf. Accessed 31 May 2019

Roth B, Grimsb E (2016) Electrical stunning of edible crabs (*Cancer pagurus*): from single experiments to commercial practice. Anim Welf 25:489–497

Rotllant G, Arnau F, Garcia JA, Garcia N, Rodriguez M, Sarda F (2002) Effect of metabisulphite treatments and freezing on melanosis inhibition in rose shrimp *Aristeus antennatus* (Risso, 1816). Food Sci Technol Int 8(4):243–247

RSPCA Australia (2017) Humane killing and processing of crustaceans for human consumption. http://www.rspca.org.au. Accessed 14 Sept 2017

Sachindra NM, Bhaskar N, Mahendrakar NS (2005) Carotenoids in different body components of Indian shrimps. J Sci Food Agric 85:167–172

Saclier M, Peczalski R, Andrieu J (2010) Effect of ultrasonically induced nucleation on ice crystals' size and shape during freezing in vials. Chem Eng Sci 65:3064–3071

Sae-leaw Y, Benjakul S (2019) Prevention of quality loss and melanosis of Pacific white shrimp by cashew leaf extracts. Food Control 95:257–266

Salin KR (2005) Live transportation of *Macrobrachium rosenbergii* (De Man) in chilled sawdus. Aquac Res 36:300–310

Salin KR, Vadhyar KJ (2001a) Effect of different chilling rates for cold anaesthetization of *Penaeus monodon* (Fabricius) on the survival, duration and sensory quality under live storage in chilled sawdust. Aquac Res 32:145–155

Salin KR, Vadhyar KJ (2001b) Cold anaesthetization and live transportation of penaeid shrimps. Fish Technol 38:71–76

Samiee K, Abdolhossein R, Keshavarz F (2017) Omega-3, 6 and 9 fatty acids composition and lipid content from liver and muscle tissues of spiny lobster (*Panulirus homarus*) in the Persian Gulf. J Food Nutr Res 5(1):27–30

Santerre CR, Cash JN, Vannorman DJ (1988) Ascorbic acid/citric acid combinations in the processing of frozen apple slices. J Food Sci 53:1713–1717

Sapers GM (1993) Browning of foods: control by sulfites, antioxidants, and other means. Food Technol 47:75–84

Sapers GM, Ziolkowski MA (1987) Comparison of erythorbic and ascorbic acids as inhibitors of enzymatic browning in apple. J Food Sci 52:1732

Sapers GM, Hicks KB, Phillips JG, Garzarella L, Pondish DL, Matulaitis RM, McCormack TJ, Sondey SM, Seib PA, Ei-Atawy YS (1989) Control of enzymatic browning in apple with ascorbic acid derivatives, polyphenol oxidase inhibitors, and complexing agents. J Food Sci 54:997–1002

Sapers GM, Garzarella L, Pilizota V (1990) Application of browning inhibitors to cut apple and potato by vacuum and pressure infiltration. J Food Sci 55:1049–1053

Sarnoski PJ (2007) Instrumental methods for determining quality of blue crab (*Callinectes sapidus*) meat. Thesis submitted to the faculty of the Virginia Polytechnic Institute and State University in partial fulfillment of the requirements for the degree of Master of Science in Food Science and Technology, 3 May 2007, Blacksburg, Virginia

Sathiya U, Valarmathi V (2018) Diversity of marine shrımp species along the coast of Nagapattinam, Tamil Nadu, India. Int J Zool Appl Biosci 3(3):410–416

Schoemaker R (1991) Transportation of live and processed seafood. INFOFISH technical handbook, vol 3. Infofish, Kuala Lumpur, Malaysia, p 17

Shakila RJ, Vasundhara TS, Rao DV (1995) Rapid quality assessment of shrimps during storage by monitoring amines. J Food Sci Technol (Mysore) 32:310–314

Shamshad SI, Kher-un-Nisa K, Riaz M, Zuberi R, Qadri RB (1990) Shelf life of shrimp (*Penaeus merguiensis*) stored at different temperatures. J Food Sci 55(5):1201–1205

Sharma A, Chen CR, Lan NV (2009) Solar-energy drying systems: a review. *Renew Sust Energ Rev*, 13:1185–1210

Sharma GP, Prasad S (2004) Effective moisture diffusivity of garlic cloves undergoing microwave-convective drying. J Food Eng 65:609–617

Shaw BG, Shewan JM (1968) Psychrophilic spoilage bacteria of fish. J Appl Microbiol 31(1):89–96

Shenouda SYK (1980) Theories of protein denaturation during frozen storage of fish flesh. Adv Food Res 26:275–311

Sherwin CM (2001) Can invertebrates suffer? Or, how robust is argument-by-analogy? Anim Welf 10:103–118

Shiekh KA, Benjakul S, Sae-leaw T (2019) Effect of chamuang (*Garcinia cowa* Roxb.) leaf extract on inhibition of melanosis and quality changes of Pacific white shrimp during refrigerated storage. Food Chem 270(1):554–556

Siddiqui MN, Chowdhury MM, Hasan MJ, Haque MN (2011) Organoleptic, biochemical and microbial changes of freshwater prawn (*Macrobrachium rossenbergii)* in different storage conditions. Bangladesh Res Publ J 5(3):234–244

Sidwell VD, Foncannon PR, Moore NS, Bonnet JC (1974) Composition of the edible portion of raw (fresh or frozen) crustaceans, finfish, and mollusks. Part I Protein, fat, moisture, ash, carbohydrate, energy value and cholesterol. Mar Fish Rev 36:21–35

Sikorski ZE, Kolakowski A (1990) Freezing of marine food. In: Sikorski ZE (ed) Seafood resources, nutritional composition and preservation. CRC Press Inc., Boca Raton, pp 112–122

Simpson BK, Gagné N, Ashie INA, Noroozi E (1997) Utilization of chitosan for preservation of raw shrimp (*Pandalus borealis*). Food Biotechnol 11(1):25–44

Skonberg DI, Perkins BL (2002) Nutrient composition of green crab (*Carcinus maenus*) leg meat and claw meat. Food Chem 72:401–404

Slattery SL, Williams DJ, Cusak A (1995) A sulphite-free treatment inhibits blackspot formation in prawns. Food Aust 47(11):509–514

Smith F, Walton G (1958) The spiny lobster industry of Florida. Florida State Board of Conservation Educational series No:11, 36 p

Solval KM, Rodezno LAE, Moncada M (2014) Evaluation of chitosan nanoparticles as a glazing material for cryogenically frozen shrimp. LWT-Food Sci Technol 57:172–180

Souty-Grosset C, Fetsner JW (2016) Taxonomy and identification. In: Longshaw M, Stebbing P (eds) Biology and ecology of crayfish. CRC Press, Boca Raton, London, New York, pp 1–31

Squires HJ (1990) Decapod crustacea of the Atlantic coast of Canada. Can Bull Fish Aquat Sci 221:532

Sreenath PG, Abhilash S, Ravishankar CN, Srinivasa Gopal TK (2008) Standardization of process parameters for ready-to-eat shrimp curry in tin-free steel cans. J Food Process Preserv 32:247–269

Sriket P, Benjakul S, Visessanguan W, Kijroongrojana K (2007) Comparative studies on chemical composition and thermal properties of black tiger shrimp (*Penaeus monodon*) and white shrimp (*Penaeus vannamei*) meats. Food Chem 103:1199–1207

Stanek M, Borejszo Z, Dabrowski J, Janicki B (2011) Fat and cholesterol content and fatty acid profiles in edible tissues of spiny-cheek crayfish, *Orconectes limosus* (Raf.) from Lake Gopło (Poland). Fish Aquat Life 19(4):241–248

Stevenson JR (1985) Dynamics of the integument. In: Bliss DE, Mantel LH (eds) The biology of the Crustacea: integument, pigments and hormonal processes. Academic, New York, pp 1–44

Strømmen I, Bredesen AM, Eikevik T et al (2000) Refrigeration, air conditioning and heat pump systems for the 21st century. Bull Int Inst Refrig 2:2–18

Su G, Ramaswamy HS, Zhu S, Yu Y, Hu F, Xu M (2014) Thermal characterization and ice crystal analysis in pressure shift freezing of different muscle (shrimp and porcine liver) versus conventional freezing method. Innov Food Sci Emerg Technol 26:40–50

Subramanian TA (2007) Effect of processing on bacterial population of cuttlefish and crab and determination of bacterial spoilage and rancidity development on frozen storage. J Food Process Preserv 31:13–31

Suklim K, Flick GJ, Bourne DW, Granata LA, Eifert J, Williams R, Popham D, Wittman R (2008) Microbiology, physical and sensory quality of vacuum packaged fresh blue crab meat (*Callinectes sapidus*) treated with high hydrostatic pressure. Food Prot Trends 28:96–106

Sun H, Lv H, Yuan G, Fang X (2014) Effect of grape seed extracts on the melanosis and quality of Pacific white shrimp (*Litopenaeus vannamei*) during Iced Storage. Food Sci Technol Res 20(3):671–677

Swastawati F, Suzuki T, Dewi EN, Winarni T (2000) The effect of liquid smoke on the quality and omega-3 fatty acids content of tuna fish (*Euthynnus affinis*). J Coast Dev 3:573–579

Tam LN, Thuy P, Uyen X, Vu THA (2018) Effect of treatment on quality of black tiger shrimp (*Penaeus monodon*). Vietnam J Chem 56(6):773–777

Tantratian S, Setjintanin J, Sanguandeekul R (2011) Improving the quality of cold storage black tiger prawn (*Penaeus monodon*) usingozonated water. In: 2nd international conference on biotechnology and food science IPCBEE, vol 7. IACSIT Press, Singapore, pp 105–109

Tapaneyasin R, Devahastın S, Tansakul A (2005) Drying methods and quality of shrimp dried in a jet-spouted bed dryer. J Food Process Eng 28:35–52

Tavares M (1998) True crabs. In: Carpenter EK, Niem H (eds) The living marine resources of the western central pacific. Cephalopods, Crustaceans, Holothurians and Sharks. FAO fisheries identification guide for fishery purposes. FAO, Rome, pp 327–343

Tavares M (2002) Shrimps. The living marine Resources of the Western Central Pacific. I. In: Carpenter EK, Niem H (eds) Introduction, molluscs, crustaceans, hagfishes, sharks, batoid fishes, and chimaeras. FAO fisheries identification guide for fishery purposes. FAO, Rome, pp 253–291

Taylor PG (1993) The application of phosphates in the processing of the pacific shrimp, or what's so different about this use? In: Proceedings of the annual tropical and subtropical fisheries technological conference of the Americas, Williamsburg, Virginia, USA, pp 72–77

Thepnuan RA (2007) In Studies on some enzymes and the effect of modified atmosphere packaging and some chemicals on quality changes of black tiger and white shrimps during refrigerated storage. Master of Science Thesis, Prince of Songkla University, Hat Yai, Thailand

Thomas FB, Thomas SD (1983) Technical operations manual for the blue crab industry. Sea Grant Publication UNC-SG-83-02. 44 pp, US

Thorarinsdottir KA, Arason S, Bogason SG, Kristbergsson K (2004) The effects of various salt concentrations during brine curing of cod (*Gadus morhua*). Int J Food Sci Technol 39:79–89

Tirawanichakul S, Na Phatthalung W, Tirawanıchakul Y (2008) Drying strategy of shrimp using hot air convection and hybrid infrared adiation/hot air convection. Walailak J Sci Technol 5(1):77–100

Tirloni E, Stella S, Gennari M, Colombo F, Bernardi C (2016) American lobsters (*Homarus americanus*) not surviving during air transport: evaluation of microbial spoilage. Ital J Food Saf 5(2):75–80

Toktas B (2018) Investigation of the inhibitory effect of some reducing agents on enzymatic browning (melanosis) in different shrimp species. MSc. Thesis in Aquaculture Engineering. Akdeniz University, Institute of Science, Antalya, Turkey, 79 p

Tripathy C (2013) Effect of delay in icing on the quality of *Littopenaeus vannamei* during chilled storage. Master of Fishery Science in Fish Processing Technology Department of Fish Processing Technology, College of Fishery Science, Sri Venkateswara Veterinary University, Muthukur, Andhra Prades

Tsape K, Sinanoglou VJ, Miniadis-Meimaroglou S (2010) Comparative analysis of the fatty acid and sterol profiles of widely consumed Mediterranean crustacean species. Food Chem 122:292–299

Tseng YC, Xiong Y, Webster CD, Thompson KR, Muzinic L (2002) Quality changes in Australian red claw crayfish, *Cherax quadricarinatus*, Stored at 0°C. J Appl Aquac 12(4):53–66

Tsironi T, Dermesonlouoglou E, Giannakourou M, Taoukis P (2009) Shelf life modelling of frozen shrimp at variable temperature conditions. LWT-Food Sci Technol 42:664–671

Tureli C, Celik M, Erdem U (2000) Comparison of meat composition and yield of blue crab (*Callinectes sapidus* Rathbun, 1896) and sand crab (*Portunus pelagicus* Linne, 1758) caught in İskenderun Bay, North-East Mediterranean. Turk J Vet Anim Sci 24:195–203

U.S. Dept. of Agriculture (USDA) (2019) Food composition databases. https://www.nal.usda.gov/usda-food-composition-database. Accessed 4 June 2019.

Ulikowski D, Chybowski L, Traczuk P, Ulikowska E (2017) A new design of crayfish traps reduces escaping and improves opportunities for long-term catching. Turk J Fish Aquat Sci 17:363–369

Vakili S, Ardakani SAY (2018) Antioxidant effect of orange peel extract on chemical quality, sensory properties, and black spots of farmed white shrimp. J Nutr Food Secur 3(1):19–26

Vanderzant C, Cobb BF, Thompson CA, Parker JC (1973) Microbial flora, chemical characteristics and shelf life of four species of pond reared shrimp. J Milk Food Technol 36:443–449

Varlet V, Prost C, Serot T (2007) Volatile aldehydes in smoked fish: analysis methods, occurrence and mechanisms of formation. Food Chem 105:1536–1556

Varlık C, Baygar T, Özden Ö, Erkan N, Metin S (2000) Sensory evaluation and determination of some physical and chemical characteristics of shrimp during cold storage. Turk J Vet Anim Sci 24:181–185

Venugopal V (2006) Seafood processing: adding value through quick freezing, retortable packaging and cook-chilling. CRC Press, Boca Raton, 524 p

Venugopal V (2020) Nutritional value of shrimp the popular shellfish. EC Nutr Special Issue 02:01–09

Vijayan PK, Balachandran KK (1981) Studies on blue discolouration in canned body meat of crab (*Scylla serrata*). Fish Technol 18(2):117–122

VijayaVenkataRaman S, Iniyan S, Goic R (2012) A review of solar drying technologies. Renew Sust Energ Rev 16:2652–2670

Villamiel M, Del Casillo MD, Corzo N (2006) Browning reactions. In: Hui YH (ed) Food biochemistry and food processing. Blackwell Publishing, West Sacramento, pp 71–100

Visciano P, Perugini M, Conte F, Amorena M (2008) Polycyclic aromatic hydrocarbons in farmed rainbow trout (*Oncorhynchus mykiss*) processed by traditional flue gassmoking and by liquid smoke flavourings. Food Chem Toxicol 46:1409–1413

Wakjira M (2011) A distance course of module on fisheries and aquaculture (Biol 421). In: Habtamu T (ed). Jimma University, Ethiopia, 154 p

Wang J, Lin T, Li JB, Liao C, Pan YJ, Zhao Y (2014) Effect of acidic electrolyzed water ice on quality of shrimp in dark condition. Food Control 35:207–212

Wang M, Wang J, Sun XH, Pan YJ, Zhao Y (2015) Preliminary mechanism of acidic electrolyzed water ice on improving the quality and safety of shrimp. Food Chem 176:333–341

Weineck K, Ray AJ, Fleckenstein LJ, Medley M, Dzubuk N, Piana E, Cooper RL (2018) Physiological changes as a measure of crustacean welfare under different standardized stunning techniques: cooling and electroshock. Animals (Basel) 8(9):158

Wheaton FW, Lawson TB (1985) Other preservation methods. In: Processing aquatic food products. Willey, USA, pp 294–304

Wikipedia (2019a) Crayfish. https://en.wikipedia.org/wiki/Crayfish. Accessed 11 Sept 2019

Wikipedia (2019b) *Astacus astacus*. https://en.wikipedia.org/wiki/Astacus_astacus. Accessed 11 Sept 2019

Wikipedia (2019c) *Orconectes limosus*. https://en.wikipedia.org/wiki/Orconectes_limosus. Accessed 16 Sept 2019

Wikipedia (2019d) *Procambarus clarkii*. https://en.wikipedia.org/wiki/Procambarus_clarkii. Accessed 16 Sept 2019

Wisaiproma N, Kasayapananda N, Pratinthonga N, Songprakorpa R, Thepaa S, Deeto S (2018) The study of shrimp drying by greenhouse drying combined with low humidity air. Int J Smart Grid Clean Energy 7(4):303–313

Wood R, Peat A, Galeano D (2007) Australian Fisheries Statistics 2006. Australian Bureau of Agricultural and Resource Economics, Canberra, Australia

Work TM, Bushway AA, Work RW, Pelletier RC (1997) The effect of cryogenic freezing on the quality of soft and hard shell lobsters (*Homarus americanus*). J Aquat Food Prod Technol 6(1):53–64

Wu TH, Bechtel PJ (2008) Ammonia, Dimethylamine, Trimethylamine, and Trimethylamine oxide from raw and processed fish by-products. J Aquat Food Prod Technol 17:27–38

Wu XF, Zhang M, Adhikari B, Sun J (2017) Recent developments in novel freezing and thawing technologies applied to foods. Crit Rev Food Sci Nutr 57(17):3620–3631

Yan Z, Sousa-Gallagher JM, Oliveira ARF (2008) Shrinkage and porosity of banana, pineapple and mango slices during air-drying. J Food Eng 83(3):430–440

Yanar Y, Celik M (2006) Seasonal amino acid profiles and mineral contents of green tiger shrimp (*Penaeus semisulcatus* De Haan, 1844) and speckled shrimp (*Metapenaeus monoceros* Fabricus, 1789) from the Eastern Mediterranean. Food Chem 94:33–36

Yanar Y, Celik M, Yanar M (2004) Seasonal changes in total carotenoidcontents of wild marine shrimps (*Penaeus semisulcatus* and *Metapenaeus monoceros*) inhabitating the Eastern Mediterranean. Food Chem 88:267–269

Yanar Y, Gocer M, Yanar M, Kucukgulmez A (2011) Differences in nutritional composition between cultured and wild green tiger shrimp (*Penaeus semisulcatus*). Ital J Food Sci 23:1–6

Yang H, Li J, Yu Y, Zheng W, Zhu S, Hu Y (2010) High pressure shelling of shrimp and quality improvement. Paper presented at the American Society of Agricultural and Biological Engineers Annual International Meeting 2010, ASABE 2010

Yatmaz HA, Gokoglu N (2016) Effects of plant extract-sulphide combinations on melanosis Inhibition and Quality in Shrimp (*Aristeus Antennatus*). Int J Food Prop 19:359–370

Yerlikaya P (2002) Mediterranean of chemical composition of blue crab (*Callinectes sapidus*) meat and its quality changes during frozen storage. MSc Thesis, Akdeniz University Food Engineering Department

Yerlikaya P, Gokoglu N (2004) A research note qualıty changes of blue crab (*Callinectes sapidus*) meat durıng frozen storage. J Food Qual 27:83–89

Yerlikaya P, Topuz OK, Buyukbenli HA, Gokoglu N (2013) Fatty acid profiles of different shrimp species: effects of depth of catching. J Aquat Food Prod Technol 22:290–297

Yue S (2014) An HSUS report: the welfare of farmed fish at slaughter. HSUS Reports: farm industry impacts on animals. Farm Animal Welfare: Science and Research. Animal Studies Repository. Humane Society Institute for Science and Policy. The Humane Society of the United States

Zaglol N, Eltadawy F (2009) Study on chemical quality and nutrition value of freshwater crayfish (*Procambarus clarkii*). J Arab Aquacult Soc 4(1):1–10

Zamorano JP, Martínez-Alvarez O, Montero P, del Carmen Gomez-Guillen M (2009) Characterisation and tissue distribution of polyphenol oxidase of deepwater pink shrimp (*Parapenaeus longirostris*). Food Chem 112(1):104–111

Zeng QZ, Thorarinsdotti KA, Olafsdottir G (2005) Quality changes of shrimp (*Pandalus borealis*) stored under different cooling conditions. J Food Sci 70(7):459–446

Zhao L, Chen J, Zhao X, Wang Z (2014) Effects of different freezing rates on the quality changes in cooked crayfish (*Procambarus clarkii*) meat during frozen storage (-20°C). Adv Mater Res 1033–1036:673–676

Zhu S, Le Bail A, Ramaswamy H, Chapleau N (2004) Characterization of ice crystals in pork muscle formed by pressure-shit freezing as compared with classical freezing methods. J Food Sci 69(4):190–197

Zhu S, Zhang C, Wu H, Jie J, Zeng M, Liu Z, Wang C, Yang H (2017) Spoilage of refrigerated (4°C) *Litopenaeus vannamei*: cooperation between Shewanella species and contribution of cyclo-(L-Pro-L-Leu)-dependent quorum sensing. Int J Food Sci Technol 52:1517–1526

Zotti M, Del Coco L, De Pascali SA, Migoni D, Vizzini S, Mancinelli G, Fanizzi FP (2016) Comparative analysis of the proximate and elemental composition of the blue crab *Callinectes sapidus*, the warty crab *Eriphia verrucosa*, and the edible crab *Cancer pagurus*. Heliyon 2:1–15

Chapter 3
Molluscan Shellfish

3.1 Bivalves

Bivalves are animals having two symmetrical shells belonging to the pelecypoda class of molluscs and collected in the families of Mytilidae, Ostreide, Pectinidae, Cardinidae, Donaeidae, Veneridae and Solenidae. They are found in marine and freshwater environments. They have not an obvious head. The body is placed in two shells joined by a hinge on the back. The shells are lined by a specialized tissue called the mantle. In both shell, there is a shell-bound mantle which secretes the shell, and a mantle cavity inside of the mantle which opens to the outer surface. Two large lamellar gills are placed on the mantle cavity. Internal organs and foot are on the abdominal side. The soft body inside of the shell includes a muscular foot, gills (or ctenidia, used for respiration and feeding), muscles, a digestive system, nerves, a three-chambered heart, and an open circulatory system (with sinuses). According to the bivalve type, the foot has elastic fibrous (byssus) structure that can penetrate and gouge the soft medium or adhere to the hard environment. Some species, such as the hard-shelled clam have siphons. The abdominal siphon takes the water and transmits it to the mantle cavity. The back siphon takes the water out of the mantle cavity and throws it out (Atay 1997; Mikkelsen and Henne 2011).

Bivalves are an important part of marine biomass. Bivalves also account for a large percentage of the total mollusc diversity in temperate regions. Marine bivalves exploit a wide range of habitats, from intertidal rocks to subtidal sands and deep-sea muds. Some bivalve groups burrow deeply into the sediments, and still others bore into rock or wood. Both wild and cultivated bivalves have some functions in the ecosystem. They play a role in water quality management. Due to their ability to filter water, they purify water from particles and make water transparent. Thus, they promote the production of phytoplankton by providing better penetration of light. Direct ammonia excretion and mineralisation of bio-deposits, produced by the bivalves, act as a source of inorganic nutrients. Through uptake and assimilation of phytoplankton, the bivalves accumulate nutrients in their tissue, and harvesting of

N. Gökoğlu, *Shellfish Processing and Preservation*,
https://doi.org/10.1007/978-3-030-60303-8_3

the product removes the accumulated nutrients from the ecosystem (Smaal et al. 2019).

Bivalves are appreciated by consumers because of their delicious and nutritious nature. Bivalves are an important source of protein. They are rich in vitamins and essential minerals, contain low fats and are health foods in terms of omega-3 fatty acids. On the other hand, because they contain environmental contaminants such as mercury and biotoxins, they pose a risk to consumer health. In this respect, regular monitoring programs are required to ensure food safety (Wijsman et al. 2019).

Marine bivalves are grouped into four major groups: mussels, oysters, clams, and scallops. Clams and oysters are the major species groups that contribute 38% and 33%, respectively, to the global production. Scallops account for 17% and mussels for 13% of the global production.

Bivalves are considered an important source of animal protein for humans. For a long time, commercial fisheries for mussels, arks, oysters, scallops, cockles, venus clams, and razor clams is done worldwide (Coan and Valentich-Scot 2006). Pearl production has been reported to be particularly important in some countries in the Indo-Pacific (Landman et al. 2001).

Mussels, clams, scallops, and oysters are the most traded species among the bivalve mollusc species in the world trade, and the majority of these originate from aquaculture. It is reported that China is the largest exporter country in 2016 and Chile comes second. China also has a significant domestic consumption. The European Union is the largest market for bivalves. Demand for bivalves has been reported to increase in recent years (FAO 2018).

It is reported that the total market value of the marine bivalve is approximately US $ 23 billion annually and its real economic value is higher due to secondary products and services such as packaging, transport, production of finished products and retail sales. Bivalve aquaculture and wild catch have shown a steady increase from 5 to 16 million tons per year over the period 1995–2015. Bivalve aquaculture nowadays dominates over wild catch almost ninefold, and this figure still increases. Marine bivalves account for about 14% of the global marine production (tonnes) in this period. Most of the marine bivalve production (89%) comes from aquaculture, with a total economic value of 20.6 billion US$ per year. Only 11% of the marine bivalve production comes from the wild fishery. Bivalve production in Asia is reported to increase year by year. Clams and oysters constitute a significant part of this increase. In 2015, scallop and mussel production was reported as 2.3 and 1.1 tons, respectively. The largest producer in Asia is shown as China. China is followed by Japan, Korea, and Thailand. This increase in production in China is attributed to the increase in demand and living standards. Japan is the leading producer of wild catching in Asia (Wijsman et al. 2019).

Bivalve mollusc fishing is done using active or passive catching tools. As an active catching tool, it is usually used bottom dredging tools working with the help of man or boat power. Passive fishing is performed by collecting from the natural environment via divers. In shore areas catching bivalve, active bottom dredging fishing gears (such as mechanical and hydraulic dredges) (especially at depths of 2–20 m) adversely affect benthic ecosystems and cause serious damage to creatures in the bottom (Colakoğlu and Tokac 2017). The type of dredge that is used varies depending on whether the sea bottom consists of mud, sand, pebble, rocks, or boulders. The dredge consists of a heavy metal frame, about 3.5 m wide, attached to a bag made from steel rings (10–12 cm diameter) with interconnecting chains as reinforcement. On smooth terrain the bag may last several trips, but on rough ground it may not even last one trip. Divers are limited to working in shallow waters (<30 m depth) because of safety and economic considerations. Usually teams of two to three divers operate from a small boat with an outboard engine. Each diver systematically covers lucrative spots within a given area, the choice invariably based on a prior working knowledge of the fishing ground (Gosling 2015).

3.1.1 Mussels

3.1.1.1 General Information About Mussels

Mussel is the common name used for members of several families of bivalve molluscs, from saltwater and freshwater habitats. The word "mussel" is frequently used to mean the bivalves of the marine family Mytilidae, most of which live on exposed shores in the intertidal zone, attached by means of their strong byssal threads ("beard") to a firm substrate. The common name "mussel" is also used for many freshwater bivalves, including the freshwater pearl mussels. Freshwater mussel species inhabit lakes, ponds, rivers, creeks, canals, and they are classified in a different subclass of bivalves, despite some very superficial similarities in appearance (Wikipedia 2019a).

Mussels are widely used as bio-indicators to monitor the health of aquatic environments in both fresh water and the marine environments. They are particularly useful since they are distributed worldwide, and they are sessile. These characteristics ensure that they are representative of the environment where they are sampled or placed. Their population status or structure, physiology, behaviour, or the level of contamination with elements or compounds can indicate the status of the ecosystem (Wikipedia 2019a).

Biology of Mussels

The mussels are composed of two shells, clamped with very strong muscles, triangular front, oval and bilaterally symmetrical. The shell consists of anterior edge, posterior edge, ventral edge, and dorsal edge. The front edge is very short, and the shells are interlocked here. The exterior of the shell is in various shades of purple and brown, while the interior is luster of mother of pearl. On the shells, there are growth lines in the form of small elliptical circles. There is byssus that provides clamping of mussels to dock walls, rocks, and all kinds of hard grounds. There are about 150 byssus fibres in a medium size mussel (Alpbaz 1993).

Mussels are organisms having separate sexes and fertilization takes place in the water column. Eggs produced in ovaries and sperms produced in testicles are sprayed from body to water. The genital opening where the sperm leaves the gonads opens into the water outlet cavity above the gills and the sperm leave from the body into this water. However, during spawning, since water outlet of the body is completely closed and a large part of the water inlet is closed, the eggs are collected in the water inlet cavity and thrown away 30–50 cm away by a quick contraction of the adductor muscle (Sirin 2012).

Mussels are creatures that are fed by filtering water. Both marine and freshwater mussels are filter feeders; they feed on plankton and other microscopic sea creatures which are free-floating in seawater. It can filter all kinds of organic and inorganic particles with a size of 2–100 μm. A mussel with an average length of 7–8 cm can drain 10–15 L of water per hour. However, in the filtration rate of mussels, mussel size, particle size, particle density, particle type, water temperature, water flow play an important role. When the nutrition of mussels is low, growth slows or stops. Meat yield decreases and ripening of gonads is not complete (Sirin 2012).

Distribution of Mussels

Mussels settle on a wide variety of substrates, for example rock, stones, pebbles, shell, cement, and wood, once the substrate is firm enough to provide a secure anchorage. In the case of Perna, mangrove mudflats represent a major habitat for tropical species. Early *M. edulis* spat either attach to filamentous algae, from which they eventually migrate onto adult mussel beds, or else they settle directly onto adult beds. On rocky shores mussel beds provide a habitat for a few but abundant macro invertebrate species. Upper distribution limits for mussels are usually governed by physical factors, primarily temperature, while predators are mainly responsible for setting lower limits (Gosling 2015).

Marine mussels are abundant in the low and mid intertidal zone in temperate seas globally. Other species of marine mussel live in tropical intertidal areas, but not in the same huge numbers as in temperate zones. The South African white mussel exceptionally does not bind itself to rocks but burrows into sandy beaches extending two tubes above the sand surface for ingestion of food and water and exhausting wastes. Freshwater mussels inhabit permanent lakes, rivers, canals, and streams throughout the world except in the Polar Regions. They require a constant source of cool, clean water. They prefer water with a substantial mineral content, using calcium carbonate to build their shells (Wikipedia 2019a).

Production and Trade of Mussels

World production of mussel reached 2.14 million tonnes in 2016, with a 35% increase since 2007. Most of the production is farmed (94%), wild caught mussels accounting for 6% of the volumes. The largest world producer of mussels is China with 879,000 tonnes in 2016. EU-28 is the second-largest world producer (522,000 tonnes in 2016) with production declining since 2007 by 6%. China and the EU-28 accounted for 67% of global mussel production in 2016. Chile, Thailand, New Zealand and Republic of Korea are the four next largest producers (respectively 313,607 tonnes, 115,000 tonnes, 94,000 tonnes and 64,000 tonnes in 2016). The production in each other country is below 30,000 tonnes (EUMOFA 2019). Within Europe, where mussels have been cultivated for centuries, Spain remained the industry leader. Aquaculture of mussels in North America began in the 1970s. In the US, the northeast and northwest have significant mussel aquaculture operations, where *Mytilus edulis* (blue mussel) is most grown. While the mussel industry in the US has increased, in North America, 80% of cultured mussels are produced in Prince Edward Island in Canada (Wikipedia 2019a).

The international trade of mussels was stable in terms of volume in 2017, while the value of mussels increased in all importing countries. The EU-28 is the main importer of mussels, accounting for about 70% of total imports. Following the production increase in 2017, Chile expanded mussel exports by 20% in the first 9 months, to reach 66,000 tonnes. While the traditional markets Spain and the United States of America continue to be strong buyers, Chile accessed the Chinese market (FAO-Globefish 2018a).

Some Important Species of Mussels

There are hundreds of species of mussel but only about a dozen or so are fished commercially. Commercially most relevant marine mussel species belong to the two genera of *Mytilus* and *Perna*. *Mytilus* species occur in temperate waters of Europe, Asia, and America, whereas Perna species are cultured in warmer waters such as Thailand, the Philippines, China, and New Zealand. Within the genus *Mytilus*, the marine mollusc *M. edulis* is commonly known as blue or black mussel due to the colour of its shell. It is mostly cultured in Canada, USA, Europe, and Africa. Another common edible mussel, *M. galloprovincialis*, originates from the Mediterranean Sea (Gosling 1992). Concerning the genus *Perna*, major aquaculture mussel species include *P. viridis*, the Asian green mussel, and *P. canaliculus*, the green-lipped mussel which is endemic to New Zealand. *P. canaliculus* is an integral dietary part of the indigenous Maori culture and is the basis of an important aquaculture and processing industry serving both export and domestic markets (Wakimoto et al. 2011).

Mytilus edulis **(Linnaeus 1758) (Blue Mussel)**
The blue mussel *Mytilus edulis* has a wide distribution in the northern hemisphere, occurring in European waters from Spitsbergen to western France, and on the Atlantic coast of North America from the Canadian Maritimes southward to North Carolina (Gosling 2015).

Shell is solid, equivalve, in equilateral, beaks at the anterior end approximately triangular in outline. Hinge line is without teeth but with 3–12 small crenulations and is under the umbones. Margin is smooth. Pallial line is wide. Anterior adductor scar is very small, posterior is large. External ligament is much concealed, extending more than half-way from the beaks to the highest point of the shell. Sculpture of fine has concentric lines. Colour is purple, blue, and sometimes brown, occasionally with prominent dark brown to purple radial markings. Periostracum is almost black, dark brown, or olive. Interior is pearl-white with a wide border of purple or dark blue.

Mytilus edulis has been harvested for centuries. Blue mussel shells have been found in kitchen middens dated at 6000 B.C. Until the nineteenth century, blue mussels were harvested from wild beds in most European countries for food, fish bait and as a fertilizer. Blue mussels are widely distributed in European waters, extending from the White Sea, Russia as far as south as the Atlantic coast of Southern France. *Mytilus edulis* has a wide distributional pattern, mainly due to its abilities to withstand wide fluctuations in salinity, desiccation, temperature, and oxygen tension. Therefore, this species occupies a broad variety of microhabitats, expanding its zonational range from the high intertidal to subtidal regions and its salinity range from estuarine areas to fully oceanic seawaters. Highly tolerant of a wide range of environmental conditions, the blue mussel is euryhaline and occurs in marine as well as in brackish waters (Baltic) down to 4‰, although it does not thrive in salinities of less than 15‰ and its growth rate is reduced below 18‰. Blue mussels are also eurythermal, even standing freezing conditions for several months. The species is well acclimated for a 5–20 °C temperature range, with an upper sustained thermal tolerance limit of about 29 °C for adults. Its climatic regime varies from mild, subtropical locations to frequently frozen habitats. *M. edulis* typically occurs in intertidal habitats, although this distribution appears mostly controlled by biological factors (predation, food competition) rather than by its capacity to survive subtidally, as demonstrated by offshore mussel culture using longlines. When predators are lacking, *M. edulis* subtidal aggregations can reach a 1.2 m thickness and individuals attain large sizes in a relatively short period of time. Although blue mussels can live up to 18–24 years, most cultured mussels are produced in less than 2 years. In the wild, *M. edulis* settles in patches of open spaces, quickly building a dense population referred to as ‘mussel beds’. Although showing a seasonal pattern, the reproductive cycle of *M. edulis* can exhibit considerable temporal and spatial variation. Gonads are usually ripe by early spring in European waters; mussels commonly show a significant loss of condition following spawning. Rapid gametogenesis leads to fully ripe gonads again in summer. Although directly driven by food availability and temperature, reproductive cycles in *M. edulis* may vary latitudinally, both in terms of onset and duration. High fecundity and a mobile free-living larval phase are two characteristics that have contributed to the development of mussel culture; in fact, the natural abundance of *M. edulis* larvae has been the key for such development. Exports mainly comprise processed products, which provide added value, and may be a major driving factor for *M. edulis* culture development. Import prices tend to be lower compared to export prices, reflecting the role of processing in this market, and facilitating international trades when local production cannot sustain the demand (FAO 2019a) (Fig. 3.1).

Fig. 3.1 *Mytilus edulis*

***Mytilus galloprovincialis* (Lamarck 1819) (Mediterranean Mussel)**
Mytilus galloprovincialis is the second most common type of *Mytilus* in the world after *M. edulis*. This species, which is seen extensively in the Mediterranean basin, is obtained from natural deposits or through culture production. Black mussels are the most cultivated species in the world and their production constitutes 95% of the cultural fishery worldwide. In the production of mussels, in which China and the EU are the two largest producers, Chile and New Zealand follow these countries (Colakoglu and Colakoğlu 2013).

In Europe, *Mytilus galloprovincialis* is found in the Mediterranean Sea and the Black Sea, and on the Atlantic coasts, in Portugal, north to France and the British Isles and Norway. In the northern Pacific the species is found along the coast of California, where it was introduced from Europe by human activity in the early twentieth century, and also in the Puget Sound region of Washington State, where it has been subject to aquaculture. *Mytilus galloprovincialis* is also present as a native lineage in parts of the Southern Hemisphere (Wikipedia 2019b).

M. galloprovincialis lives on hard substrates from the intertidal zone to depths of 40 m. It is found along coasts and rocky shores, and in sheltered harbours and estuaries. *M. galloprovincialis* is cultivated in Albania, Bulgaria, China, Egypt, the Federal Republic of Yugoslavia, France, Greece, Italy, Morocco, Portugal, the Russian Federation, Spain, South Africa, Turkey, Ukraine, Canada, and in the Scottish lakes (Cabi 2019a).

Mytilus galloprovincialis is dark blue or brown to almost black. The two shells are equal and nearly quadrangular. The outside is black-violet coloured; on one side the rim of the shell ends with a pointed and slightly bent umbo while the other side is rounded, although shell shape varies by region. It also tends to grow larger than its cousins, up to 15 cm, although typically only 5–8 cm (Global Invasive Species Database 2019).

***Perna viridis* (Linnaeus, 1758) (Asian Green Mussel)**
Perna viridis, known as the Asian green mussel, is an economically important mussel, a bivalve belonging to the family Mytilidae. It is native in the Asia-Pacific region but has been introduced in the Caribbean, and in the waters around Japan, North America, and South America. *Perna viridis* ranges from 80 to 100 mm in

length and may occasionally reach 165 mm. Its shell ends in a downward-pointing beak. The smooth periostracum is dark green, becoming increasingly brownish towards its point of attachment (umbo), where it is lighter. The Asian green mussel is found in the coastal waters of the Indo-Pacific region. However, the mussels are introduced to other areas as invasive species via boat hulls and water ballasts. The mussels live in waters that are 11–32 °C with a wide-ranging salinity of about 18–33 ppt. *P. viridis* is harvested in the Indo-Pacific region as a food source due to its fast growth. However, it can harbour deadly Saxitoxin produced by the dinoflagellates that it feeds upon. It can also be used as a bio monitor to indicate pollution caused by heavy metals, organochlorides, and petroleum products (Wikipedia 2019c).

The green mussel *Perna viridis* has the most widespread distribution, extending from India eastwards to south-east Asia, the China coast, Indonesia, Philippines, and Samoa. The species is a recent invader to the Caribbean Sea, including the subtropical south-eastern United States region (Gosling 2015).

Perna perna **(Linneaeus, 1758) (Brown Mussel)**
Perna perna, the brown mussel, is an economically important mussel, a bivalve mollusc belonging to the family Mytilidae. It is harvested as a food source but is also known to harbour toxins and cause damage to marine structures. It is native to the waters of Africa, Europe, and South America and was introduced in the waters of North America. *Perna perna* is usually 90 mm long although it can reach sizes of up to 120 mm. The mussel is easily recognized by its brown colour, but its identifying characteristic is the "divided posterior retractor mussel scar". Its pitted resillal ridge also differentiates the mussel from other bivalves. The brown mussel is native to the tropical and sub-tropical regions of the Atlantic Ocean. It is found in waters off the west coast of Africa and the coast of South America up to the Caribbean. It is accidentally introduced as an invasive species to the coast of Texas via the boat hulls and water ballasts of ships from Venezuela. Its distribution includes Chile, Peru, and South Africa (Wikipedia 2019d).

Nutritive Composition of Mussels

Mussels are an important source of nutrients containing essential nutrients and are preferred as an inexpensive protein source for human consumption, especially in countries with low socioeconomic levels. Since shellfish are low in fat, especially low in saturated fat, containing omega-3 fatty acids, excellent protein sources and particularly good sources of iron, zinc, copper and vitamin B_{12}, increasing the intake of various shellfish should be encouraged to ensure a healthy diet (Dong 2001) (Tables 3.1 and 3.2). Mussels are an environmentally sustainable way of producing dietary protein and long-chain omega-3 fatty acids. Mussels are not only a sustainable source of omega-3 fatty acids, but also a sustainable source of essential amino acids. The protein content in mussels varies between 12.6 and 24.0 g/100 g mussels, depending on the variety. Mussels contain a range of vitamins and minerals found in other meat-based sources of protein such as B-vitamins and trace minerals. Mussels are relatively high in EPA and DHA (Carboni et al. 2019).

Factors such as water temperature, nutrient availability and reproductive cycle, and season can influence biochemical composition of bivalve molluscs. There are many studies that examine the lipid content of mussels seasonally. The total lipid content of *P. viridis* was found higher in spring than in the other seasons (Li et al. 2007). Similar results were reported for *M. edulis* (Lin et al. 2003). In a study comparing the lipid contents of mussels obtained from the Black Sea coast of Bulgaria, total lipid content of cultured mussels (2.95%) was found to be higher than that of wild mussels (2.05%) (Stancheva et al. 2017). Similar results for *M. galloprovincialis* were reported from the Romanian Black Sea coast (Sirbu et al. 2011). Contrary results have been reported from the Mediterranean coast of Greece (Zlatanos 2008), from the Adriatic Sea (Ventrella et al. 2008) and from Italy (Prato et al. 2010). A previous study conducted on *M. galloprovincialis*, mussels seasonally harvested in two sites in Italy showed differences in biochemical composition depending on the different environmental conditions. As the result, it has been reported that the sessile habitat of mussels revealed the fact that their chemical composition strongly depended on the presence of phytoplankton sources and hence the harvest season. Moreover, it has been found that mussels from both sites were characterised by low lipid contents, especially the lipid fraction contained elevated levels of n-3 polyunsaturated fatty acids, cholesterol, plant sterols and carotenoids and low levels of n-6 polyunsaturated fatty acids (Orban et al. 2002). In another study, Baker and Hornbach (2001) found significant seasonal differences in the chemical composition of *Actinonaias ligementina* and *Amblema plicata*. Various environmental factors such as temperature, food availability, plankton composition, and physiological factors have contributed to these differences. Mladineo et al. (2007) reported a

Table 3.1 Proximate compositions of some mussel species (g/100 g)

Species	Moisture	Protein	Fat	Ash	Carbohydrate	Reference
Mytilus galloprovincialis	77.45–78.65	16.80–17.40	1.85–2.51	–	2.55–2.73	Merdzhanova et al. (2016)
	73.35–76.20	17.40–19.92	1.40–2.89	–	2.0–2.73	Merdzhanova et al. (2017)
	78.77	10.44	1.33	2.76	6.63	Azpeitia et al. (2017)
	85.71–85.86	8.98–9.81	1.75–2.18	0.73–0.76	1.57–1.87	Sirbu et al. (2011)
	79.48–83.41	7–10	0.99–2.08	1.12–1.70		Dernekbası et al. (2015)
	77.09–81.2	13.02–17.63	1.48–1.67	1.50–1.52		Stratev et al. (2017)
Unio terminalis	80.36	11.87	2.55	1.68	–	Ersoy and Sereflisan (2010)
Potamida littoralis	81.69	11.97	1.05	1.61	–	Ersoy and Sereflisan (2010)

Table 3.2 Fatty acid composition of some mussel species (mg/100 g)

	SFA	MUFA	PUFA	n-3	EPA + DHA	Reference
Mytilus galloprovincialis	31.25–33.19	13.0–16.25	52.50–54.90	33.05–35.95	31.53–33.65	Merdzhanova et al. (2016)
	25.60–31.0	16.0–17.91	53.0–56.5	35.25–40.06	34.32–39.3	Merdzhanova et al. (2017)
	25.66–29.19	15.06–19.81	50.67–55.80	28.45–35.61	22.38–31.93	Dernekbası et al. (2015)
	20.9–35.6	12.2–22.5	49.4–60.3	33.1–46.0	27.1–51.1	Ventrella et al. (2008)
	49.29–53.71	33.19–39.63	7.55–11.16	1.93–5.2	1.42–1.96	Prato et al. (2010)
	40.75–44.57	24.87–26.46	30.56–32.79	19.02–27.6	16.96–23	Stancheva et al. (2017).
	35.87–46.31	10.19–21.11	40.92–52.54	36.03–45.98	32.88–56.25	Stratev et al. (2017)
Unio terminalis	32.13	19.60	37.08	22.57	19.12	Ersoy and Sereflisan (2010)
Unio elongatulus	28.19–34.49	35.32–43.98	28.63–32.23	7.7–8.71	2.13–4.07	Ekin et al. (2011)
Potamida littoralis	30.21	22.84	32.41	18.89	14.77	Ersoy and Sereflisan (2010)
P. viridis	24.87–34.04	17.33–31.61	37.93–48.38	32.40–40.70	22.0–35.84	Li et al. (2007)

dramatic increase in protein but a decrease in lipids between June and July in horse-bearded mussel (*Modius barbatus*). *M. galloprovincialis* was found to contain low protein due to ovulation in the months when water temperature was highest (Stratev et al. 2017).

Food availability in the environment is also an important factor affecting the composition of mussels. The food quality is a determinative factor for the growth of mussels, and microalgae are the major fatty acid source for bivalve molluscs. Since bivalves are generally regarded as herbivores, it is assumed that phytoplankton is the main components of their diet and FA profiles. Phytoplankton, algae, and other plants are at the base of the marine food chain and they can synthesize these unsaturated PUFAs in high quantities. In a study which investigated fatty acid compositions of freshwater mussel (*Unio elongatulus*) collected from four different locations, it was found that mussel collected from the river which is rich in fauna and flora and feeding activity is more intense contained higher n3 fatty acids (Ekin et al. 2011).

The development of gamete in mussels affects the chemical composition. During the maturation of gonads, an increase in lipid and protein content is observed. This increase continues until spawning. Increases in the levels of polyunsaturated fatty acids in bivalves occur during periods of high primary productivity. Gametes contain high levels of lipid reserves, and lipid levels vary according to reproductive activity. In studies, lipid levels of *Mytilus platensis*, *Mytilus galloprovincialis* (De Moreno et al. 1980), *Perna perna* (Narváez et al. 2008) and *Mytilus edulis* (Dare and Edwards 1975; Zandee et al. 1980) significantly decreased with spawning.

Lipid classes found in mussel oil have been reported to include sterol esters, triglycerides, free fatty acids (saturated and unsaturated), carotenoids, sterols, and polar lipids (Sukumaran et al. 2010).

3.1.1.2 Postharvest Quality Changes in Mussels

Microorganisms play an important role in the degradation of seafood. Bivalve molluscs are filter-feeding organisms, and they can accumulate pathogenic bacteria and viruses. Contamination of seafood by microorganisms can be at various stages such as processing, storage, and distribution. Water, facilities, equipment, and personnel are sources of contamination. The processing step is particularly important due to the potential high microbial load on the surface. The number of microorganisms present in the product determines whether the contamination can cause microbial degradation or disease from seawater (Odeyemi et al. 2018).

The quality of seafood depends on the characteristics of the individual (such as species, sex and age), external factors (such as the region, food and water quality, temperature and salinity) and post-harvest variable factors (harvesting method, transport and storage). Season is an effective factor on the quality of bivalves. The chemical composition of bivalves may change with the season. Especially the reproductive cycle is an important factor in this. They are in the waiting phase where they use their stored energy in winter. They use the energy they have stored as glycogen for both live and gametogenesis. Since gonads develop in this phase, there is an energy requirement for the gametes. When the spawning season starts, the glycogen content is usually minimum in winter and maximum in autumn (Gosling 2015). Information on the effect of seasonal variation on chemical composition is given above in the section "Nutritive composition of bivalves". This change significantly affects the quality of the mussel.

Despite the change in environmental factors, a reaction called stress occurs in the living. Stress disrupts the physiological balance of the organism. Bivalves react to this stress differently. For example, mussels close their valves in the case of stress and oxygen intake decreases. Such variability will condition both the handling and storage of bivalves. Therefore, it is very important to understand the results of the stress. Depending on the degree and the duration of the stress period, the animal will become more fragile and therefore there will be occurred differences in metabolism, by decreasing the quality of the bivalve. Thus, stress negatively affects the final quality of seafood, which causes negative consequences for the industry. Glycogen is reported as an indicator of stress in fish as its content decreases with increasing stress (Lopes Da Costa 2018).

The fishing and harvesting methods of bivalves are an effective factor on the product quality. The methods vary according to the type of bivalve and the environment in which the bivalve lives. Among the methods, catching by hand causes the least stress effect on the living. Dredging and trolling methods cause environmental effects that destroy the seabed. This strenuous activity leads mussels to exhausting and to death (Lopes Da Costa 2018).

Transport and storage are the most stressful stage for bivalves. Transport is carried out by wet, semi-dry or dry methods. Since wet tanks need to be ventilated and this process is expensive, dry transport can be applied. Bivalves can stay in a place outside the water without losing quality. However, the quality of the end product may be affected according to the duration of the bivalves under conditions other than their natural environment. In addition to the lack of oxygen during dry transport and storage, there is also a hazard exposing of bivalve to drying. Drying also reduces the effectiveness of the respiratory system (Lopes Da Costa 2018).

Mussels have high degree of perishability. They are very sensitive for post-harvest handling. Shellfish deteriorate more quickly than fish because they contain a larger amount of amino acids, on the other hand, mussels harvested from contaminated waters can be heavily contaminated with bacteria, and thirdly, mussels are bulky when their shells are not opened, so they are laborious and expensive to transport (Legaspi 1979). Quality of mussels means total number of microorganisms, absence of pathogen, virus, toxin, heavy metal and parasite and presence of good sensory properties. Deterioration of seafood usually occurs in two ways, autolysis, and bacterial degradation. Autolysis occurs after the breakdown of tissue components by enzymes present in the tissues of seafood. This breakdown causes significant changes in the odour, flavour, and texture of the product. Bacterial degradation follows autolysis.

3.1.1.3 Post-harvest Handling of Mussels

Post-harvest handling of mussels is more difficult than that of fish, not only because of its higher degree of perishability, but also because of adverse effects of its habitat. Due to their high perishability, various studies have been conducted to keep mussels alive longer or to increase shelf life. For this purpose, various processing and preservation techniques have been developed.

On the other hand, to minimize adverse effects of their habitat and the health risks, the water quality in shellfish-growing environments should be monitored. The quality, value and safety of molluscan shellfish can be compromised by pollution of marine environments. In terms of public health, the growing area of mussels should be "approved". Shellfish-growing areas may be classified as 'approved' if the area is not contaminated with faecal material, pathogenic microorganisms, poisonous or deleterious substances, to the extent that consumption of the shellfish might be hazardous (Su and Liu 2014). Marketing and distribution problems are brought about by the bulky nature of shellfish, particularly if they are to be marketed with shell on. Modern processing technology makes it possible to process fish and shellfish in such a manner by which the above-mentioned problems are minimized, if not completely solved. Other fish processing methods, like pickling, drying, smoking, canning, bottling, or converting shellfish into powder, have also partially solved the problem of marketing and distribution (Legaspi 1979).

To maintain post-harvest quality in shellfish and limit the growth of human pathogens, prolonged exposure to high temperatures should be avoided. Shellfish harvested for raw consumption need to be cooled to 10 °C within a certain time to control the growth of *Vibrio parahaemolyticus* or *Vibrio vulnificus*.

Keeping and Processing Mussels Alive

In terms of quality and safety, it is best to transport mussels alive. Bivalve species such as oysters, mussels, and hard-shell clams can survive for extended periods out of water and can be traded for human consumption as live animals. Mussels will be exposed to some level of stress during all or part of the trade chain. Capture, de-clumping, fluctuating temperatures, sunlight and other bright lights, wind or drafts, handling and physical damage, poor water quality during holding, conditioning and purification are factors causing the stress. Mussels are generally able to recover from such stresses, however if any or a combination of those stresses are sufficiently intense, then poor quality (broken shells, gaping, unpleasant smell) or dead mussels will result. Thus, transport systems need to ensure mussels are held in conditions that keep stress to a minimum (Barrento et al. 2013). Particular supply chains may differ from harvest to point of sale, although the following procedures are generally typical: rope grown mussels are removed from long lines and are washed, and de-clumped from each other and fouling organisms, before being "graded" according to size. Depending on the bacterial loading of the original water body and national legislation, mussels may also be "depurated" (immersed in clean seawater, for up to 42 h, to pro-mote removal of bacteria), and then de-byssed. Onward transportation may be up to 48 h duration. On arrival, mussels can be sold immediately, maintained in cold, damp storage, or immersed again ("re-watered" or "re-conditioned") to improve quality prior to sale (Barrento and Powell 2016).

Shellstock shellfish should be transported in clean storage bins with effective drainage and be shipped on pallets in a truck that is properly maintained to prevent contamination, deterioration, and decomposition. If the mussels are pre-chilled in the refrigerator, or if the ambient temperature and travel time are sufficient to allow the growth of bacteria and the degradation of mussels, they should be transported in trucks with a refrigeration system equipped with an automatic temperature control capable of maintaining the air temperature at 7.2 °C or less. Any ice used in the transportation of shellfish should be made on-site from potable water in a commercial ice machine. All containers used to transport shellstock shellfish should be constructed from safe materials and allow for easy cleaning. They should be cleaned with potable water, detergents, and sanitizers after use (Su and Liu 2014).

Depuration of Mussel

Mussels, oysters, and clams are filter feeders and accumulate pathogenic bacteria. Depuration is defined as a process utilizing the natural water-filtering mechanism of shellfish to release contaminants from the digestive tract into clean and unpolluted water. Depuration is intended to reduce the number of pathogenic organisms that may be present in shellfish harvested from moderately polluted (restricted) waters to such levels that the shellfish will be acceptable for human consumption without further processing. The process is not intended for shellfish from heavily polluted (prohibited) waters nor to reduce the levels of poisonous or deleterious substances that the shellfish may have accumulated from their environment (NSSP 2017). This treatment can also remove most of the sand in the stomach. Depuration of mussels is necessity before marketing due to increasing of pollution of seawater. Those most widely subjected to the process include oysters, mussels, and clams. It is essential to employ depuration in order to remove or reduce the bacteria load causing typhoid and viruses causing gastroenteritis (non-*typhi Salmonellae* and *Campylobacter*) and infectious hepatitis accumulated in bivalves from the marine environment before human consumption (Singh and Nagalakshimi 2013). Some species such as cockles, scallops and razor clams pose specific challenges to depuration, for example the mobility of scallops makes them difficult to contain in baskets and to prevent them stirring up settled detritus. While depuration may be the only mitigation strategy for those species eaten raw, such as oysters, many other species of bivalves are lightly cooked before eating and depuration will provide an additional safeguard (FAO 2008). In the process of filter-feeding bivalve molluscs may also concentrate and retain human pathogens derived from sewage contamination of growing waters. The hazards posed by bioaccumulation are compounded by the traditional consumption of molluscan shellfish raw or after minimal heat treatment and by consumption of the entire animal, including the viscera. Depuration processes are applied to shellfish such as oysters, mussels, scallops, and clams that are sold alive. Shellfish, which are commercially marketed and consumed as food in many parts of the world, are supplied to the market by reaching the hygienic conditions determined by the depuration process.

Purification, also known as depuration, involves the transfer of the molluscs from the harvesting area to enclosed systems with flowing clean water with the appropriate salinity and temperature. Here, the molluscs continue filtration and normal digestive activity and over a period of about 48 h, they purge themselves of most bacterial contamination present, although under certain conditions, longer depuration periods may be required for adequate purification (Martinez et al. 2009).

Depuration starts by placing shellfish into tanks containing clean sea water. Thus, the animals continue their normal feeding activity (filtration) and, after a period, remove microorganisms that cause diseases through the gills and the intestine. By removing the wastes accumulated at the bottom of the tanks in certain periods, it is ensured that microbial hazards removed from the animal body do not re-enter the organism. After depuration process, the bivalve products should remain alive while packaging and the quality of the product must be maintained. Resumption of filtration activity requires that the animals be not subjected to undue stress prior to the depuration process. It means that the harvesting method and subsequent handling should not shock the animals too much and that they should not be exposed to temperature extremes. Once placed in the system, the physiological conditions should be such as to maximise the activity of the animals.

The sea water to be used in the depuration process must be disinfected during the entry into the tanks and after being used in the tanks without being returned to the system. Sea water used in depuration is disinfected with chlorine, UV light, ozone and iodophors. Chlorine was one of the earliest means used to disinfect seawater for depuration. When used with seawater of low to moderate sediment and organic loads, it is an effective bactericide. However, there are concerns with its effectiveness against viruses (FAO 2008). In a study, three types of depuration methods were tried for mussels (*P. viridis*) collected manually from natural beds. In the first one, the mussels were washed thoroughly and filtered and then kept in filtered sea water for 24 h. The second was kept in filtered and aerated water for 24 and 48 h. In the third, the mussels were kept in filtered seawater and washed in filtered seawater thoroughly after 2 h of chlorination at levels of 2.0, 2.5 and 3.0 ppm. According to the results the bacterial load of the mussels was lowered effectively, either by washing in filtered sea water for 24 h or by keeping them in aerated sea water for 48 h. Chlorination yielded the best results (Malarvanan and Edward 2002). Ren and Su (2006) reported reduces in *V. parahaemolyticus* and V. *vulnificus* by application of 30 ppm chlorine. The advantage of ultraviolet light treatment is that it does not change the chemical and physical properties of water while it purifies water. UV light has been used for decontaminating seawater for shellfish purification. In one study, it was reported that the levels of coliforms decreased in sea water, which was subjected to UV treatment for 72 h (Vasconcelos and Lee 1972), and in another study, it was reported that 2 h of UV treatment decreased the *V. parahaemolyticus* population (Greenberg et al. 1982). Ozone has been shown to be effective in inactivating both bacteria and viruses by affecting their genetic material (RNA or DNA). A decrease in *Escherichia coli* and *V. parahaemolyticus* populations was found in mussels after being depurated in sea water containing 0.139 ppm ozone for 44 h (Croci et al. 2002). A 2-log reduction in *V. vulnificus* has been reported after 24 h of depuration in seawater containing 0.6–3.1 ppm ozone (Schneider et al. 1991). In another study, a decrease in *E. coli* and coliforms was reported after ozone application (Maffei et al. 2009).

Shucking of Mussels

Shucking is the processing step that removes the edible portion of the mollusc from the shell. It is usually done by hand, mechanically or through heat shock with steam or hot water. This step may expose the product to microbiological or physical contamination. Physical removal of shellfish meat from the shell will often expose the product to dirt, mud and detritus that should be removed before further processing through washing (FAO 2008).

Bivalve molluscs, such as clams, oysters, scallops and mussels, have two main shells or valves which are joined together by a hinge and held shut by adductor muscles which function to open and close the valves which, when open, pump in sea water containing food. One class of bivalve, which includes oysters and scallops, contains one adductor muscle and is known as monomyrian. Another class, which includes clams and mussels, contain two adductor muscles and is known as dimyrian (Hanks Jr and Grieb Jr 1971). The muscles are strong enough to close the valves of the shell when they contract, and they are what enable the animal to close its valves tightly when necessary, such as when the bivalve is exposed to the air by low water levels, or when it is attacked by a predator. When the adductor muscles relax, the valves of the shell are automatically pulled open to some extent by a ligament, which joins the valves together and which is usually located on the hinge line between the umbos of the shell. The resiliency of the ligament is what causes the valves of the bivalve mollusc to open when the adductor muscles relax (Wikipedia 2019f).

Heat shocking of molluscan shellfish is defined as a process that uses any form of heat treatment such as hot water, water-saturated air, or dry heat for a short time to facilitate detachment of meat from the shells. Heat treatment is not a cooking process, but a method of removing only meat (FAO/WHO 2000). Opening the shells with heat shock and removing the meat from the shell is done by steaming or dipping into boiling water. During this process, there is a loss of moisture in the meat, protein degradation and a decrease in the juiciness of the meat which causes the tissue to become rubber-like. At the same time, the original shape and structure of mussels are lost during cooking. The mussels should first be washed and then either steamed for about 4 min at 115.5 °C or for about 6 min in boiling water at 100 °C or steam at atmosphere pressure; some processors prefer boiling to steaming since the meats shrink less. Mussel meats immersed in boiling water are sterilized after about 2½ min; additional processing time should be just sufficient to cook the meats and no more, since overcooking causes excessive shrinkage of the meats. The liquor produced can be kept if required for use in bottling or canning. The mussels should be quickly cooled by water spray to prevent toughening of the meat (Waterman 1963). In the previous study, live green mussel (*P. viridis*) was subjected to heat treatment by three different methods (hot water baths, water-saturated air, or dry heat). Various time temperature combinations were used for this. The most effective method for shucking was water bath followed by water-saturated air or dry heat. Hot water bath treatment at 100 °C gave 100% open mussels (Azanza and Ventura 2005).

High pressure (HP) processing is a useful technology for shucking the raw meat from the rigid shell of crustaceans and molluscs without cooking, thus meat removal significantly more efficient without changing the size and shape of the meat retaining nutritional qualities. In a study, mussel meat treated at 100 MPa, was found difficult to be shucked and remained attached to the shells by the adductor muscle and at the edges of the mantle. In mussels treated at 200 MPa, the meat was easily removed, but the adductor muscles remained on the shells. The meat got easily detached from the shell, when subjected to 300 and 400 MPa pressure treatment. Application of HP increased the hardness of mussel tissue. The author stated that this increase in the texture may be due to the protein-protein interactions which results in tissue elasticity and hardness. Furthermore, colour values significantly increased after HP application (Bindu et al. 2015).

3.1.1.4 Mussels Processing and Preservation

Chilling/Refrigerating of Mussels

The principle of the refrigerated preservation of foods is to reduce, and maintain, the temperature of the food such that it stops, or significantly reduces, the rate at which detrimental changes occur in the food. These changes can be microbiological, biochemical, and/or physical. An efficient and effective cold chain is designed to provide the best conditions for slowing, or preventing, these changes for as long as is practical. Effective refrigeration produces safe food with a long, high-quality shelf life (James and James 2014). Mussels are sold either as a live, fresh product or processed into a variety of chilled or frozen products. Due to perishable nature of mussels, shucked mussels can rapidly deteriorate. Therefore, it must be stored properly.

Chilling or refrigerating provides short term storage. Erkan (2005) determined a shelf life of 4 days for shucked mussels (*M. galloprovincialis*) stored at 4 °C. In another study, mussel (*M. edulis*) stored in insulated boxes in direct contact with ice was kept in sensorially acceptable for 8 days (Chinnamma et al. 1970). *P. viridis* remained acceptable for 3 days at 7 °C (Zamir et al. 1999). In a study, applicability of chilling to keep mussels (*Perna viridis*) alive for a longer period has been investigated. In that study, different packaging and transport methods were applied. Used packaging methods were traditional way using sacks, styrofoam box and styrofoam box with two kg block of ice inside a metal tray. All the mussels packed with ice-free methods died after 3 days, whereas no single death was observed in 3 days (Yap and Orano 1980).

Chilling or refrigeration is generally used to store products that have been processed or preserved by other methods such as using preservatives, packaging under modified atmosphere and vacuum, application of high pressure etc. Thus, it increases the efficiency of preservation and the shelf life of the product. In various studies, treatment with organic acids or their salts was used to increase the quality and the shelf life of mussels. Masniyom and Benjama (2007) used lactic, acetic, and citric acids for extension of green mussel (*P. viridis*) during storage at 4 °C. After immersion into lactic, acetic and citric acid solutions the mussel was packed in polyethylene bags and stored at 4 °C. Lactic acid dipped samples, particularly with 0.2 M, showed the greater acceptability than did those dipped in other acids throughout the storage of 27 days. The control sample had the acceptability only for 6 days of storage. Mediterranean mussel (*Mytilus galloprovincialis*) meat (shucked mussels) was stored at 5 ± 2 °C in plastic pouch with water. Sodium lactate and potassium sorbate were added to in the water. At the end of study, it was found that while potassium sorbate significantly retarded the growth of mesophilic aerobic bacteria, psychrotrophic bacteria, and especially lactic acid bacteria, sodium lactate did not affect. Authors have suggested that potassium sorbate can be used to increase the shelf-life of Mediterranean mussel at 5 ± 2 °C in plastic pouches with water, up 8 days (Vasakou et al. 2003). Shucked green mussel (*P. viridis*) was pre-treated with lactic (2%) and citric (2%) acids and stored at 3 °C. It was found that lactic acid pre-treatment is effective in to control sensory, microbiological, and physicochemical spoilage of green mussel during chilled storage. Lactic acid treated samples were acceptable up to 15 days. Although the retention effect of citric acid has been determined, it is stated that it is not suitable for commercial use because it causes more weight loss in the product (Arcales and Nacional 2018). It was reported that the shelf life of mussels (*M. galloprovincialis*) increased 2–3 times compared to the control group at 4 °C when lactic acid was applied with chitosan (Terzi et al. 2013).

Modified atmosphere packaging (MAP) is another preservation method applied to increase shelf life and cold storage stability of mussels. It has been demonstrated in various studies that a longer shelf life can be achieved when the mussels are stored in cold storage after being packaged in different atmospheres. Goulas et al. (2005) packed cultured mussels (*Mytilus galloprovincialis*) under MAP using three different gas mixture (M1: 50%/50% (CO_2/N_2), M2: 80%/20% (CO_2/N_2) and M3: 40%/30%/30% ($CO_2/N_2/O_2$)) and investigated the quality changes of mussel meat during stored at 4 °C by comparing with mussels packed aerobically and under vacuum. Microbiological results revealed that the M2 and VP delayed microbial growth compared with that of air-packaged samples. Total volatile basic nitrogen and trimethylamine nitrogen values of mussels packed under modified atmosphere remained lower than acceptable limit values in 15 days of storage. However, aerobic, and vacuum-packed mussels exceeded the limit values during this time. In another study, cultured mussels (*M. galloprovincialis*) were packed under different atmosphere (MP1: 60% CO_2/20% N_2/20% O_2; MP_2 = 60% CO_2/40% N_2) and stored at 3 °C. Mussels packed in air and under vacuum were used to control the results. According to sensory, biochemical, and total viable count analyses, it was found that the MAP including oxygen (MAP1) extended the shelf life of refrigerated mussel flesh as compared to MAP with the same CO_2 content but without oxygen. In addition, VP extended the shelf life of mussels but at a lower degree than MAP with oxygen (Goulas 2008). Similar results have been reported for wild mussels (*M. galloprovincialis*) packed under MAP (M1 = 50%/50% CO_2/N_2: M2 = 80%/20% CO_2/N_2: M3 = 65%/35%CO_2/N_2) (Caglak et al. 2008). An alternative process, an integrated cooking and vacuum-cooling system, to cook shucked mussel (*Perna perna*) to avoid cross contamination and reduce exudation inside modified atmosphere package was developed. Mussels were stored for 25 days at 3 °C under modified atmosphere (MAP) of 50% CO_2/50% N_2. This alternative method with the modified atmosphere was efficient to control exudation and microbiota of mussels stored over 25 days (Lima et al. 2017). Furthermore, it was found that the coating with alginate-based and green tea (2.5%) and vitamin C (1.2%) incorporated before packaging in a modified atmosphere was more effective to increase the shelf life of cooked green mussels (*Perna viridis*) at 4 °C. Mussels packed under MAP containing gas mixture of 60% CO_2: 20% N_2: 20% O_2 were observed to have a shelf life of more than 28 days (Teerawut et al. 2017).

Freezing of Mussels

Shellfish keep their shells tightly closed. If the shell is open, should be tapped it lightly. The shellfish should respond by snapping the shell closed. If it does not respond, it is dead. The dead ones should be discarded immediately. They should be kept solidly frozen. Should not be thawed and re-frozen them. Repeated freezing and thawing reduce the quality of the shellfish and allows spoiling. They should be thawed only enough for the number of servings needed and thawed in the refrigerator on the day to be used.

The mussel-freezing process starts with cooked and shucked meat. It is frozen as blocks which are laid out on trays. Mussel meats are proper to either the plate freezing or the air blast freezing process, since the meats are small and provide good surface contact with the cooling medium (Waterman 1963). The freezing process is usually carried out in blast freezers at −40 °C for a long enough time to obtain a temperature of −20 °C in the geometrical centre of the mussel meat block. The process normally takes approximately 50 min. The meats may be packed in a variety of containers before freezing, for example waxed cartons or polythene bags, or may be frozen unwrapped and packed or glazed afterwards. The unwrapped meat is glazed to mitigate the water loss by evaporation. Finally, glazed and not necessarily packed meats are stored in a cold room at a temperature of −20 °C. Their shelf-life under these conditions should be 8–9 months. Thawed meats after 8–9 months storage will then be in excellent condition with flavour and texture equal to fresh (Waterman 1963; Almonacid et al. 2015).

Thawing of the meats in 10° salt solution, or soaking the thawed meats in a similar solution, may enhance the flavour, the salt acting as a condiment; the use of salt during cooking or subsequent rinsing of the meats before freezing may however accelerate the development of off flavours in the frozen meats during cold storage (Waterman 1963).

Mussels should be pre-cooled for effective and rapid freezing. This will also affect product quality. In a study examining the effect of pre-cooling with ice on the quality and shelf life of frozen mussels (*M. edulis*), freshly frozen mussels remained acceptable for 40 weeks, while the mussels frozen after ice storage of 8 days were kept acceptable for only15 weeks (Chinnamma 1973).

The quality and shelf life of frozen seafood varies depending on the biological differences and pre-freezing processes. If the product is initially of poor quality, its shelf life is also significantly reduced. In addition, storage conditions and storage time greatly affect the quality of the final product (Gokoglu 2002). In various studies, different shelf life periods were determined for mussels frozen under different freezing and storage conditions. Gokoglu et al. (2000) have determined a shelf life of 4 months for mussels (*M. galloprovincialis*) frozen and stored at −20 °C. During the storage, trimethylamine (TMA-N) total volatile bases (TVB-N) and pH values increased, and sensory acceptability decreased.

Some researchers have used some preservatives to maintain the quality and improve storage stability of frozen mussels. Kaba and Erkoyuncu (2005) have treated mussels with sodium tripolyphosphate before freezing and determined that it was effective in reducing weight loss. In another study, polyphosphates reduced the amount of expressible moisture of the muscles (Paredi et al. 1996). Ablett et al. (1986) have investigated the usefulness of ascorbic acid and chelating agents in retarding sensory deterioration and progression of oxidative rancidity in frozen cooked mussels. Ascorbic acid with and without chelating agents was found effective in preventing of oxidative rancidity of mussels (*M. edulis*) at −12 °C for 20 weeks. However, these agents showed no effect in preventing sensory deterioration of mussel meats under the same conditions. Storage at −30 °C maintained sensory quality and demonstrated less progression of lipid oxidation compared with mussels held at −12 °C.

Thermal Processing of Mussels

Thermal processing is a method of preserving food by heating in sealed containers to eliminate microbial pathogens at a given time and temperature. Thermal processing includes packing the material with oil or brine within a metal can, bottle, or pouch, sealing the container completely and heating to kill most microorganisms in the products.

Sterilisation and pasteurization are two main temperature categories employed in thermal processing. In addition, the thermal treatment can be divided into categories depending on the temperature, method, equipment, fish species, packaging method or microbial target to be applied.

Sterilization is a classic thermal method aiming inactivation of all pathogenic bacteria and their spores and performing at temperatures ranging from 110 to 135 °C. Mussel meat is suitable for canning in oil, brine, and sauces (Skipnes 2014). After depuration, the whole live mussels are either heated in open containers or steamed in the autoclave until the meat has a structure in which it can be easily shucked. Shucked mussels are boiled for 5 min in brine after washing thoroughly. Boiled meats are filled in cans, filling fluid such as oil or brine added, exhausted and heat process in steam is done. Once mussel meats are packed into the open cans, hot liquid brine, consisting of an aqueous solution of 1–3% salt and 0.1% citric acid is added to each can at a temperature of 75 °C (Almonacid et al. 2012).

The canning process of mussels requires two types of heat treatment. The first is the pre-cooking stage. This is done in hot water or steam at about 100 °C. As for the latter, after the filling of the mussels in cans, thermal sterilization is applied in pressurized retorts at 115–125 °C (Almonacid et al. 2015). The primary purpose of the precooking step is to separate the mussel shells from their valves, facilitate the manual removal of mussel meats.

According to Holdsworth and Simpson (2007), the thermal sterilization of filled and sealed cans is carried out at a specific time to ensure fo = 6 min lethality at 110–130 °C. After heat treatment, cooling with water at 19–21 °C is carried out. Finally, the cans are stored at room temperature. Under these conditions, they have a shelf life of 4 months.

The pasteurization process aims to inactivate vegetative cells and does not aim to inactivate the spores of all pathogenic bacteria. This process is generally applied to acidic foods and chilled foods.

Shucked mussels (*M. galloprovincialis*) were placed in a thermostable polyethylene-polyamide pouch with lemon juice, apple vinegar, onion, salt and black pepper., then pasteurized at 70 °C for 8 min. Pasteurized samples were stored in a refrigerator (4 ± 1 °C). According to the results the pasteurization process reduced the microbial load of the mussels. Authors stated that the pasteurized mussel stored at 4 °C could be safely consumed for 9 days (Tosun et al. 2018).

Canning of mussels is performed filling them into proper packaging material. For this purpose, different materials such as tin cans, retort pouches and bottles are used. Tin plate cans made of steel is the most used container for canning. Since tin containers affect the sensory properties of the product, aluminium containers and tin free containers have been developed over time. The use of rigid containers polymer coated Tin-free Steel cans as alternative to conventional cans is one of the recent developments. Although metal cans have the advantages of being available in different sizes, superior strength, high speed production, easy filling and closing, they also have the disadvantages such as heavy weight and difficulty to re-closure and disposal (Bindu et al. 2014).

Retort pouches are used as an alternative to traditional packaging materials such as cans in the thermal processing of mussels. When compared to conventional containers retort, pouches have many advantages such as they require less space when they stored as empty, can be easily transported, opened easily, provide fast heat penetration during heat treatment and thus shorten the processing time which maintains the nutritional and sensory properties of the product. Other advantages are light weight, cost effectiveness, ease of opening and reheating (Bindu et al. 2014). Live mussels were vacuum-packed in retort pouches after washing, shucking, removing the gut, cleaning by washing, mixing with salt and seasonings, and frying in the electric fryer for 3 min at 170–180 °C. The retort temperature was maintained at 121 °C. After processing the pouches to the required F0 value, they were cooled rapidly to 55°C. The results showed that a Fo value of 9.8 and a cook value of 90 min were optimal for processing fried mussels in a retort a pouch. The product remained in good condition after a storage period of 12 months at room temperature (Bindu et al. 2004). Chopped mussel meat packaged in retort pouches was processed in a laboratory-scale water immersion retort, adapted for processing under overpressure conditions. Retort temperature effects on product yield and on cook value were evaluated by setting the F0 at 7 min. The effects of different pre-treatments (salting and marination) on the characteristics of mussels were evaluated. Processing in retort pouches showed good performance. The salting pre-treatment of mussel meat increased product yield after processing (Tribuzi et al. 2015).

Mussels packed in glass are usually pickled either in brine or vinegar solution. To produce bottled mussel meat, the cooked, cleaned mussel meats are brined for up to 3 h in a 10° salt solution, drained and allowed to stand for 3 days in a vinegar and salt solution. The meats are then packed into glass jars and covered with spiced vinegar that has been diluted with an equal quantity of water. The jars are then sealed and sterilized. Mussel liquor obtained during extraction of the cooked meats from the shell may be used instead of water to dilute the solution before adding to the bottled meats (Waterman 1963). Similarly, white wine or wine vinegar may be used as alternatives to distilled vinegar. Extracted mussel (*P. viridis*) meat was filled sterilized into bottles and brine solution was added to remaining space. Closed bottles subjected to wet steam sterilization in pressure cooker for 1 h at 110 °C. It was concluded mussel meat can be exposed to heat process in bottles to produce ready to eat value added product and can be stored in room temperature up to 6 months (Jayasooriya et al. 2014).

Thermal processing seems to be a good alternative to obtain a shelf-stable product with good nutritional value. However, changes in the composition of mussels can be observed during the canning process. Heat treatments cause loss of water from the mussel meat tissue because of protein denaturation, along with other trace nutrients. This loss of water translates directly to loss of mass in the mussel meats and substantial yield loss in the manufacturing process (Almonacid et al. 2012). Almonacid et al. (2015) reported that the effects of the process on mussel composition are mainly caused by two mechanisms. One of these mechanisms is the material exchange between mussel meat and processing medium (water, steam, brine, air) and the other is thermal inactivation of high temperature. Firstly, water loss and increase in concentration are observed in the pre-cooking stage. On the other hand, it is reported that a factor affecting water loss during heat treatment may be salt concentration. Salt has been reported to increase the water retention capacity as well as the solubility of proteins in muscle tissue by increasing the gaps between the filaments in the myofibrils (Hamm 1986). Moreover, process variables such as precooking time, brine salt concentration, and retort temperature had effect on the ultimate drained weight in the final mussel product (canned mussel meats) (Almonacid et al. 2012).

Due to the increase in this dry matter, the other components of the mussel increase. Reduction of moisture content with an increase of protein, fat and ash content was observed after thermal processing of green mussel (*P. viridis*) (Biji et al. 2015). Thermal process has led to a reduction of amino acid and fatty acid content in mussel samples (Biji et al. 2015).

The thermal processing of food may cause severe quality deterioration, such as degradation in colour and texture, nutrient loss, cook loss (weight loss) and area shrinkage, rendering the products reducing consumer acceptance. The most significant changes in mussel during heating result from protein denaturation. Protein denaturation reduces water holding capacity, shrinks muscle fibres, and causes connective tissue degradation, subsequently leading to a harder and more compact tissue texture. There are structural differences between shellfish and mammalian muscle. Therefore, the effect of thermal processing would be more detrimental to shellfish. Paramyosin which forms the cores of the thick filaments in the adductor muscle of invertebrates is covered by a cortical layer of myosin which is heat stable. However, there is less connective tissue in invertebrate whole body compared to vertebrate muscle also resulting in relatively higher textural changes resulting from protein denaturation. Mussel meat composition is different from finfish and other meat and contains less connective tissue and collagen leading to a greater compression force during the heating treatments, because of proportionately less gelatine formation (Ovissipour et al. 2013).

Sous-vide cooking means cooking food under controlled temperatures in vacuumed plastic bags. There is a very strictly controlled temperature application in sous-vide cooking. This term includes cooking under conventional vacuum and at low temperature. This technique is also described as long-term cooking at low temperature. The meat is cooked at low temperature for a long time and the juicy flavour and softness are provided (Christensen et al. 2012; Mortensen et al. 2012). Sous-vide technology can be applied to many products from vegetables to soups, meat and meat products and various sauces. There are few studies on sous-vide cooking of seafood. Only one study on sous-vide cooking of mussels has been found (Bongiorno et al. 2018). In this study, freshly harvested mussels (*M. galloprovincialis*) were packaged under vacuum in polyamide/polypropylene bags and heat-treated in a steam oven at 85 °C for 10 min and cooled immediately to 3 °C. The sous vide cooking resulted in being able to preserve the quality of mussels and extend their shelf-life to 21 storage days. Moreover, with the addition of brine a shelf-life extension of up to 30 days was possible, in comparison to mussels subjected to conventional cooking. Further studies are needed to cook mussels with this method. It should be examined particularly for microbial safety.

Marinating of Mussels

Marinating is a semi-preservative method for fish based on the treatment of muscles with marinade solutions containing salt, sugar, spices, oil and acids etc. Marinating solutions are also used to prevent the growth of microorganisms and to inhibit the activity of enzymes and thus marinating reduces the bacterial and enzymatic activity, contributes to the improvement of sensory qualities of the product and ensures its extended but limited shelf life (Arason et al. 2014). Although there are several studies on the processing of fish by marination, there is little published information on the production of marinated mussels. There are only a few studies on this subject (Fig. 3.2).

In a study, a shelf life of 15 days at 4 °C was determined for mussels (*M. galloprovincialis*) marinated using acid, salt, and vegetable oil (Cherifi and Sadok 2016). Green mussels were marinated using acetic acid and salt after depuration, shucking, washing and blanching processes. This product, which was stored at room temperature, was determined to have a shelf life of 16 weeks (Unnikrishnan-Nair et al. 1989). Some marinades are produced with addition of different ingredients to improve the quality and extend the shelf life. In a previous study, mussels (*M. galloprovincialis*) which are boiled in 100 °C water for 5 min, removed from shells and byssus, were marinated. The marination solution consisted of citric acid (3.5%) and NaCl (2%). The products were packaged in plastic bags. Different ingredients such as carrageenan LM, konjac flour E425 i, phosphate, yeast extract, xanthan gum and maltodextrin were added into the marination solution to decrease weight loss and

Fig. 3.2 Marinated mussel

enhance sensory properties. The best results were obtained with solution containing 0.2% carrageenan. Weight increase and good sensory responses were determined with this solution (Guldas and Hecer 2012). In another study, mussel (*P. perna*) marinate was prepared using two kinds of marinate solution. In the first group, in addition to acetic acid and salt (NaCl), soy oil, sorbic acid (maximum 0.1 g/100 g), spices, onion and green pepper were added to the marinate solution. The second group was prepared by adding mustard together with this mixture. All marinades were stored at 4 °C. The data from microbiological, physicochemical, and sensory analysis showed that the marinated mussel was stable for 50 days at 4 °C, with both formulations (Aveiro et al. 2007).

Drying and Smoking of Mussels

Drying is the removal of water content to safe levels that can slow down the actions of enzymes, bacteria, yeasts, and moulds. The effects of heating on the activity of microorganisms and enzymes are also important when food is dried. Research on the drying process in the food industry has three goals: economic considerations, environmental concerns, and product quality aspects (Nguyen et al. 2014). Like other dried seafood, dried shellfish has been commonly produced and consumed in Asia due to its convenient for storage and transportation. Drying of mussel meat increases the value of this product and extends the shelf life. Tribuzi and Laurindo (2015) dried the cooked mussels (*P. perna*) in different methods and examined the effects of these methods on the rehydration capacity of the product. For this purpose, the live mussels were cooked at 100 °C for 5 min, cooled and separated from their shells and dried by three different methods (freeze drying, vacuum drying and air drying). In this study, it has been shown that it is possible to obtain good quality product in shorter time than in air and vacuum drying by freeze drying method under working conditions.

Natural sun drying has been used since ancient times for agricultural products. Sun drying is the most convenient and the cheapest processing technique to preserve fish and fish products particularly in tropical and subtropical countries, where solar radiation is abundant, inexhaustible, and environmentally friend (Nguyen et al. 2014). The products are dried by spreading in the open air. The products are placed on various materials such as trays, hangers, frames, etc. For this purpose, bamboo frames are mostly preferred in tropical regions (Gokoglu 2002). However, a major problem of open sun drying is the loss of quality due to contamination with dust and excreta from birds and animals (Nguyen et al. 2014). Therefore, other drying techniques have been developed and various studies have been conducted on the use of these techniques in drying mussels.

Recently innovative drying methods have been tried to dry mussel meat in some studies. Spray drying is a unique technique which can produce powder of special particle size. In a study, to produce dried powder from freshwater mussels, after washing the mussels were deodorized by two different methods. In the first method, 0.3% sodium chloride, 1% citric acid and 0.1% yeast powder, in the second method 1% ginger and 0.5% cooking wine were used. The deodorized mussels were placed in the meat grinder and the small pieces were transferred to the ultrafine grinder with 20% water for the subsequent pulverization process and dried in a spray dryer (Zhang et al. 2013). Freeze drying is an innovative drying process that uses ice sublimation as the main drying mechanism. This differs from conventional drying methods in that it is based on capillary movement and evaporation of liquid water for drying. In freeze-drying, water in the products is removed by sublimation at low temperature and low pressure below the triple point. First, the water in the products is frozen in the pre-freezing step and then removed by sublimation of ice in the primary drying step (Nguyen et al. 2014). McCarron et al. (2007) have tried freeze-drying process of mussel (*M. edulis*) with the aim of stabilizing shellfish toxins. In this study, they used mussels naturally contaminated with domoic acid, okadaic acid, dinophysistoxin-2, and azaspiracid-1, -2 and -3. Wet materials were used for control purposes. The freeze-dried material containing domoic acid was stable over the whole duration at all temperatures, while in the wet material domoic acid degraded to some extent at all temperatures except −20 °C. In freeze-dried and wet materials containing lipophilic toxins, okadaic acid, dinophysistoxin-2, azaspiracid-1 and azaspiracid-2 were stable over the whole duration at all conditions, while concentrations of azaspiracid-3 changed significantly in both materials at some storage temperatures. In another (Kipcak 2017) study, microwave drying have been used for *M. edulis*. In this study, several microwave power levels (90, 180, 360, 600 and 800 W) were applied to mussel (*Mytilus edulis*) to determine their effect on drying kinetics, rehydration characteristics and energy consumptions. At the end of study, it was found that for microwave power levels of 90, 180 and 360 W optimal drying times were found to be 16, 5 and 2 min, while they were 80 and 60 for 600 and 800 W respectively. The microwave power level of 360 W was found to be the most effective, considering the minimum energy consumption.

Dried products which are rich in lipids and unsaturated fatty acids lose their flavour and oxidize easily during storage. Mussel (*M. edulis*) meat boiled and later separated from the shells were sun dried and packaged in vacuum packs and air packs. Packed samples were stored at room temperature and the changes of their lipid and fatty acids were investigated for 1 year. During the storage, while free fatty acids (FFA) and saturated fatty acids increased greatly, phospholipids and polyunsaturated fatty acids decreased. Packaging methods had a great influence on the oxidation of mussel and vacuum packaging showed best results (Changhu et al. 1998).

Smoking is a process that gives taste and flavour to fish because of the penetration of the smoke obtained by the burning of the plant materials most often wood. Fish and shellfish have been preserved for many years using smoke. It has been used as an inexpensive alternative preservation method to prevent post-harvest losses, especially in less developed countries. It has become a popular product preferred in developed countries due to its different flavour and texture (Patterson 2004).

Although smoking process generally is used for fish, mussel meats are traditionally and commercially smoked in various parts of the world. Whole mussels are usually smoked with oak wood. Smoked mussels are delicious as stews or chowders or eaten with a splash of lemon. However, there are a limited number of studies conducted on smoking of mussels. It is seen that there is a need to carry out studies on proper smoking method, process parameters, appropriate wood selection, and packaging method and storage conditions for mussel meat. In addition, changes in product quality during storage should be examined in detail. A few studies on this subject are summarized below (Patterson 2004).

In a study, the cooked, cleaned mussel meats were soaked in brine 50% at room temperature for 5 min to improve taste and consistency. The brined mussel meats were drained for 5 min. The meats were dipped in edible vegetable oil and placed on racks in a semi-controlled mechanical smoking kiln that was electrically heated. Smoke was produced by burning wood (beech) sawdust under forced airflow. The mussel meats were hot smoked at 82 °C for 30 min and were cooled at room temperature. Smoked mussels were packaged in polystyrene boxes wrapped with stretch film and stored at 4 °C. In these conditions, a shelf life of 12 days was determined for mussel meat (Turan et al. 2008).

In another study, live mussels were washed and depurated by immersion in water chlorinated for 2 h. The mussel then was shucked. The meat was blanched in 5% brine. After 5 min blanching in boiling brine, the meat was drained and smoked at 80–90 °C for 30 min in a conventional vertical smoke kiln. The smoke was generated by burning coconut husk and saw dust. Smoked meat was dried to 10% moisture level in a dryer. The authors stated that under these conditions the product will have a shelf life of more than 6 months (Muraleedharan et al. 1979).

Canned smoked mussel meat is popular on the international market because of its characteristic flavour. Shelf life of mussels can be prolonged by canning after being smoked. There are a few scientific studies on this subject. Sengor et al. (2004) conducted such a study. In this study, block-frozen mussel meat was used. Defrosted mussels were dipped into brine for 1 h. After draining, mussel meats were dried at 30 °C for 15 min and smoked with oak wood sawdust smoke for 45 min in a semi-control mechanical smoking kiln. Following the smoking process mussel meat was canned. They were sterilized at 120 ± 1 °C for F0 5.2 ± 0.1 min. The authors reported that microbial control was guaranteed by the technological process. In another study, after depuration, live mussels (*Perna indica*) were washed, boiled for 10–15 min, and shucked. The meat was blanched in brine for 5 min. After blanching, the drained meat was spread on trays and dried for 1.5 to 30 min to facilitate uniform and better absorption of smoke. Smoking was done in a conventional smoking kiln where sawdust generated the smoke. Smoked meats were placed into lacquered cans and filled with heated sunflower, ground nut and cotton seed oil. The filled cans were sterilized at 121 °C for 15 min. During the storage period of 9 months, mussel meat was found to be acceptable sensorially (Patterson 2001).

Liquid smoke is also used to smoke seafood. Liquid smoking extract is prepared by the dry distillation of wood and then concentrated. Concentrated smoke is dissolved in a solvent, such as water of oil and can be used directly on products (Arason et al. 2014). The use of liquid smoke has several advantages such as providing more rigid flavour control, lowering costs, giving less environmental damage, better in quality and prolonging their preservation, compared traditional smoking procedures (Dimitriadou et al. 2008). Farmed mussels (*M. galloprovincialis*) were brined in 0, 10 and 20% sodium chloride solution for 5 min. Then, mussel samples were placed on drying racks for 2 min in order to be partially dried. The smoking process was conducted using a 10-L electrical commercial steamer added 20, 50 and 80 mL of liquid smoke diluted in 1 L of tap water (2, 5 and 8%). Thereafter, the smoking process took place by the production of smoky steam from the liquid smoke solution. Samples were placed at one layer and processed for 6 min. The applied steam pressure conditions were: 1, 1.5 and 2 bar and the consequent temperatures were 100, 111 and 120 C, respectively. After each process, the liquid smoke solution was discarded, and a new solution was used. After cooling, the samples were stored at 4 °C. The optimal hedonic conditions of the mussel products were achieved at smoke concentrations 3.8–8%, brine level from 8.5 to 13.5% at pressure 1 bar and from 11.5 to 16.5% at pressure 1.5 bar. It was concluded that to produce high quality smoked mussel products can be processed by immersing mussel samples in approximately 13% brine concentration for 5 min and can be steamed in approximately 6.5% liquid smoke at 1.5 bar pressure (Petridis et al. 2012). In a study comparing the use of liquid smoke in the smoking of mussels (*M. galloprovincialis*) with traditional method, it was noticed an unflavoured taste formation in smoked mussel treated with liquid smoke. Therefore, it has been reported that practically liquid smoke is not suitable for this product. It was determined that the aromatic properties and colour were pleasant in the black mussel, which has been given the aroma of smoke with the traditional method when the oven conditions were checked. Therefore, it has been reported that practically liquid smoke is not suitable for this product. When the oven conditions were checked, the aromatic properties and colour were found to be pleasant in the black mussels which were given the aroma by the traditional method (Sengor et al. 2003).

3.1.2 Clams

3.1.2.1 General Information About Clams

Clam is a common name for several kinds of bivalve molluscs. The word is often applied only to those that are edible and live as infauna, spending most of their lives halfway buried in the sand of the seafloor. Clams have two shells of equal size connected by two adductor muscles and have a powerful burrowing foot. Clams in the culinary sense do not live attached to a substrate (whereas oysters and mussels do) and do not live near the bottom (whereas scallops do). In culinary usage, clams are commonly eaten marine bivalves (Wikipedia 2019e).

Clam production and aquaculture are important economic resources in a worldwide context (Fiz da Costa 2013). Clams are an important food item, especially in Mediterranean countries. In Italy, Christmas Eve is the day of the year with highest seafood consumption, including clams and anchovies that are generally used in appetizers (FAO-Globefish 2019).

Although the amount of production is lower than other bivalve species, clams have an important place in world trade. International trade of clam was stable during the first 9 months of 2017 compared with the same period in 2016. Around 170,000 tonnes were exported, and 180,000 tonnes were imported. The main importing countries continued to be Japan and the Republic of Korea, both buying mainly from China. China is in the main clam exporter globally, accounting for about 70% of world clam exports. Most clams are traded in live or fresh form (FAO-Globefish 2018a). FAO reported the amount of capture production to be 591,265 tonnes for all of clams, cockles, and ark shells in 2016 (FAO 2016).

There are several bivalve families in the Clam category. This book will include families with economic importance in terms of fisheries and aquaculture.

Biology of Clams

Clams are a very diverse group of bivalves in that there is notable variation in the shape, size, thickness, colour, and degree of sculpturing of the shell from one species to the next. The one feature that all clams have in common is that they burrow into the seabed. Consequently, both shell and body display modifications necessary for this type of existence. The quahog clam *M. mercenaria* has a thick, triangular shell. It is grey or brown with a sculpturing of numerous shallow concentric rings that run around the shell, parallel to the hinge. Annual rings are clearly visible on the shell exterior and thus ageing in this species, and indeed in many of the other commercially important clam species, is an easy task. There are three conspicuous teeth on each valve and each tooth fits into a corresponding socket on the opposing valve. This ensures an intimate fit when the valves are closed. The shell interior is marked by an anterior and posterior adductor muscle scar, a distinct pallial line, and a short pallial sinus. The depth of the pallial sinus is a very reliable indicator of the length of the siphons, and thus the burrowing depth of a clam species. *M. mercenaria* and the softshell clam *Mya arenaria* can reach a shell length of 150 mm, while the surf clam *Spisula solidissima* grows as large as 220 mm. The palourde *Ruditapes decussata* and the Manila clam *Ruditapes philippinarum* are much smaller, with a maximum shell length of about 75 mm (Gosling 2015).

Distribution of Clams

Venerid clams are the target species of bivalve fishing all over the world due to their worldwide distribution. In general, more varieties of venerid species are harvested in the Pacific and Indian oceans than in the Atlantic and the Mediterranean. In different geographical areas, various species are exploited due to their wide distribution area, while other species with more limited distribution are harvested only in more restricted areas (Gaspar et al. 2013).

It is reported that giant clams (family Tridacnidae) lived along shallow shorelines and on reefs from South Africa to beyond French Polynesia and Japan to Australia (Othman et al. 2010). Giant clams are found throughout the tropical Indo-Pacific, generally inhabiting the shallow water of coral reefs (Gosling 2015).

Although Asian clam (*C. fluminea*) is a freshwater species native to southern and eastern Asia (Russia, Thailand, Philippines, China, Hong Kong, Taiwan, Korea and Japan) and Africa (Britton and Morton 1979), it is now found in freshwater and salt water throughout the USA, including all five Gulf states and northern Mexico, and much of Europe (Cabi 2019b).

The European clam fisheries are dominated by several species in the Atlantic coast such as the European clam (cross-cut carpet-shell clam), *Ruditapes decussatus*, the palourde (in French), *Venerupis corrugata*, and the Manila clam, *Ruditapes philippinarum*. In the Mediterranean, the dominant species is the stripped venus clam, *Chamelea gallina* which presents fisheries with relevant socio-economic importance particularly in the Adriatic Sea (Scarcella and Cabanelas 2016).

Manila clam *Ruditapes philippinarum*, indigenous to the western Pacific, was introduced first onto the West coast of North America and from there into western Europe. The hard clam *Mercenaria mercenaria* is distributed on the Atlantic coast of North America from the Gulf of St. Lawrence to Florida and is particularly abundant from Maine to Virginia. The softshell clam *Mya arenaria* occurs in many temperate and subarctic areas along the north-east and north-west coast of the Atlantic Ocean and the north-east coast of the Pacific Ocean. Other clams that also have a broad geographic range in tropical waters are species of *Arca, Mactra, Meretrix* and *Paphia* (Gosling 2015).

Fishing/Harvesting of Clams

Although some venerids fisheries occur in the subtidal zone of coastal areas, the majority takes place in intertidal flats and shallow subtidal areas of estuaries, coastal lagoons, sheltered bays, and mangroves. In these areas, the most common harvesting methods are hand picking, signing, digging, or raking. In hand picking, harvesters get down on their hands and knees or squat down and then sweep their hands through the sediment to dislodge the clams. Signing and digging consists in walking along the shore looking for siphon holes. When a hole is found, the harvester picks the clam from the sediment with his hands or digs the clam out of the sediment using simple tools such as spoons, knives, rakes, forks, shovels or grubber hoes. Some harvesters use a clam gun to collect clams that are buried deeper into the sediment. This tool comprises a 10–15 cm diameter tube with a handle and a small air vent at the closed upper end. When a hole is found, the tube is placed over it and pushed down with an up-and-down, rocking, or twisting movement. When the clam is enclosed, the air vent is blocked with a finger and the core of sand pulled up and dropped on the beach to collect the clam. Raking is a technique that involves dragging a rake through the bottom until a scraping is felt. By that time, the harvester pushes the rake into the sediment, pulls it towards him and upwards to collect the clam. In shallow subtidal areas, treading is a widely used method that consists in probing the bottom with the foot until a clam is felt. When a clam is found, the harvester simply bends down and picks it up. Hand dredging and bull raking are also two common techniques used in shallow areas (ranging between 0.3 and 1.5 m depth). Clam kicking is another harvesting method employed in the low intertidal. In this method, the fisherman uses the propeller backwash to dislodge clams out of the substrate, throwing them on the sediment surface. Then, clams are picked up by hand, or using hand dredges or bull rakes. In sheltered areas and coastal zones ranging from 1 to 50 m depth, clams can be harvested by divers (Gaspar et al. 2013).

Some Important Species of Clams

Veneridae (Venus Shells)

The Veneridae or venerids, common name: venus clams, are a very large family of minute to large, saltwater clams, marine bivalve molluscs. Over 500 living species of venerid bivalves are known, most of which are edible, and many of which are exploited as food sources. Many of the most important edible species are commonly known simply as "clams". Venerids make up a significant proportion of the world fishery of edible bivalves. The family includes some species that are important commercially, such as the hard clam, *Mercenaria mercenaria* (Wikipedia 2019f).

Venus shells or Veneridae include the most important representatives of clam species in the world. For example, *Venerupis philippinarum* or Manila clam, which has been introduced to various parts of the world since the beginning of the twentieth century, is by far the most cultured clam species. The hard clam *Mercenaria*

mercenaria Linnaeus, 1758 is an important recreational and commercial species harvested in the United States (Fiz da Costa 2013).

Venerids have rounded or oval solid shells with the umbones (projections) in turned towards the anterior end. Three or four cardinal teeth are on each valve. The siphons are short and united, except at the tip, and are not very long. The foot is large (Wikipedia 2019g).

Over 800 species have been identified in the Veneridae family, of which about 20 species are collected from natural beds or cultured to increase production and meet market demands. *Mercenaria mercenaria*, *Venerupis decussate*, *Venerupis philippinarum*, *Cyclina sinensis*, *Chamelea gallina*, *Venerupis corrugate*, *Ameghinomya antiqua*, *Leukoma staminea*, *Meretrix lusoria*, *Polititapes aureus* are commercially important venus clam species (Arias-Pérez et al. 2013).

Mactridae (Surf Clams)

The Mactridae are bivalves found in shallow waters, to a depth of 50 m, and sheltered areas. They are active burrowers of sandy to muddy bottoms. *Mactra luzonica*, *Mactra maculate*, *Macta mera*, *Mactra achatina*, *Mactra cuneata*, *Mactra violacea*, *Meropesta capillacea*, *Meropesta pellucida* are important species of this family.

Donacidae (Wedge Shells)

They are quick shallow burrowers living in sandy bottoms. They are both suspension and deposit feeding animals. *Donax cuneatus*, *Donax faba*, *Donax deltoids*, *Donax incarnates*, *Donax scortum* are some species of this family.

Myidae (Softshell Clams)

Mya contains most species of this family. *M. arenaria*, *M. baxteri*, *M. pseudoarenaria* and *M. truncata* are among the well-researched species.

Solenidae (Razor Clams)

Solenidae is an ecologically and economically important family of marine bivalves called razor clams. They are filter feeding animals which adapted to switch and deep borrowing in soft bottoms with their powerful foot. Intertidal species of Solenidae are sometimes actively exploited, in the Philippines, Indonesia and Malaysia. Pharella acuditens, *Pharella javanica*, *Sliqua winteriana*, *Solen Lamarckii*, *Solen rosemaculatus* are some species of this family.

Saeedi and Costello (2013) reported that no species of any of the families are reported from Antarctica. Otherwise the Mactridae and Veneridae were cosmopolitan. The Donacidae, Myidae, Phardidae and Solenidae were absent from New Zealand and the Pacific islands. The Myidae has the most restricted.

Nutritive Composition of Clams

Clam meats are also an important source of protein as in mussels. As with other shellfish products, they have low fat content. Clams are also an excellent source of minerals.

There are several studies on the chemical composition of different clam species. In one of these studies, in foot, mantle and viscera of Asian hard clam (*Meretrix lusoria*), moisture content was 76.23, 84.22 and 80.89%; protein content 12.75, 9.09 and, 9.61%; fat content 1.58, 3.53 and 6.58; ash content 1.23, 1.94 and 2.58% and carbohydrate content 7.89, 1.20 and 0.32% were reported (Karnjanapratum et al. 2013).

In another study, moisture, protein, lipid, and ash contents of venerid clam (*Chamellia gallina*) were reported to be 67%, 10.12%, 2.57%, and 1.66%, respectively. The mean concentrations of elements in tissues were found as boron 2.37–4.24 mg/kg, chromium 0–0.76 mg/kg, Cobalt 0–0.43 mg/kg, copper 0.71–5.30 mg/kg, Manganese 0.30–5.94 mg/kg, Zinc 13.08–77.76 mg/kg, Nickel 0–1.22 mg/kg, iron 2.46–114.22 mg/kg, Aluminium 1.23–75.49 mg/kg, lead 0.18–3.24 mg/kg, Barium 0.66–15.97 mg/kg, and Cadmium 0.04–0.69 mg/kg (Arik-Colakoglu et al. 2011).

Orban et al. (2006) have reported that protein (8.55–10.75 g/100 g), total lipid (0.73–1.59 g/100 g), glycogen (2.25–4.96 g/100 g) and non-protein nitrogen (0.54–0.78 g/100 g) contents of venus clam (*C. gallina*) changed seasonally and there was a relationship between proximate composition and condition index. They found a prevalence of polyunsaturated fatty acids (PUFA, 41.6–48.1% of total fatty acids) over the saturated (SFA, 29.1–39.3% of total fatty acids) and monounsaturated ones (MUFA, 14.2–23.4% of total fatty acids) throughout the year.

Seasonal variation in proximate composition is also reported for venerid clam (*Chamellia gallina*) and wedge clam (*Donax trunculus*) collected from the Marmara Sea. 81.43–88.75% moisture, was 5.91–9.48% protein, 0.45–1.38% fat, 0.59–1.57% ash and 2.19–4.87% carbohydrate for *C. gallina*, while for *D. trunculus* the moisture was as 80.14–87.33, protein as 6.94–11.24, fat as 2.36–4.55, ash as 2.59–5.68 and carbohydrate as 0.96–4.48 were reported (Ozden et al. 2009).

The biochemical composition of *Ruditapes decussatus* has been found to change seasonally due to the reproductive cycle (Anibal et al. 2011). In the study of the authors, it was determined that gametogenesis started in January and that ovulation lasted from June to September and resting period lasted from October to December. Therefore, it was determined that the nutritional values of clams were higher in summer than in winter.

Mean moisture, protein, fat, and ash contents of venus clam (*C. gallina*) collected from Black Sea coasts of Turkey were reported as 85.76%, 8.49%%, 0.54% and 2.27% respectively (Olmez et al. 2003). The moisture, protein, fat, and ash contents of clam (*Mactra chinensis*) muscle collected from the coasts of the Japan Sea were reported as 80.16%, 14.55%, 1.89%, 0.553%, respectively. Na (1134–1527 mg 100 g^{-1}) and K (1088–2293 mg 100 g^{-1}) were found to be the main macro elements in clam muscles. The major microelements present in the tissues of the clams were determined as Fe, Zn, Cr, Ni and Mn. Cu, Cr, Mo, and Al were the other detected elements. The tissues of *M. chinensis* were characterized by the highest content of Fe, Ni, Al, and Mn (Tabakaeva et al. 2018).

Hirimuthogoda et al. (2016) determined the moisture, protein, fat, and ash contents of clam (*Geloina vexans*) collected from two different regions in Siri Lanka as 83.08–90.98, 4.41–11.98, 0.56–1.09 and 1.03–3.13, respectively. In the same study, the authors determined total saturated, total monounsaturated, total unsaturated, total n-3 polyunsaturated fatty acid levels as 34–36%, 19.2–20.3%, 24–24.3% and 17.3–18.8%, respectively.

In a study, in clam (*Meretrix casta*) collected from Cuddalore and Parangipettai Coast, South East Coast of India, the total protein content varied between 45.67% and 30.021%., the carbohydrate concentrations between 4.21 and 15.67% and lipid content 5.63 and 1.11% (Srilatha et al. 2013). In this study, it was reported that the total essential amino acids were recorded 48.67% and 47.21% and non-essential amino acids were recorded 43.87% and 39.4% in *M. casta*. Moreover, this study showed that a total of 20 amino acids which exhibit high levels of lysine followed by phenylalanine and histidine in the tissue of *M. casta*. The result revealed in this study showed that, M. *casta* meat is a potential source for food value due to high quality protein, as well as balanced essential amino acids.

3.1.2.2 Post-harvest Handling of Clams

After the harvest, surf clams are placed in metal-mesh cages. These are mechanically unloaded at the dock and stored, if necessary, in a refrigerated room. They are then dumped onto a conveyor belt. The seafood industry which moves the clams through a gas flame to open the shell. Some processors use a steam or hot water bath to open the shells. Then the clams are carried to a table where the meats are extracted from the shell by shuckers, and the viscera, referred to as the sandbag, are removed. These two steps are often still done by hand, but mechanical shakers and eviscerators are also being used. The meats are then washed and cut or diced for packing and freezing. The surf clam meats are frozen in sheets, called blankets, for sale and shipment to food processors and restaurants. Surf clams are usually served in chowders, soups, or fried clam strips. The smaller hard clams are usually sold live and, in the shell, and are consumed raw, steamed, or as specialty items such as clams casino. Chowder clams are either sold in the shell or fresh shucked for further processing. Surf clam processors occasionally process chowder clams in the same manner as surf clams. Soft clams are sold in the shell for steamed clams or as a shucked and frozen product used in restaurants as breaded and fried clams. They also are sold fresh shucked in cans or frozen. The Manila clam is sold and processed like hard clam. It seldom reaches chowder size, so most are sold and used like little neck and cherry stone hard clams (Castagna 1990).

Keeping and Processing Clams Alive

Bivalve, including clams, often are presented to the consumer in the live state; however, they are often processed for the live market. Bivalves can survive out of the water for extended periods, which allows them to go through the processing chain.

Clams can be stored wet or dry to keep alive. Freshly harvested clams (*Galatea paradoxa*) obtained from the Cross River, Nigeria were subjected to live storage for 7 days in or outside of water. It was determined that 57% of wet clams and 35% of dry clams died in 7 days. There was no death on day 1 in wet storage, but no death on day 1–3 in dry storage. Dry storage was found to be effective in reducing post-harvest mortality and weight loss and would increase profitability in the clam industry (Ekanem and Achinewhu 2006).

Depuration of Clams

Depuration is a process in which shellfish are placed in tanks with clean seawater and allowed to continue their natural filter feeding activities by cleaning themselves from sewage contaminants. Many factors such as, water quality, oxygenation and flow rates of water, temperature, water to bivalve ratio, salinity, removal, and deposition, of faecal material, purification and system design affect the depuration process.

In a study based on microbiological indices before and after depuration, the survival rate of depurated clams (*C. gallina*) and meat yields were investigated. After landing, clams harvested from offshore natural beds transported to the depuration plant. Depuration was performed monthly for 1 year. At the end of the study, *Escherichia coli* decreased by 62% and fecal coliforms decreased by 54%. The authors reported that after 24 h of depuration, except for August, the faecal coliform count of all samples was below the legal limits and that *E. coli* was above the limits in December and January. While no decrease in *E. coli* in August was reported, there was a decrease in other samples. Between March and September, Salmonella spp. and *Vibrio parahaemolyticus* were not detected, while *Vibrio alginolyticus* was detected. All these results showed that the effects of depuration conditions on microbiological quality were variable. On the other hand, meat yield and survival rate of *C. gallina* were not affected by depuration process (Maffei et al. 2009).

In another study aimed to determine the depuration times of *Donax trunculus* and *Tapes decussatus*, clams were contaminated with *Escherichia coli*, *Salmonella enterica* subsp. *enterica* serovar *Typhimurium* and *Vibrio parahaemolyticus*. Clams harvested from natural beds in Çanakkale Strait and Marmara Sea (Turkey) were depurated in tanks having filtration and ozone system. As a result of the study, it was determined that the bacteria load of both clams decreased by 40% in the first 12 h. Depuration time for all bacteria was determined as 66 s for *T. decussatus*. For *D. trunculus*, the situation was slightly different. Depuration time for *S. typhimurium* was found to be 66 h and for E. coli was 78 h. However, *V. parahaemolyticus* was detected at 1.7 Log10 cfu/g even after 72 h. Therefore, it has been proposed that

V. parahaemolyticus should be considered when classifying bivalve production beds (Arik-Colakoglu et al. 2014). Other researchers have reported that 4-day depuration is sufficient for clams containing *Salmonella* in the category B limits, while the C limits may be reduced by 50% over the same period (El-Shenawy 2004).

In a study to determine the depuration capacity of clams, *Venus gallina* were contaminated with *Vibrio parahaemolyticus* and *Vibrio vulnificus* to determine accumulation capacities in weak, medium, and highly contaminated waters. All clams were exposed to contamination for 72 h. Depuration trials were performed in close-circuit seawater-disinfection system that uses filtration, Ultraviolet and ozone. Most of the depuration trials with *V. parahaemolyticus* showed a decrease in initial bacterial loads after36–48 h, but in subsequent periods, the trend remained stationary. In *V. vulnificus* tests, clams showed a scarce depuration capacity instead (Barile et al. 2018).

Depuration is not only effective on microbial contamination. The effects of depuration on heavy metals accumulation were also investigated. El-Shenawy (2004) investigated clams (*Ruditapes decussatus*) harvested from two sites Timsah Lake, Ismilia, Egypt and found that the bioaccumulation of metals varied strongly according to the sampling site. After 48 h of depuration, it was found that Fe, Ni, Co, Cu, and Mn reduced significantly compared with the initial concentrations in clam tissue at the two stations.

Shucking of Clams

Most clams are shucked by hand. The hard clam is held in the palm of the hand with the shell hinge against the palm. A strong, slender knife is inserted between the valves and the shell is pried open, the abductor muscles cut, and the meat removed from the shell. Other clams are also shucked by hand (Hackney 1990).

Surf clams, ocean quahogs, and occasionally large hard clams are shucked mechanically. The clams are either placed on a flat conveyor belt and passed through an open gas flame or are steamed at high temperature in pressure vessels. The heating, which often partially cooks the meat, greatly weakens the muscle shell bond, and the meat is removed from the shell by tumbling. The meat and shells are separated in a brine tank. Surf clams, ocean quahog and large hard clams are often frozen after mechanical shucking. After the clams are removed from the shell, the viscera are removed. The meats are then washed and cut or diced for packaging and freezing (Hackney 1990).

Freshly harvested clams (*Galatea paradoxa)* from the Cross River, Nigeria, were subjected to heat treatment (steam and water at 60, 70, 80, 90, 100 °C) for 1–6 min after 24 h depurations. The effects of level of heat treatment on opening were evaluated. Some chemical shucking aids (NaOH, $NaHCO_3$, Na_2CO_3, NaCl) in 60 °C water was also used. According to the results boiling water was found most effective in opening the clams. Shucking was achieved in 1 min as 100%. Steam was least effective. For 100% opening needed 6 min in case of steam. In general, shucking aids reduced opening time. The shucking aids at 60 °C increased the incidence of opening to 100% in 2–3 min (Ekanem and Achinewhu 2000).

3.1.2.3 Clam Processing and Preservation

Chilling/Refrigerating of Clams

Chilling by ice and mechanical refrigeration are the most common means of retarding microbial and biochemical spoilage. Temperature control is the most critical factor for providing a good product. When clams are harvested during the warm summer months, on-deck temperatures can exceed 30 °C. For optimum quality and shelf life, bivalves should be cooled and stored at temperatures less than 10 °C. Processing shellfish for the live market is usually limited to washing, sorting, and packing. Hard clams (*Mercenaria mercenaria*) often are sold in the shell by size, with smaller clams being higher priced. The clams are sorted either manually or mechanically, and then washed and boxed for shipment. The Manila clam (*Tapes philippinarium*) is handled in a manner similar to the *Mercenaria* clam. The soft-shell clam (*Mya arenaria*) also is washed and boxed before sale. The clams are often placed in clean water for a short time to remove grit and sand from the intestinal area (Hackney 1990). Live clams are generally displayed on the ice which keeps them cold and extends their longevity. It should be noted that these products cannot be overwrapped, or they will suffocate (Haby and Coale Jr 1990).

The shelf life of refrigerated clam is relatively short. It is susceptible to microbiological and chemical deterioration and hence, the consumption of this product is expected to be shortly after purchase. The extension of product shelf life is very important to increase the market and satisfy consumer demands (Vongsawasdi et al. 2011).

Clams stored in direct contact with ice and non-contacted in packaged form were found to remain in good condition for 2 and 4 days, respectively (Basu and Gupta 1984).

The temperatures of the harvested clams should be kept under control within 20 h after harvest in summer. The product should be kept at a temperature of 7.2 °C or less and should not be out of temperature control for more than 2 h. The Food and Drug Administration (FDA) states that in accordance with the interstate transport regulations, it is necessary to allow shippers to cool their products to a temperature of 10 °C or less. Clam transporters prefer to transport the product quickly, usually within 2 h of receipt, until the product has cooled sufficiently (Brenton et al. 2001).

In a study, the live clams were transferred to the laboratory. After washing, they were depurated. The clams were steam shucked at 110 °C for 10 min in a retort. The shells were opened, and the shucked meat was collected manually, packed in two different material polyester polyethylene laminate (PEST-PE) and low-density polyethylene (LDPE) and stored in a box with flake ice at 2 °C. At the end of study, it was found that the shelf life of black clam was 22 days in PEST-PE and 16 days in LDPE (Sreedevi et al. 2018).

Freezing of Clams

Although only a very small part of the commercial harvest of clams may be frozen, only the best quality raw material should be selected for this process. Frozen storage life is limited to only 4–6 months at 0 °F (−18 °C), rancidity and toughening of the flesh being the major limiting factors. For preparation of clams before freezing are washed, and then are opened. The meats are washed, drained, and packed (Banks et al. 1977).

Many species of clams are harvested commercially. Only a very small part of the commercial harvest is frozen. Clams are not usually frozen in the shell, but small amounts are frozen as meats. They are shucked by cutting the adductor muscle and scraping out the meat. Meats that are frozen are usually chopped or minced for use in chowders (Kolbe and Kramer 2007).

Marinating of Clams

Clams are also pickled but to a lesser extent than mussels and oysters. Pickled cockles are common in France. The meat is dipped in 3% salt brine, drained, and covered for 3 days with a 3% vinegar solution containing 3% salt. The cockles are then drained, packed, and covered with spiced vinegar (Hackney 1990).

Canning of Clams

The clams are washed and scalded. The meats are shaken out of the shells and split along one side to remove sand and mud. They are washed a second time and the siphon, body side walls, and stomach are removed. The remainder is chopped and packed into cans. Brine or clam juice is added, then the cans are closed and retorted under pressure. They are removed from the steamer and sorted by size, small or large, and by colour, light or dark. Hand shucked clams also are used. The clams are packed into "C" enamelled cans. Clams shrink considerably during processing, so "fill weights" are greater than "drain weights." Dark discoloration is a problem sometimes encountered with canned clams (Hackney 1990).

3.1.3 Oysters

3.1.3.1 General Information About Oysters

Oyster is the common name for several different families of salt-water bivalve molluscs that live in marine or brackish habitats. In some species the valves are highly calcified, and many are somewhat irregular in shape. Many, but not all, oysters are in the superfamily Ostreoidea (Wikipedia 2019g).

They are one of the best known and most widely cultivated marine animals. Oysters are highly esteemed sea food and considered a delicacy in USA, Europe, Japan.

Biology of Oysters

Oysters differ from other bivalves in having a highly irregular shell form. The shape of the shell is typically dictated by environmental constraints, and they can grow over or around adjacent objects, including other oysters (NRC 2004).

Oysters are filter feeders, drawing water in over their gills through the beating of cilia. Suspended plankton and particles are trapped in the mucus of a gill, and from there are transported to the mouth, where they are eaten, digested, and expelled as faeces or pseudofeces. Oysters feed most actively at temperatures above 10 °C. An oyster can filter up to 5 L of water per hour. While some oysters have two sexes (European oyster and Olympia oyster), their reproductive organs contain both eggs and sperm. Because of this, it is technically possible for an oyster to fertilize its own eggs. The gonads surround the digestive organs, and are made up of sex cells, branching tubules, and connective tissue (Wikipedia 2019h).

Oysters do not regulate their body temperature or the salinity of their body fluids; thus, their metabolic activity is closely tied to the temperature of their surroundings, and the salt content of their blood is the same as that of the ambient water. The ability of oysters to tolerate different environments is species specific. For instance, the European oyster, *Ostrea edulis*, grows in relatively cool, clear, water of high salinity. Crassostrea species, in contrast, are more typically inhabitants of estuaries in which they tolerate wide fluctuations in temperature, salinity, and turbidity.

Members of the genus *Crassostrea* shed their gametes directly into the water where fertilization occurs, and larval life is spent entirely in the water column. In contrast, fertilization, and partial larval development in *Ostrea* take place in the interior of the oyster's shell. Females release eggs within the shell cavity, and fertilization occurs when sperm shed by nearby male oysters get drawn into the female cavity. The larvae develop partially among the female's gill filaments, which turn dark and become gritty as the larvae produce shells and become pigmented. The female's unpleasant appearance and texture at this time are the principal reason that eating *Ostrea* species is avoided in the summer when reproduction occurs. The larvae of *Ostrea* species are eventually expelled from the female's shell cavity and complete their development in the water column. Oysters that are brooders produce smaller numbers of offspring than nonbrooders (NRC 2004).

Oyster farming has become an increasingly important global aquaculture activity, accounting for the greatest proportion of molluscan aquaculture production. Out of 100 known oysters species, only several are widely farmed. The most widely cultured species are *Crassostrea angulata* and *Ostrea edulis* in Europe. Other species are *C. iredalei* (Philippines); *C. gigas* (Japan, Korea, West coast of the United States and Canada); *C. commercialis* (Australia); *C. brasiliana* (East coast of Southern South America); *C. chilensis* (West coast of South America);

C. margaritacea (South Africa); *C. gasar* (along the central West coast of Africa). *C. gigas* has been recently introduced into France, England, Morocco, Australia, and New Zealand (Garrido-Handong 1990).

Production and Trade of Oysters

Oysters are typically produced in one of three different ways: natural, managed and cultivated. Natural oysters grow and reproduce without human intervention and are often available for harvest by anyone with the appropriate licenses and permits. In contrast, managed oysters are supervised by harvesters who scrape the oyster beds periodically to reduce clustering. In the case of cultivated oysters, immature oysters are transported to man-made beds where they can mature (Lutz 2012).

With a total of 135,000 tonnes in 2016, oyster catches represent 0.15% of the global fisheries production. USA and Mexico provided 81% of the world catches in 2016. Spain and Croatia account for 56% of EU catches in 2016. From 2007 to 2016, the world oyster aquaculture production increased by 27% to reach 5.6 million tonnes in 2016, driven by the development of China's harvest. China's production represented 86% of the global oyster farmed production in 2016. EU oyster farmed production decreased by 38% over the last decade. The main species produced in the EU is the Pacific cupped oyster (*Crassostrea gigas*), with 92% of harvested volumes in 2016. In 2016, France alone represented 79% of EU production (EUMOFA 2018a).

Fishing/Harvesting of Oysters

Oysters are harvested in a variety of methods. In areas where oyster reefs are exposed by low tide, they are handpicked. In shallow areas, tongs are used to harvest oysters. Tonging is one of the oldest methods of harvesting, where tongs work like a pair of post-hole diggers with handles. Oysters are also harvested by dredging from oyster boats, using metal baskets with rows of spike-like teeth. Oysters are harvested throughout the year, but the meat yield differs with the season. Oysters harvested in the winter yield roughly eight pounds of oyster meat per sack while oysters harvested in the summer on average yield six pounds per sack (Lutz 2012).

Some Important Species of Oysters

Commercially important species are classified in three major genera: *Ostrea*, *Saccostrea*, and *Crassostrea* and a few minor genera.

***Ostrea edulis* (Linnaeus, 1758) (European Flat Oyster)**

Ostrea edulis is a bivalve mollusc that has an oval or pear-shaped shell with a rough, scaly surface. The irregular shell has a distinct hooked beak, patterned with delicate foliation. The two halves (valves) of the shell are different shapes subcircular to circular and in equivalve. Left shell is deeply concave and fixed to the substratum, the right being flat with rougher edges and sitting inside the left acting as a lid. The flat oyster can grow very large (>20 cm) and become very old (>20 years) (FAO 2009) (Fig. 3.3).

Natural populations of the European flat oyster are found along the west coast of Europe and Morocco in the northeast Atlantic. This species can be found on both coasts of the United States. European flat oysters are available as fresh seafood and generally consumed on the half shell (Heinonen 2014).

Cassostrea gigas **(Thunberg 1793) (Cupped Oyster, Pacific Oyster)**
C. gigas has an elongated rough shell, which can reach a 20–30 cm size. Although highly variable, the two valves are solid but unequal in size and shape. The left valve is slightly convex, and the right valve is quite deep, and cup shaped. One valve is usually cemented to hard substrata. Shells are sculpted with large irregular, rounded radial folds. Radial ribs are on both shells starting from the umbo. Usually whitish, they show purple streaks and spots. Its inner side is white. The adductor muscle scar is kidney shaped (Héral and Deslous-Paoli 1991; Cabi 2019c).

The Pacific oyster is native to Japan, but it has been introduced to the United States and France. There have been secondary introductions to many other countries. This species occurs at depths ranging from 0 to 40 m in estuarine. Global production of this species has exceeded that of any other species and continues to expand, with major producing countries including China, Japan, Korea, the United States, France, European states, Australia, New Zealand, and South Africa. Much of the production is consumed by local markets and is only imported when there is a surplus. The preferred product form is fresh and on the half shell, while canned, frozen and vacuum-packed forms are less common (Heinonen 2014).

Crassostrea virginica **(Gmelin 1791) (Eastern Oyster)**
Oysters are sedentary with their lower valve firmly cemented to hard objects. Their flattened, distorted shells are extremely variable in shape. The shells of Crassostrea

Fig. 3.3 *Ostrea edulis*

virginica are typically broadly oval and thick and grow to about 10–15 cm in length. The lower valve is convex and upper valve flat, usually with concentric ridges and lines. Exterior colour is dirty white to grey. The interior is bright white with a deep purple or red-brown muscle scar (Eldredge and Smith 2001).

C. virginica are abundant in shallow saltwater bays, lagoons, and estuaries. Oysters favour estuaries and embayments with low salinities and are intolerant of prolonged exposure to fresh water or marine conditions. They are found in shallow water of tidal to subtidal depth of constant turbidity and salinity but can withstand a wide range of temperatures (Cabi 2019d).

The cupped Eastern oyster occurs in estuaries and marine coastal environments, and is cultured in the United States, Canada, and Mexico. This species is marketed fresh on the half shell and frozen, or incorporated into value added products, such as soups and chowders (Heinonen 2014).

Saccostrea commercialis **(Iredale and Roughley, 1933)**
The Sydney rock oyster occurs in intertidal estuarine habitats, as well as on natural subtidal dredge beds. The main producer country of this species is Australia. Sydney rock oysters are cultured for the half shell market and are sold in supermarkets and restaurants. The composition of oyster meat varies, depending on oyster type, breeding area, harvest season, feeding conditions at farms, spawning season, and oyster size (Heinonen 2014).

Nutritive Composition of Oysters

The oyster contains vitamins A, B_1, B_2, C, D and E and a serving of 15 g of its meat has the same amount of vitamin C as 3 g of lemon juice. Some important essential amino acids (lysine, histidine, tyrosine) are also found in oyster meat. It is abundant in minerals (Hasanspahić 2011).

It is reported that the main component of oyster meat is water, which is tightly bound to proteins in the structure and is not easily released, even under high pressure, and is an index of freshness. The composition of oysters with high protein content and low-fat levels is similar to the nutritional composition of other molluscs (Table 3.3).

Carbohydrates that can be ignored in fish are found in significant amounts in oysters (Asha et al. 2014). Energy stored in the form of glycogen prior to gametogensis is utilized during periods of high metabolic demand during the production of gametes in oysters. Oysters begin to form and store glycogen after spawning and glycogen content reaches a maximum a few months before the next spawning. It is used as an energy source during the rapid proliferation of sex cells and by the end of the reproductive cycle it is at a minimum (Linehan et al. 1999). It has been reported that the glycogen content of the oyster changes seasonally (Berthelin et al. 2000). In a study, it was determined that the glycogen level of *C. gigas* was at the maximum level in April before the gonad volume increased, and the lowest level in August in relation to spawning (Perdue and Erickson 1984). Linehan et al. (1999)

Table 3.3 Proximate composition of oyster species (g/100 g)

Species	Moisture	Protein	Fat	Ash	Carbohydrate	Reference
Unknown	73.37	13.31	0.53	11.87	0.92	Kiin-Kabari et al. (2017)
Crassostrea madrasensis	82.64	9.41	3.25	1.01	3.2	Asha et al. (2014)
C. gasar	79.7–81.05	12.87–13.1	2.16–2.21	2.08–2.13	2.43–3.75	Akinjogunla et al. (2017)
Crassostrea rhizophorae	82.5–82.8	11.00–13.0	2.5–2.7	1.5–1.7	0.6–2.0	Lira et al. (2013)
	81.1–83.0	9.3–10.2	1.5–2.0	2.8–3.7	2.7–4.4	Martino and Cruz (2004)
Pinctada radiata	79.02–5.53	6.59–16	0.43–1.09	0.46–2.7	–	Gokoglu et al. (2006)

determined that he oysters began to store glycogen from December to a maximum in May. Glycogen levels then began to decrease and reached a minimum from August to December. In another study, an inverse relationship was found between glycogen levels and lipid levels of *Crassostrea gigas*. In winter, lipid concentrations were the lowest and the highest glycogen levels. During the maturing period (spring), while lipids accumulate in gonads, glycogen reaches minimum values in maturity. In addition, the protein content at the maturing stage was found to be high and decreased at the beginning of spawning (late summer). This change in protein level has been associated with nutrient availability (Dridi et al. 2007). The smell, the appearance and the taste of oysters vary according to their physiological state and the season in which they are eaten. At the time when they are best for consumption, oysters are called seasonal. In winter they create large quantities of glycogen, which makes them fatty and light in colour and gives a savoury smell and a sweet taste. In late spring and summer, they grow thin, tough and dark, due to reproduction. Therefore, the best time to eat oysters is in February, and the caloric value is highest from January to April (Hasanspahić 2011).

Proteins are the main structural material of gonads during gametogenetic development. There is a change in proteins with the development of gonad. Proteins are also used as energy reserves where glycogen levels are not sufficient (Berthelin et al. 2000).

The lipid reserves of mollusc mainly depend on the environmental influences on metabolic activities, the nutritional value of the food supply and the stage of gonadal development. The oysters collected from the wild habitats exhibited higher lipid content as compared to those cultured ones. Some researchers have reported significant variations of lipid content of oysters seasonally (Salaskar and Nayak 2011; Subasinghe et al. 2019).

Oyster meat is rich in both macro elements and trace metals (Table 3.4). Oysters are reported as excellent sources of iron (Fe), copper (Cu), zinc (Zn), and selenium (Se) (Noel et al. 2012). On the other hand, oysters also serve as bioindicators for monitoring environmental contaminants. As they are filter feeders, they can accumulate elements in their bodies.

The fatty acid compositions of some oyster species reported by some researchers are shown in Table 3.5. The fatty acid compositions of some oyster species reported by some researchers are shown in Table 3.5. Many factors impact the fatty acid composition, such as diet, geographic locations of catch and seasons of the year, which may be related to water temperature (Chen 2011). Fatty acids composition of oyster also changes seasonally. Temperature is one of the variables that greatly influence fatty acid composition. As temperature decreases, the level of unsaturation tends to increase, in order to maintain the freezing point below that of the surrounding water and to ensure membrane fluidity and general body flexibility (Lira et al. 2013). The sum of SFA was found significantly lower in winter than in the other seasons. Monounsaturated fatty acids (MUFA) were significantly lower in spring, than in the other seasons. As for PUFA, it was lower in autumn than in the other seasons, while the sum of the concentration of the n-3 fatty acids was lower in summer than in the other seasons. The n-3/n-6 ratio was higher in spring and winter than in summer and autumn (Martino and Cruz 2004). The growth conditions of the oyster and especially the presence of nutrients in the environment in which it grows affect their fatty acid composition. The availability of phytoplankton resources affects the fatty acid composition. Due to availability of non-phytoplanktonic organic material in the cultured condition resulted in the accumulation of higher proportion of SFAs (Freites et al. 2002). While the total MUFA content of the cultured oyster was found to be higher than that obtained from the natural environment, total PUFA content was found to be higher in the wild oyster (Chakraborty et al. 2016a). The fatty acids (FA) profile of oysters is dependent on the FA composition

Table 3.4 Macro and micro elements in some oyster species (mg/100 g)

	Macro elements					References
	Na	K	Ca	Mg	P	
C. madrasensis	1170	975	309	270	–	Asha et al. (2014)
Unknown	263.2	56.73	52.53	55.76	286.22	Kiin-Kabari et al. (2017)
C. gasar	46.5–57.1	–	91–96.6	12.7–20.03	29.7–33.53	Akinjogunla et al. (2017)
Pinctada radiata	494.3–672.0	126.1–191.7	36.33–87.23	57.14–150	1.94–157.58	Gokoglu et al. (2006)

Micro elements

	Mn	Cu	Zn	Fe	Cr	Se	
C. Madrasensis	0.81	14.7	95.5	33.3	nd	2.4	Asha et al. (2014)
Unknown	–	–	96.56	16.52	–	–	Kiin-Kabari et al. (2017)
C. gasar	0.09–0.12	–	–	–	–	–	Akinjogunla et al. (2017)
Pinctada radiata	0.06–0.49	0.06–0.22	23.85–87.45	0.96–6.92	–	–	Gokoglu et al. (2006).

Table 3.5 Fatty acid composition of oysters (% of total FA)

Species	SFA	MUFA	PUFA	n-3	EPA + DHA	Reference
C. Madrasensis	32.19	13.71	44.77	36.78	34.85	Asha et al. (2014)
C. rhizophorae	17.10–28.9	8.6–23.7	15.7–27.2	11.3–21.1	7.1–17.0	Martino and Cruz (2004)
Ostrea edulis	16.8–33.7	9.0–17.5	49.4–72.6	36.2–58.9	24–38.4	Abad et al. (1995)

of the diet. It was reported that the diet contained 100% seaweed resulted in poor in highly unsaturated FA. Such a diet is unable to meet nutritional requirements of oysters. On the other hand, microalgae containing diets showed higher contents of PUFA, namely EPA and DHA (Rato et al. 2019). Significant differences were found in the fatty acid composition of oysters fed with different microalgae diets (Pennarun et al. 2003).

3.1.3.2 Postharvest Quality Changes in Oysters

Since bivalve molluscs are susceptible to microbial deterioration, their shelf life is considerably short, which limits their distribution and trade. In addition to being highly perishable, oysters can be harmful to public health as they can carry pathogenic bacteria related to outbreaks. The safety of raw oysters for consumption depends upon their initial degree of contamination, mainly due to the quality of seawater from which they are extracted or cultured, rather than to postharvest storage conditions. Natural spoilage flora and microbial pathogens may grow during and post harvesting conditions and affect composition and texture of oysters and cause health risks for consumers. The initial number and type of microorganisms present at the oysters affect the rate of deterioration. In addition, unfavourable conditions during transport and storage may cause re-contamination and rapid deterioration. When common conditions of oysters change to unusual environments, significant physiological changes in oysters may occur. Thus, in the new conditions, there is a decrease in total carbohydrate and ATP levels, causing biochemical changes in the muscle. Therefore, the occurrence of significant biochemical changes during harvesting, transport and storage can lead to a significant negative impact on final quality (Montanhini and Neto 2015).

As in fish, after death, oysters pass through rigor mortis autolysis and bacterial spoilage. Post-harvest changes in oysters depend on some factors such as capture method, handling, processing, and storage (Songsaeng 2010). Spoilage of oysters occurs by microbial growth. It has been reported that the microbial flora of mollusc shellfish varies significantly depending on the quality of the water from which these fish are taken and the quality of the wash water and other factors (Jay 2000). In a study where the number of faecal coliforms in water and oyster (*Crassostrea iredalei*) samples collected from two different oyster growth areas exceeded microbiological limits, it was concluded that these results were obtained because these regions were surrounded by residential houses and fishponds (Sorio and Peralta 2018). It has been also reported that the microflora of the oysters contains Pseudomonas, *Vibrio*, *Serratia*, *Clostridium*, *Proteus*, *Bacillus*, *Escherichia*, *Lactobacillus*, *Flavobacterium*, *Enterobacter* and *Micrococcus* (Jay 2000). Cao et al. (2009a) detected gram-negative bacteria as dominant in oyster (*C. gigas*), and they reported that 22% and 20% of these were *Pseudomonas* and *Vibrionaceae*, respectively. *Shewanella*, *Alcaligenes*, *Enterobacteriaceae*, *Moraxella*, *Acinetobacter*, *Flavobacterium*, *Corynebacterium*, *Staphylococcus*, *Micrococcus*, Lactic acid bacteria and *Bacillus* were also detected as minor organisms. After death, the microbial flora of oysters may increase depending on environmental conditions. Indeed, some researchers have reported increases in the total viable count (TVC) (Cruz-Romero et al. 2008; Hu et al. 2008; Cao et al. 2009a). In oyster samples collected from seafood commercial establishments and the cultivation area, *Vibrio cholerae*, *V. parahaemolyticus* and *Salmonella* spp. were not detected. The counts of coliforms at 35 and 45 °C indicated that samples obtained from both the cultivation area and place of sale were contaminated. *E. coli* was detected in 4 (9%) samples collected in the cultivation area and in 16 (35.5%) samples obtained from commercial establishments. It was commented that inadequate storage of oysters in the period between collection and sale may contribute to reaching high numbers (Pereira et al. 2006).

A measure of pH decrease is considered a better test of spoilage in oysters and other molluscan shellfish than volatile nitrogen bases (Jay 2000). Although pH decreases due to lactic acid formation from glycogen after death, pH also increases with the increase of basic compounds in the later period. When pH drops, the net surface charge on muscle proteins is reduced. In this case, loss in water holding capacity is seen with partial denaturation. This also causes toughening in the texture (Songsaeng 2010).

During spoilage of oysters, total volatile basic nitrogen (TVB-N) including trimethylamine (TMA), dimethylamine (DMA), ammonia and other volatile basic nitrogen increase, and these compounds are considered as spoilage indicators. Studies have reported increases in TVB-N contents during storage of oysters (Balasundari et al. 1997; Lopez-Caballero et al. 2000; Cao et al. 2009b; Songsaeng 2010; Songsaeng et al. 2010). It is reported that total volatile basic nitrogen (TVBN) can be used as freshness indicator for raw Eastern oysters, with the acceptability of 11 mg/100 g (Zhang et al. 2017).

Due to their high content of unsaturated fatty acids, oysters have post-mortem changes in lipids. The hydrolysis and oxidation events in lipids decrease the quality of oyster. Changes in oyster lipids can be observed during refrigerated or frozen storage. Indeed, some researchers have reported increases in peroxide (PV) and thiborbutiric acid reactive substances (TBARs) during storage of oysters (Balasundari et al. 1997; Jeong et al. 1990).

3.1.3.3 Post-harvest Handling

Processors typically sell oysters as fresh raw shucked, processed half shell or as other value-added products such as smoked, cooked, canned, and breaded oysters. Oysters generally reach consumers live in the shell, as fresh, frozen, or canned product, or further processed such as frozen and breaded (Lutz 2012).

Keeping and Processing Oysters Alive

Since oysters are preferably consumed in half shell, they should be presented to the consumer alive. Therefore, the survival of these bivalves after harvesting is very important for industry and linked to the final quality of the oysters. Oysters are usually sold as a live product that can be stored for several weeks before consumption In a live form, oyster quality is not only affected by microbial change as seen in other forms of nonviable seafood, but also by the influence of oyster defence and physiological systems that control microbial growth and maintain animal health (Fernandez-Piquer et al. 2012).

Oysters can survive out of water for weeks if carefully handled and kept moist and cool but, since they rapidly lose liquor from within, particularly if the edge of the shell is damaged, they should reach the inland wholesaler within 3 days of harvesting to be in prime condition. They should be carried and stored with the cupped half of the shell downwards and kept moist by covering with a damp cloth. They should be packed in a manner that protects them from mechanical damage, and should be kept at all times at a temperature between 1 and 10 °C. Commercial packaging ranges from a simple barrel, box or sack to a specially designed container with separate compartments for individual oysters, the degree of sophistication depending on the value of the product, the journey time, and the market for which it is destined (Stroud 1981).

Wild oysters (*O. edulis*) were stored alive in polypropylene boxes at different temperatures (3, 6, and 9 °C) during different days (3, 6, 9 and 12 days). Oysters were placed into the boxes with a wet sheet of newsprint placed over the animals to keep them wet, with any drainage. The oysters were placed tightly together to keep them under some pressure during storage. Oysters were transferred from the farm to storage room within 1 h. The survival rate of oysters was considered high with only 6 deaths in 168 animals stored at different temperatures until 12 days (Lopes Da Costa 2018).

Oysters (*O. edulis*) supplied from a farm were kept in an aquarium at 9 °C. A group oyster was packed in polystyrene boxes with drainage holes in the bottom. The other group packed on ice was placed in a cold storage chamber at 5 °C. The third group oysters were packed with wood wool moistened with seawater instead of ice. For control, a group oysters were placed in the tank containing seawater at 9 °C. This work showed low mortality during the 23 days of storage (Aaraas et al. 2006).

Pacific oysters (*Crassostrea gigas*) were stored in air (out of water) for 20 weeks, and the survivors re-immersed in the sea water and investigated for a few months. Survival rate was found to be 52–80% for oysters sprinkled with water during storage at 7 °C. Re-immersion affected survival positively (Seaman 1991).

Depuration of Oyster

Depuration is an effective post-harvest treatment for the removal of most bacterial species from oysters. Depuration has been used for reducing sewage-associated bacteria, such as coliforms and *E. coli* in shellfish. The efficacy of a depuration process is largely dependent on the biological activities of oysters and the nature of the microorganism. Temperature is the most critical factor affecting the pumping rate. The volume of water pumped by oysters is regarded as a predictor of biological activity of oysters (Shen et al. 2019).

Shen et al. (2019) reported that the rate of water to oyster is effective in *V. parahaemolyticus* depuration in oysters. They achieved a reduction of >3.00 log with depuration with a water to oyster (*Crassostrea gigas*) ratio of 2:1 for 2 days for *V. parahaemolyticus*.

The combination of chlorine and UV radiation has been reported to increase disinfection efficiency and reduce the time required for depuration. Two disinfection methods are thought to have a synergistic effect (Koivunen and Heinonen-Tanski 2004). Although chlorine is used as a disinfectant in depuration tanks, it can be toxic to bivalves. On the other hand, UV radiation can eliminate bacteria and viruses without toxic effects (Correa et al. 2012). Chlorine has also been suggested to cause organoleptic changes in shellfish meat (Lee et al. 2008).

Viruses are mainly concentrated in the digestive gland of the bivalves. It has been reported that if depuration of shellfish is carried out appropriately, bacteria can often be reduced to undetectable levels, but viruses can be reduced at lower rates because they are more resistant and persistent than bacteria (Dore and Lees 1995). The oysters were contaminated with hepatitis A virus (HAV) or human adenovirus type 5 (HAdV5) by Correa et al. (2012). Oysters which are harvested after 48, 72 and 96 h and infected with viruses, were placed into a closed system depuration tank recirculating seawater. A sterilising system consisting one 18 W ultraviolet (UV) low-pressure tube, a sand filter and a refrigeration system was used. The temperature of the depuration system was maintained at 19–20 °C during all of the experiments to prevent the molluscs from spawning. In conclusion, after 96 h of UV treatment, the depuration system studied in this work purified oysters that were artificially contaminated with HAdV5 and HAV.

Populations of fungi and bacteria showed marked reduction after depuration of Mangrove oysters collected from Benya lagoon, located at Elmina in the Central Region of Ghana (Obodai et al. 2010).

It has been reported that the presence of *V. parahaemo*lyticus in marine environments is associated with water temperature and that *V. parahaemolyticus* can rarely be detected in oysters or environmental samples until water temperatures rise to

15 °C. Therefore, depuration at ambient temperatures has been reported to be ineffective in reducing oyster *Vibrio* contamination. For this purpose, the efficacy of refrigerated-seawater depuration for reducing *Vibrio parahaemolyticus* levels in Pacific oyster (*Crassostrea gigas*) was investigated. Raw Pacific oysters were inoculated with a mixed culture of five clinical strains of *V. parahaemolyticus* and depurated with refrigerated seawater (5 °C) in a laboratory-scale recirculation system equipped with a 15-W gamma UV sterilizer. It was concluded that the depuration of refrigerated sea water for 96 h showed a reduction in *V. parahaemolyticus* populations of >3.0 log MPN/g in oysters collected in winter, however, it was concluded that 144 h of depuration at 5 °C was required to achieve 3-log reduction in summer collected oysters (Su et al. 2010).

Depuration was also applied to heavy metals. Oysters (*Crassostrea gigas*) contaminated with heavy metals were transplanted to cleaner waters. The oysters kept in the new environment were sampled at regular intervals and heavy metal levels were examined. According to the results of the study, Cd, Cu, Hg, Ag and Zn were eliminated by *C. gigas*, while As, Cr, Pb, MN, Ni and Se showed no significant changes. It has been reported that the difference in heavy metal depuration is related to the bonding and deposition characteristics of each metal (Okazaki and Panietz 1981).

Shucking of Oysters

Oysters are composed of two calcareous valves or shells wherein the soft body of the oyster lies. The left valve is usually cupped and has a projection at the anterior region that curves upward called the umbo. The oyster normally rests on the left valve. The right valve is usually less cupped and flatter in appearance. The valves are joined by a resilient lamellar ligament at the anterior margin of the oyster. This ligament is biased in the open position and counteracts the force of the adductor muscle which closes the valves. The ligament goes into tension when the oyster is closed and into compression when the oyster is in the relaxed state and open. Adductor muscle is fast acting and provides protection for the oyster. Sessile intertidal bivalves such as oysters frequently remain closed for long periods of time. Thus, to the side of the translucent muscle, there is a smaller, crescent-shaped muscle with smooth muscle fibres and opaque in colour that holds the valves closed for extended periods of time. Both muscle types are attached to the inside of the shell on both the left and right valves. The strength of this opaque muscle is what makes it difficult to open an oyster. In the mid-1800s, the oyster knife was invented. The oyster knife is still the standard today as it allows for inexpensive access to the oyster. It is good for prying the shells loose at the hinge and severing the adductor muscle. Some shuckers use a hammer in conjunction with the knife to crack the bill of the oyster so the knife can be inserted more easily. As there is no heat involved in the process, the final product also remains in a raw state. Shucking by knife is labour intensive and as labour costs increase and qualified shuckers become harder to find, profit margins may be reduced (Martin and Hall 2006). During shucking by hand, a sharp knife is inserted between the shells, cutting the adductor muscle, and opening the oyster. This requires a skilled workforce, as not only is the process hazardous but inexpert handling can damage the oyster meat, reducing the quality and appearance of the finished product (Cruz-Romero et al. 2007).

Because of the difficulties of manual shucking, other techniques have been developed for shucking the oysters. Some devices and machines shucking the oysters mechanically have been developed.

Because of being a filter feeder and accumulating bacteria and viruses from the water it inhabits, it is one of the dominant bivalves in seafood-borne diseases. They are generally consumed whole and raw. Therefore, they create risks for consumers. In recent years, a process has been developed a method eliminating pathogenic bacteria. It has been reported that HP treatment provides a significant advantage in detaching the oyster adductor muscle from the shell. HP treatment at 241 MPa for 2 min caused detachment of adductor muscle in 88% of oysters (*Crassostrea gigas*), while treatment at 310 MPa, with immediate pressure release, resulted in 100% efficiency of shucking (He et al. 2002). In a study comparing the effects of high pressure (260 Mpa for 3 min) and heat treatment (50 °C for 10 min and 75 °C for 8 min) applied to the oysters, it was determined that the yield was higher with HP. Because, adductor muscle of oyster was not cut, so that the oyster retained all its moisture (Cruz-Romero et al. 2007). Moreover, recovered oyster meat has been reported to have a good shape and appearance (Lopez-Caballero et al. 2000). HP processed oysters can be shucked with minimal effort and skill, as the consumer only needs to remove the surrounding band to open the processed oyster. In addition to reduced labour cost and risks, and the safety of the product is improved (Cruz-Romero et al. 2007). The process calls for 100–800 MPa of hydrostatic pressure to destroy pathogenic organisms and separate the adductor muscle from the shell. HP treatment makes the seafood safe for eating in raw condition due to the inactivation of microorganisms. *Vibrio* species have been reported to be susceptible at pressures of 200–300 MPa (Berlin et al. 1999). HP treatment of 345 MPa for 90 s reduced *Vibrio parahaemolyticus* in Pacific oysters (*Crassostrea gigas*) (Calik et al. 2002). Moreover, Hydrostatic pressure has been reported to have no significant effect on the sensory and nutritional properties of oysters. The biggest drawback of this method is that it is expensive and requires high investment expenses. HP application to Pacific oysters (*Crassostrea gigas*) with pressure between 207 and 310 MPa at 0, 1, and 2 min reduced initial microbial load by 2–3 logs and counts remained at a reduced level through the storage study (He et al. 2002).

Fitting of oysters with heat-shrinkable plastic bands before treatment holds the shells together and reduces the loss of interval fluid. Oysters treated in this way do not gape but can be shucked with minimal effort and skill (Murchie et al. 2005).

3.1.3.4 Oyster Processing and Preservation

Chilling/Refrigerating of Oysters

Chilling is defined as a unit operation that reduces the temperature of a food to −1 to 8 °C to reduce biochemical and microbiological changes and to extent shelf life (Songsaeng 2010).

Storage conditions, especially storage temperature, affect the quality of oysters. Other factors such as type of product, microbial and enzymatic inactivation degree, hygienic conditions, packaging are also effective on shelf life of chilled oyster. Quality changes of shell-on white-scar oyster (*Crassostrea belcheri*) packed in air and stored at ambient (30 + 2 °C) and chilled (4 + 2 °C) temperatures were compared with shell-on oysters packed in 2.5% brine and stored at ambient and chilled temperatures. Lower changes in chemical and microbiological qualities were found in chilled shell-on oysters when compared with those stored at ambient temperature. Shell-on oyster packed in 2.5% brine showed a slower decrease in pH but increased in total volatile basic nitrogen (TVB-N), total viable count (TVC) and psychrotrophic bacteria at a faster rate than those stored in air. According to microbiological and sensory analyses results, it was found that oysters (shell-on) can be kept at chilled temperature in air and in brine (4%) for 9 and 10 days respectively (Songsaeng et al. 2010). It was reported that the total viable count (TVC) increased from 6.6×10^4 to 5.5×10^6 in 14 days at 4 °C in oyster (*C. gigas*), reaching 9.7×10^8 at 7 °C (Hu et al. 2008).

Oysters (*Crassostrea gigas*) shucked and wrapped in plastic bag were stored at 0, 5 and 10 °C. During the storage, the oysters were investigated for sensory, chemical, and microbiological quality. The shelf-life of oysters stored at 10 °C, 5 °C., and 0 °C was found to be 6–7 days, 10–11 days, and 17–18 days, respectively (Cao et al. 2009a, b).

Oysters (*O. edulis*) were stored at different temperatures (5, 7 and 15 °C) for 11 days and sensory changes were investigated. Oysters stored at 5 and 7 °C showed changes on the seventh, and those stored at 15 °C on the fifth day (Hasanspahić 2011).

Effects of temperature and period of postharvest storage on the microbiological quality and shelf life of raw mangrove oysters, *Crassostrea brasiliana* were investigated and it was reported that raw mangrove oysters remained in safe microbiological conditions for consumption up to 8 days after harvesting, regardless of temperature. It was also stated that their shelf life may be extended to 15 days if they are stored at temperatures not exceeding 15 °C (Montanhini and Neto 2015).

An on-board rapid chilling unit for small vessels was designed to control *vibrio* during warmer months. *Vibrio* levels did not change significantly with icing application, while gaping increased (Thomas 2016). In another study, it was reported for oysters (*Crassostrea virginica*) which were kept at 54.5 °C for 15 min after dipping slurry ice (Thomas et al. 2016). Melody et al. (2008) reported significant gaping in oysters (*Crassostrea virginica*) exposed to the iced after 7- and 14-days post-harvest.

Freezing of Oysters

The oysters for freezing should be alive. Freezing causes the shells of oysters to open and makes subsequent shucking easier. Oysters can also be frozen in the half shell. Oysters should never be put into a cold store to freeze. Cold stores are designed only to hold already frozen products at the required low temperature. Frozen whole oysters packed in polyethylene bags can be kept in good condition for 6 months in a cold store at −30 °C. The liquor within the shell acts as a glaze to protect the meat from dehydration (Stroud 1981).

Unless there is a market for frozen whole oysters, or freezing is to be used as means of opening the shell, it is more economical to freeze the shucked meats. Misshapen whole oysters that are unacceptable for the live trade can also be utilized in this way. Meats frozen either individually or in blocks can yield an excellent thawed product. The shucked meats are thoroughly washed to remove sand, grit, and other shell debris, inspected to remove any ragged or broken meats, drained to remove excess water, and frozen individually in an air blast freezer before packing. Typically, the frozen meats are glazed and packed under vacuum in bags made of a flexible film that has a high resistance to the passage of water vapour and oxygen. An alternative procedure is to pack the meats under vacuum in a rigid container before freezing them; containers of metal, plastics or waxed cardboard with inner liners have been used, sometimes with shaped pockets for individual meats. The strained shell liquor is also sometimes added to the packed meats before freezing them; this has the added advantage that any remaining air spaces in the pack are filled (Stroud 1981).

Freezing temperature is important in frozen storage (Hatano et al. 1990; Wang et al. 2007). Faster ATP degradation was reported in oysters frozen at −20 °C than those frozen at −30 °C (Wang et al. 2007).

It is reported that frozen storage inhibits growth of bacteria and capable of achieving certain degrees of reductions of *V. parahaemolyticus* in oyster meat. Muntada-Garriga et al. (1995) determined that the viable cells of *V. parahaemolyticus* in oyster meat frozen and stored at different temperatures (−18 and −24 °C) can be inactivated and stated that the time of total inactivation depends on the initial number of micro-organisms and incubation temperature. Other researchers have also reported considerable reductions of *V. parahaemolyticus* in Pacific oysters (*C. gigas*) homogenates during frozen storage at −10, −20 and −30 °C and inactivation of *V. parahaemolyticus* is more effective in oysters stored at −10 °C. Similarly, Chae (2007) reported *V. parahaemoliticus* reductions and more effective reductions at −10 °C in frozen stored oysters at −10, −23 and −30 °C (Liu et al. 2009). Moreover, it was reported that blast freezing or heat shocking could be used as an effective post-harvest process to reduce *V. parahaemolyticus* in oysters (*C. virginica*) which exposure to cold air at −29 °C and stored at −20 °C (Manley 2008).

The quality of frozen oysters is influenced by many factors, such as storage temperature and time, packaging, and thawing conditions. Freshness of raw oysters is another factor affecting the quality of frozen oyster. Shucked and then frozen oysters (*C. gigas*) immediately after harvest and periodically frozen oysters after keeping at 5 °C showed differences in quality during frozen storage. Frozen oysters, usually stored at −20 °C, maintained their commercial quality over 1 year, but this was not achieved with samples frozen after keeping at 5 °C for 3 days (Lee et al. 2017).

During frozen storage slightly rancid off-flavours can develop and start to acquire a greenish yellow discoloration. Unacceptable taste and coloration of oysters during frozen storage has been reported (Hatano et al. 1990). Lipid oxidation was determined in Japanese oyster (*C. gigas*) during frozen storage. Lipid oxidation was found to be faster in oysters stored at −20 °C than stored at −35 °C. However, the use of dibutylhydroxytoluene or natural vitamin E solution was effective in inhibiting oxidation (Jeong et al. 1990).

Thermal Processing of Oysters

Heating

This process was originally developed to facilitate the shucking process by relaxing the oyster's adductor muscle. This process also reduces pathogens in the oyster. In the heat shock post-harvest process, oyster shell-stock is subjected to 65 °C for 5 min, then immediately cooled in an ice slush for 10 min (Manley 2008).

Pasteurization

Pasteurization is usually used for *Vibrio* elimination in oysters. A heat treatment above 50 °C has been reported to be sufficient to reduce *V. vulnificus* to undetectable levels (Cook and Ruple 1992). Heat treatment of 50 °C for 5 min has been found effective in reducing the numbers of *V. vulnifcus* and *V. parahaemolyticus* by 99.9%. In case of treatment for 10 min they have been reduced to non-detectable levels. This pasteurization process also increased the shelf life by reducing the number of aerobic bacteria by 99.99% (Andrews et al. 2000). Pasteurisation further has the advantage of increasing the shucking yield of oysters.

Pasteurization further has the advantage of increasing the shucking yield and shelf life of oysters. The shelf life is extended from 1 week to untreated oysters to 3 weeks under refrigeration. Another advantage of this type of process is its versatility. The process can be adjusted to a small-or large-scale to fit the needs of an individual processing plant. However, the disadvantage of this process is that the processing temperature and time can change the sensory properties of the oysters (Espinoza 2013). Andrews et al. (2003) reported that temperatures above 52.5 °C affect the texture and taste of oyster due to protein degradation. It is reported that every 10 °C increase in temperature increases protein denaturation by approximately 600-fold, and the cutting strength of oyster meat is significantly reduced when the temperature rises from 57.5 to 70.0 °C (Lekjing et al. 2017). After thermal processing, It was reported that the plum and adductor of oyster is shrunken, colour changes to yellowish green, texture becomes softer, occurs cooked meat/sweet odour, and sweet in taste (Songsaeng 2010). Chai et al. (1991) stated that heating for 8 min at 75–76 °C could reduce the microbial population in oyster meat and provide the optimum physical and sensory quality for whole oysters.

3.1.4 Scallop

3.1.4.1 General Information About Scallops

The scallops are very valuable benthic marine molluscs belonging to the Bivalve class and the Pectinidae family (Fisher 2000). The common term 'scallop' may be used to designate members of the superfamily Pectinoidea, which includes the extant families Pectinidae, Entoliidae, Spondylidae, and Propeamussiidae (Beninger and Pennec 2006).

Biology of Scallops

Scallops are a cosmopolitan family of bivalves which are found in all the world's oceans, although never in fresh water. They are one of very few groups of bivalves to be primarily "free-living", with many species capable of rapidly swimming short distances and even of migrating some distance across the ocean floor. A small minority of scallop species live cemented to rocky substrates as adults, while others attach themselves to stationary or rooted objects such as sea grass at some point in their lives by means of a filament they secrete called a byssal thread. The majority of species, however, live recumbent on sandy substrates, and when they sense the presence of a predator such as a starfish, they may attempt to escape by swimming swiftly but erratically through the water using jet propulsion created by repeatedly clapping their shells together. Scallops have a well-developed nervous system, and unlike most other bivalves all scallops have a ring of numerous simple eyes situated around the edge of their mantles (Wikipedia 2019h).

The family Pectinidae displays one of the greatest degrees of anatomical differentiation within the bivalve. Organs are quite distinct and easily located (Beninger and Pennec 2006).

The shell of a scallop consists of two sides or valves, a left valve and a right one, divided by a plane of symmetry. With the hinge of the two valves oriented towards the top, one side corresponds to the animal's morphological anterior or front, the other is the posterior or rear, the hinge is the dorsal or back/top region, and the bottom corresponds to the ventral or underside. Scallops have a single central adductor muscle; thus, the inside of their shells has a characteristic central scar, marking the point of attachment for this muscle (Wikipedia 2019i).

The adductor muscle of scallops is larger and more developed than those of oysters, because scallops are active swimmers; some species of scallops are known to move en masse from one area to another. In scallops, the shell shape tends to be highly regular, and is commonly used as an archetypal form of a seashell (Wikipedia 2019i).

Scallops have many small eyes arranged along the edge of their mantles. These eyes represent an innovation among molluscs, relying on a concave, parabolic mirror of guanine crystals to focus and retro-reflect light instead of a lens as found in many other eye types (Wikipedia 2019i).

Scallops are filter feeders and eat plankton. Unlike many other bivalves, they lack siphons. Water moves over a filtering structure, where food particles become trapped in mucus (Wikipedia 2019i).

In the Pectinidae, gametes are released into the sea and fertilisation occurs externally. Sperm penetration takes place while the egg is at the metaphase I stage of meiosis (Cragg 2006). Spermatozoa and ova are released freely into the water during mating season, and fertilized ova sink to the bottom. After several weeks, the immature scallops hatch and the larvae, miniature transparent versions of the adults called "spat", drift in the plankton until settling to the bottom again (an event called spatfall) to grow, usually attaching by means of byssal threads (Wikipedia 2019i).

Distribution of Scallops

There are some 400 known living species in the bivalve family Pectinidae, commonly known as scallops. They occur in all the seas of the world from polar regions to the tropics. Most of the commercially important species occur in the inshore waters of the continental shelves but scallops are found in waters of all depths from the intertidal zone down to some 7000 m (Brand 2006).

Scallops inhabit all the oceans of the world, with the largest number of species living in the Indo-Pacific region. Most species live in relatively shallow waters from the low tide line to 100 m, while others prefer much deeper water. Although some species only live in very narrow environments, most are opportunistic and can live under a wide variety of conditions. Scallops can be found living within, upon, or under rocks, coral, rubble, sea grass, kelp, sand, or mud. Most scallops begin their lives as byssally attached juveniles, ability that some retain throughout their lives while others grow into free living adults (Wikipedia 2019i).

Fishing/Harvesting of Scallops

Bottom-cultured scallops are harvested either by hand (SCUBA divers) or dredge. Scallops cultured in suspension are harvested by different means. For example, the harvest of *P. yessoensis* from suspension culture employs the use of vessels outfitted with winches to lift longlines and associated nets (Heinonen 2013). For wild scallops, dredge harvest techniques are used. Dredges are steel-framed structures with or without a cutting bar on the leading edge that drags above the surface of the substrate and collect scallops in an attached steel-ring bag (Mercaldo-Allen and Goldberg 2011).

Hand harvest techniques and those associated with suspended scallop culture are believed to have no significant impacts on habitat (Heinonen 2013). The impacts of dredges on seafloor habitat have been reported as reducing of habitat complexity and species diversity, causing the shifts in community structure, loss of vertical structure, and reducing productivity or biomass (Mercaldo-Allen and Goldberg 2011).

Production and Trade of Scallops

Scallop trade dropped sharply in the first three quarters of 2017, mainly due to the lower supply from Peru. The United States of America and China are the main scallop exporters and importers. Peru had three consecutive years of disastrous scallop production. As a result, Peruvian scallop exports, which in the past amounted to around 9000 tonnes for the review period, were down to only 2400 tonnes in the first 9 months of 2016 and 2017. The sanitary office of Peru closed off the main production area for exports of live scallops to the EU28 in 2017. There are signs that the situation will improve in 2018, because many juvenile scallops have recently been found in the water, so there should be plenty of scallops around for the harvesting season in November 2018 (FAO-Globefish 2018a).

China is by far the main scallop producer worldwide, with 1.9 million tonnes production per year, which represents over 90% of the world scallop production. Most of this production is consumed domestically and therefore trade is limited. In early 2018, some scallop farms experienced problems with high mortalities due an increase in water temperature, a decrease in rainfall and a major reduction in plankton abundance (FAO-Globefish 2019).

Some Important Species of Scallop

Many of the species of scallops grow large enough to have commercial interest. The most important species in Japan is *Patinopecten yessoensis*, otherwise known as the giant "ezo scallop". Other important species cultured in Japan are the "fan shell scallop" *Pecten albicans*, and the "queen scallop" *Chlamys nobilis*. In the USA the principal species of interest are the "Bay scallop", *Argopecten irradians*, "the calico scallop", *Argopecten gibbus*, and "the purple hinge rock scallop", *Hinnites multirugosus*. In Canada, "the giant scallop", *Placopecten magellanicus*, is the most important in terms of harvested quantities. The Chilean *Argopecten purpuratus*, and the Argentinean *Zygochlamys patagonica* are all commercially important species. Two valuable species are found in European waters, the King scallop, *Pecten maximus*, and the smaller species *Chlamys opercularis*, also known as Queen scallop. Other important species in Europe are the Icelandic scallop, *Chlamys islandica*, and the Greek scallop, *Chlamys glabra*. Other commercially important scallops are: *Pecten fumata* or Australian scallop, *Amusium pleuronectes* or Asian moon scallop,

Chlamys farreri, *Amusium japonicum*, and *Pecten novaezealandiae* (Lovatelli 1987). Some scallop species of commercial importance are mentioned below.

Argopecten irradians **(Lamarck, 1819) (Bay Scallop)**
The bay scallop, *A. irradians*, is geographically distributed along the Atlantic and Gulf coasts of the United States. This species of scallop used to support a large fishery on the East Coast of the United States, but since the 1950s, the fishery of the wild scallops has decreased greatly. This decrease is apparently the result of several negative influences, one of which is a reduction in sea grasses due to increased coastal development and concomitant nutrient runoff. Another possible factor is reduction of sharks from overfishing. A variety of sharks used to feed on rays, which are a main predator of bay scallops. With the shark population reduced, in some places almost eliminated, the rays have been free to feed on scallops to the point of greatly decreasing their numbers. Bay scallops are now raised in aquaculture in Florida. They were introduced into China for the 1980s and are the basis of a vibrant aquaculture industry in that country. The species was introduced for culture in China during the early 1980s and has since become one of the dominant species of scallops cultivated in that country (Wikipedia 2019i).

Chlamys farreri **(Müller, 1776) (Chinese Scallop/Zhikong Scallop)**
Chlamys farreri, also known as the Zhikong scallop, is a Pacific Asian subtropical species (Shi et al. 2013). Zhikong scallop has a wide distribution along the coasts of North China, Korea, Japan, and Eastern Russia. It is a most dominant scallop species for aquaculture, and its production has reached approximately 80% of the total scallop production in China (Zhang et al. 2011). *C. farreri* is popular seafood in East Asian countries because of its large and edible adductor muscle, which makes it a significant species in aquaculture and fishery (Shi et al. 2013).

Patinopecten yessoensis **(Jay, 1857) (Yesso Scallop)**
Patinopecten yessoensis is a cold-water species common to the north-western Pacific, including the Kuril Islands, Sakhaline (Russia), Hokkaido and northern Honshu (Japan) and the north-eastern coast of Korea. *P. yessoensis* is commonly distributed on subtidal sand flats at a water depth of 10–50 m. Most of the production of *P. yessoensis* comes from China and Japan, with the Republic of Korea and the Russian Federation contributing as minor producers. There is also a relatively small production source originating from the Pacific coast of Canada. *P. yessoensis* is one of the most highly favoured scallops on the international market and can be consumed in a variety of manners: raw, boiled, fried, dried or canned. Fresh adductor muscle is highly appreciated in sushi bars, and baked scallop meat is popular in seafood restaurants. The first aquaculture of *P. yessoensis* was initiated in Mutsu Bay on northern Honshu, in 1965, using naturally harvested *P. yessoensis* seeds. *P. yessoensis* is currently cultured in Japan, China, Korea, Russia, Taiwan, and Morocco using sawing and the long line hanging method (Cabi 2019e).

Pecten maximus/Chlamys maximus **(Linnaeus, 1758) (Great Scallop)**
Pecten maximus is native to the eastern Atlantic Ocean, from Norway to Spain, as well as the Azores and the Canaries. The species is considered a delicacy and has

been fished in large numbers. *Pecten maximus* lives on sand and gravel bottoms but it can be found in mud as well. The young molluscs live attached with their byssus to a hard substrate, but when they become adult the shells are free-swimming. They spend most of the time resting on the lower valve in self-dug depressions in the bottom, so that the upper valve is parallel to the seafloor. Sand, mud, gravel or living organism cover the upper valve so that only the margin of the shell (with all tentacles and eyes) remains visible, most active during the day. When disturbed the animal retracts with a quick movement into its valves and becomes virtually undetectable (FAO 2019b) (Fig. 3.4).

Placopecten magellanicus **(Gmelin, 1791)**
The Atlantic sea scallop, *Placopecten magellanicus*, supports one of the highest valued commercial fisheries in the United States and Canada and is found along the western North Atlantic continental shelf from Newfoundland to North Carolina (Inglis et al. 2016). It is distributed in the Northwest Atlantic Ocean from the Strait of Belle Island, Newfoundland, Canada, to Cape Hatteras, North Carolina, and commercially exploited throughout most of its range. The sea scallop resource is primarily distributed on the continental shelf in depths ranging from 18 to 110 m. The sea scallop is one of the larger scallops in both shell height and adductor muscle (meat) weight (Fisher 2000).

Nutritive Composition of Scallops

The effect of the season on the chemical composition of the scallop (*Nodipecten subnodosus*) adductor muscle was determined. The lowest values were obtained for the moisture content in summer, while the highest values were observed in winter. Autumn values were the highest for protein content. Carbohydrates peaked in summer with values higher than 10%, possibly because of the high glycogen content reported before for this season. The lowest values of carbohydrates occurred in

Fig. 3.4 *Pecten maximus*

winter, indicating low energy reserves during that season (Beltran-Lugo et al. 2006). Seasonal changes in proximate and fatty acid composition have also been reported for *Flexopecten glaber* from Ionian Sea (Prato et al. 2019). Protein and lipid contents of *Flexopecten glaber* showed were the highest during spring-summer and the lowest in the autumn-winter months. This has been reported to result from the accumulation of substitutes to be used during gamet maturation or from further muscle growth. The proximate compositions of some scallop species reported by some researchers are shown in Table 3.6.

In terms of nutrients, gonads of scallops are richer than muscles. It has been reported that the commercial scallop (*Pecten fumatus*) is a better source of long chain n-3 PUFA than muscle (Su and Dinh 2003) (Table 3.7). In the gonads of *Flexopecten glaber*, higher protein, fat, ash (Berik and Çankiriligil 2013) EPA and DHA contents (Telahiguea et al. 2010; Prato et al. 2018) were determined than those of muscles. On contrary, in a study investigating the composition of fatty acids in different body parts (adductor muscle, digestive gland, gonad, gill and mantle) of scallop (*Argopecten purpulatus*), EPA and DHA levels were highest in adductor muscle compared to other organs (Caers et al. 1999).

The reproductive cycles of the scallops also affect their nutrient storage levels. Gametogenesis requires energy, and there are different ways of obtaining energy among species. Some species store energy in different organs and tissues through feeding before gametogenesis. The onset of the oocyte development phase depends on the accumulation of nutrient reserves and transfer from digestive gland to gonads.

Table 3.6 Proximate composition of scallop adductor muscle (g/100 g)

Species	Moisture	Protein	Fat	Ash	Carbohydrate	Reference
Pecten maximus	74.6–78.2	18.0–20.0	0.4–1.3	1.4–1.5	2.7–6.8	Manthey-Karl et al. (2015)
Argopecten irradians	74.15–83.6	13.4–21.5	0.2–0.9	0.2–8.7	0.1–3.8	Webb et al. (1969)
	79.8	15.9	0.6	1.23	–	Silva and Chamul (2000)
Argopecten gibbus	76.1–81.8	13.3–17.5	0.2–1.1	1.2–1.8	0.3–3.7	Webb et al. (1969)
Nodipecten subnodosus	75.1	17.8	0.11	1.4	7.08	Jiménez-Ruiz et al. (2012)
	72.5–80.7	15.1–17.3	0.3–0.6	1.3–1.5	1.8–10.5	Beltran-Lugo et al. (2006)
Flexopecten glaber	82.1–84.8	8.2–11.9	0.7–1.2	2.2–2.8	1.2–3.3	Prato et al. (2019)
	79.3	12.39	1.9	1.30	–	Berik and Çankiriligil (2013)
	79.3–82.8	12.0–13.7	1.1–1.9	1.3–2.0	1.07–5.06	Berik et al. (2017)
Pecten irridans	80.70	15.4	0.50	1.4	1.70	Silva and Chamul (2000)

Table 3.7 Fatty acid composition of scallop adductor muscle (% of total FA)

Species	SFA	MUFA	PUFA	n-3	EPA + DHA	Reference
Pecten maximus	28.5–30.7	6.8–7.3	51.6–53.9	46.7–49.8	43.9–46.2	Manthey-Karl et al. (2015)
	20.1–31.7	3.1–9.2	34.9–59.2	47.7–65.7	43.2–55.6	Grahl-Nielsen et al. (2010)
Flexopecten glaber	48.9	16.9	34.2	29.75	24.28	Prato et al. (2018)

During gonad maturation, concentrations of lipid and protein vary (Arellano-Martínez et al. 2004).

It was reported that the tissue fatty composition of scallops is affected by both intrinsic (phylogeny, age, sex, reproductive cycle) and external factors (diet, temperature, salinity, and depth). Even among different organs of pectinids, it has been reported that the composition of fatty acids varies (Grahl-Nielsen et al. 2010).

3.1.4.2 Post-harvest Quality Changes in Scallops

After death, as in fish, scallops go through steps as rigor-mortis, post-rigor, autolysis, and bacterial degradation (Ocano-Higuera et al. 2006). These changes are mainly affected by storage temperatures. In case of temperature abuse, the freshness of the scallop is lost rapidly.

Fresh sea scallops should have whitish meat, very little liquid, and a fresh, even slightly sweetish, odour. Bay scallops should have meat that may be creamy white, light tan or pinkish; should be practically free of liquid; and should have a fresh odour (Dean 2000).

Two postmortem changes have been reported in the scallop (*Placopecten magellanicus*) muscle. One of these is the accumulation of hypoxanthine (Hx), the nucleotide degradation product, and the other is the production of octopine, the final product of glycogen degradation (Hiltz and Dyer 1970).

Indices of quality based on nucleotide degradation (hypoxanthine and K value) have received special attention for monitoring the freshness of fishery products during handling and processing (Ocano-Higuera et al. 2006; Pacheco-Aguilar et al. 2008). The concentration of major adenine nucleotides and their related compounds in post-mortem muscle correlates well with the loss of freshness in a wide range of fish. Adenosine triphosphate (ATP) degradation in invertebrates proceeds via adenosine in lieu of Inosine monophosphate (IMP). Regardless of the species and muscle type, ATP decreases rapidly within the first 24 h post-mortem. In fish muscle, ATP is metabolized as ATP (adenosine triphosphate) → ADP (adenosine diphosphate) → AMP (adenosine monophosphate) → IMP (inosine monophosphate) → HxR (inosine) → Hx (hypoxantine). Changes in Adenosine triphosphate (ATP) in scallop adductor muscle stored at 5, 0 and −3 °C were investigated and a slow decrease in ATP was observed during storage at 5 °C and a rapid decrease at −3 °C was observed. Adenosine monophosphate (AMP) deposition was observed at all temperatures with a decrease in ATP. Inosine (HxR) increased more rapidly than other temperatures at −3 °C and hypoxanthine (Hx) increased more slowly than HxR (Kawashima and Yamanaka 1992). It was reported that Hx is associated with loss of freshness and flavour, and Hx deposition in the scallop (*Nodipecten subnodosus*) adductor muscle stored in ice (0 °C) after 12 days reflects the initial stage of autolytic deterioration as well as bacterial degradation (Pacheco-Aguilar et al. 2008). Increase in Hx of adductor muscle of chilled scallop (*Chlamys tehuelchus*) (De Vidode Mattio et al. 2001). Similar increases have been reported for *Placopecten magellanicus* (Hiltz and Dyer 1973).

The K value is defined as the ratio (×100) of non-phosphorylated ATP breakdown products to the total ATP breakdown products. Despite that, K value is widely accepted as a freshness index for many fish species, it is not the same for shellfish (Ocano-Higuera et al. 2006; Pacheco-Aguilar et al. 2008). However, K value showed a linear increase in the adductor muscle of Catarina scallop (*Argopecten ventricosus*) stored in ice (0 °C) and it was stated that K value can be used as good indicator for monitoring loss of freshness during the shelf life of Catarina scallop (Ocano-Higuera et al. 2006). Similarly, a logarithmic increase in K value of adductor muscle of Pacific lions-paw scallop (*Nodipecten subnodosus*) during the storage at 0 °C (ice) (Pacheco-Aguilar et al. 2008).

Formation of octopine from arginine and pyruvic acid is one of the major biochemical processes occurring in postmortem scallop muscle (Venugopal 2006). The final anaerobic glycolysis products in molluscs are different from vertebrates. During anaerobic metabolism, octopine accumulates in the muscles of invertebrates instead of lactic acid. However, it was found that octopine scallop is an inappropriate freshness index (Wongso et al. 1998).

One of the biochemical changes that occur in fisheries after death is the formation of biogenic amines. Biogenic amines are formed by decarboxylation of amino acids. In a study investigating the biogenic amine contents of scallop (*Pecten maximus*) stored in ice and at 10 °C, agmatine was found to be the dominant amine and increased during storage and reached the limit values. Putrescine and cadaverine increased rapidly after 2 days in ice and 6 days at 10 °C storage respectively (Mackie et al. 1997). In a similar study, it was reported that agmatine, putrescine and cadaverine increased with the progression of deterioration in scallop adductor muscle stored at 5 and 15 °C and putrescine and ornithine were good indicators for the freshness of the scallop adductor muscle (Yamanaka 1989).

3.1.4.3 Post-harvest Handling of Scallops

Transport and Live Storage of Scallops

The main goal regarding live transport must be to deliver live scallops in good condition, every time, and without significant product loss (Overaa 2001). To be able to mention quality for a food product, freshness is essential, it is essential to be alive to mention freshness in case bivalves. If the bivalve is alive, it means a good price for the farmer. However, scallops are susceptible to stress if they live outside their habitat. Aerobic conditions and temperature changes have a significant effect on this. This leads to loss of quality. The scallops are more vulnerable to dry as they do not close the valves when they are out of the water. In addition, the growth of microorganisms and the accumulation of waste products cause deterioration of physical conditions and thus the death of the scallop. Therefore, the time and temperature for transport and storage must be well regulated (Lopes Da Costa 2018).

There are three methods of transporting scallops: dry, water and packaging. Temperature, moisture, oxygen, and pressure are effective factors for dry transport of scallops. Maintaining the scallops at low temperature generates low metabolism. When scallops are held at lower metabolic states, they require less oxygen. Adding moisture prevents the scallops from dehydrating and keeps the gills moist so that gas exchange can continue to take place. A high oxygen level in the transport box creates a higher oxygen gradient in the air surrounding the gills and facilitates the transport of oxygen into the scallops. Careful placement of scallops in the shipping container places a moderate amount of pressure on the shells, which keeps them closed to help avoid excessive dehydration. Oxygen, water temperature, carbon dioxide, ammonia, water flow and induction of spawning are effective factors in transporting scallops in water (Overaa 2001).

Juvenile and adult scallops are suitable for extended live shipment in insulated containers without the addition of seawater. Survival and quality are maintained if metabolic rates and other processes are depressed by cool, humid conditions. High humidity from seawater-moistened packing materials and the application of gentle mechanical pressure on the scallops will prevent desiccation of delicate mantle and gill tissues and lessen incidence of valve flapping that can cause damage to mantle tissue. Effective transport methods for scallops will take advantage of these factors (Heath 2001).

The packing methods are primarily based on the harvester's experiences. The most common method is packing the scallops in layers in styrofoam boxes with moist paper or wood fibres under and on top of the product. Some exporters use freshwater ice on top of the product to keep the temperature low. Use of freshwater ice is not proper for scallop. Because, the melting water leaks into the scallop, thus causes osmotic abnormalities and death (Overaa 2001). Styrene or styrofoam boxes with cardboard sleeves or outer boxes, with frozen gel packs for cooling are also common. However, other well-insulated containers will serve well, but are generally more expensive. The frozen gel packs are placed at the bottom of the container and insulated from the scallops with a layer of packing dampened with seawater. The scallops are positioned over the covered gel packs and are covered with another layer of damp packing material. The lid is placed on the container and taped securely shut. It is important to put some mechanical pressure on the scallops to prevent gaping and dehydration. If Styrofoam is used, the box is then placed in its cardboard housing (Heath 2001).

In general, chilling or refrigeration is effective for keep freshness and avoid the mortality of live scallop. It is reported that scallops react to stress factors differently depending on the season. It has been stated that thermal stress due to large differences in seasonal temperatures is more critical for scallop survival than differences in reproductive conditions (Chandrapavan et al. 2012). In a study investigating the effects of storage temperature and time on scallop (*Pecten maximus*) survival, it was determined that the increase in temperature and time caused the decrease in quality. It was determined that it can be stored at 3 °C for 7 days without loss of quality, it can only be stored at 6 and 9 °C for 3 days, and all scallops are dead after 3 days. In addition, it was concluded that the harvest season of the scallops influenced product quality and the best season for harvest was August (Lopes Da Costa 2018).

In a study simulating the transport stages designed in accordance with the conditions assumed necessary to introduce the scallops to the market, lion's paw scallops (*N. subnodosus*) harvested from a culture system, cleaned before transport and storage were placed in coolers made of expanded polystyrene. Before sealing the coolers, frozen refrigerant gel at −20 °C was introduced. Transportation was simulated under the most realistic conditions possible. In the first phase, the coolers with the refrigerant gel and the scallops were carried in a pick-up truck at ambient temperature for 14 h. Then a second stage of transportation was simulated, in which the sealed coolers were placed in a refrigerated room. The total simulated transportation time was 48 h. According to biochemical and microbiological analyses, the adductor muscle of scallop was found edible until day 4 under the conditions of transport and storage utilized in the study (Jiménez-Ruiz et al. 2012).

Depuration process also may affect the survival rates of scallops. In a study that aimed to simulate the supply chain under laboratory conditions and examine the survival rates of scallops (*Patinopecten yessoensis*) without and after the depuration process, the first group was packaged with ice in a polyethylene insulation box after depuration process of 48 h, the second group was packed after 24 h depuration and 24 h temporary keeping, and the third group was directly packaged with the same method. During pre-processing, no differences were found between groups 1 and 2 (survival rate always is 100%); however, a significant difference in survival of group 3 was observed, the death rarely occurred in 48 h and the mortality of group 3 was 100% in 64 h (Pan et al. 2018).

Shucking of Scallops

Scallops cannot maintain tight shell closure for prolonged periods, and they die soon after their removal from water. For this reason, generally scallops are shucked on-board ship and the meats are stored in ice until the vessel returns to port. After removal of the adductor muscle (meat), the shell and remaining visceral mass are discarded (Naidu 1987). The hand-shucking step is similar to that for oysters except that scallops do not close their shells tightly, making for easier knife entry. The soft body parts are removed, leaving only the "eye" or adductor muscle meat. In some cases, sea scallops destined for export to European markets are processed so that the roe (gonads) remain attached to the eye. In Europe, the roe is esteemed as a delicacy and may be even more desired than the meat (Hackney and Rippen 2000).

Commercial manual shucking is done by highly skilled people and requires considerable strength, persistence, ability, experience, willingness. However, manual shucking, even by skilled people, may result in damage to the edible parts with yield loss and quality reduction. Also, the efficiency of manual shucking is low (Yi et al. 2013). The recovery of meat in shucking of scallops is very important. Incomplete recovery causes economic losses. The on-board shucking of the scallops is fast, and losses are also observed during commercial rapid shucking. In scallops (*Chlamys islandica* and *Placopecten magellanicus*) shucked at sea by an experienced fisherman scallop, losses were seen depending the size. The average loss for commercial size scallop was found to be 11%. It is recommended that the process be performed more slowly as commercial fast shucking causes the losses in meat yield, but it has been reported that it should be investigated whether it will cause additional labour costs (Naidu 1987).

Because of these difficulties in manual shucking, many shucking methods have been developed, especially for oysters (Yi et al. 2013). It was reported that the invention of a machine that automatically notches and separates shells, cuts flesh and separates shells from meat for the first time in 1907 was patented (Martin and Hall 2006). Subsequently, patents have been reported in which invented devices and apparatus that mechanically shocking of scallops (Doiron 1949; Brown 1967; Wenstrom and Gorton 1985).

Bay and calico scallops usually are shucked on land. Their small size makes then uneconomical to shuck by hand on-board ship. Calico scallops are shucked mechanically. Machines that employ a shock-heat-shock method have been used. In this process, the scallops are passed through a sorter to remove trash and then are fed into a tank of water heated to 80–100 °C or through a steam tunnel. Rollers that revolve in opposite directions grip the shells and sling them with considerable force against a steel baffle slanted at a 45″ angle. They are removed by conveyor and undergo a second shock-heat-shock treatment. They are then dropped onto a vibrating screen that separates the meat and viscera from the shells. The meat then goes to an eviscerator, basically several paired rollers that grips and pulls the viscera from the meat. The meat is then washed, or it may be placed in a brine tank to remove shell fragments (Hackney and Rippen 2000).

In addition, thermal methods and freezing process are also used for shucking bivalves. However, since these products are generally consumed raw, the thermal method is not preferred in these cases. Although the process of freezing breaks the bond between the muscle and the shell, this may not always be complete.

High hydrostatic pressure (HHP) is an alternative for shucking of bivalve. HHP systems can ensure that shellfish are kept raw and that the connection between the adductor muscle and shell is completely and reliably released in high efficiency (Murchie et al. 2005). In a study conducted by Yi et al. (2013), Bay scallops (*Argopecten irradians*) were shucked with HHP in different pressures and times and compared with scallops shucked by hand using knife. It has been concluded that 100% detachment of the adductor muscle was observed when treated at 200 MPa/3 min and 350 MPa/0 min and higher yields were obtained HHP-shucked than those manual shucked samples.

3.1.4.4 Processing of Scallops

The shucked meats are washed, packed into cloth bags, and stored surrounded with ice in the hold. In hot weather, the scallops are sometimes pre-cooled in an ice slush prior to stowage. When sea scallop vessels return to port, the bagged scallops are usually transferred to a processing plant. The meats are separated and washed in a tank of chilled water, then inspected and graded by size prior to packing in plastic or metal containers. A bright orange stain, usually on one area of the meat, occasionally occurs due to the growth of pigmented bacteria during storage at sea. These scallops are removed and are rarely seen in the marketplace. Commonly the meats are treated with sodium tripolyphosphate before they are packed to reduce drip losses (Hackney and Rippen 2000).

Firstly, ice is used during post-catch handling operations. Fresh scallops should be stored on ice in a refrigerator. Pacific lions-paw scallop adductor muscle was maintained during at least 12 days of ice storage (Pacheco-Aguilar et al. 2008). Scallops are usually placed in refrigerated stores when landed. Pre-treatment in processing plants is also refrigerated storage. Additional applications such as packaging and the use of preservatives have been applied to chilled/refrigerated scallops. In a study where one group of scallops (*P. maximus*) was stored on melting ice (0 °C), the other group was stored at 4 °C after being wrapped with aluminium foil and cling film, it was found that the storage in ice was more effective and the packaged products had shorter shelf life (Ruiz-Capillas et al. 2001). In another study, potassium sorbate treated scallop (Pecten alba) was stored up to 28 days at 4 °C (Bremner and Statham 1983). In another study investigating the effects of radiation application on the shelf life of scallop (*Amusium ballati*), it was determined that scallops that were applied 0.5, 1, 1.5, and 3 kGy radiation and stored at 0 °C had a shelf life of 18, 23 and 42 days, respectively. Non-irradiated scallops had a shelf life of 13 days (Poole et al. 1990).

3.2 Cephalopods

Cephalopods are among the most complex and advanced invertebrates. They are distinguished from the rest of the Phylum Mollusca by the presence of circumoral (around the mouth) appendages commonly referred to as arms and tentacles (Leslie and Lipinski 2018). The group known as cephalopods (class Cephalopoda) is the most complex in the phylum Mollusca, and indeed, in all the invertebrate phyla (Vecchione 2002). Cephalopods include exclusively marine animals that live in all oceans of the world with the exception of the Black Sea, from the Arctic Sea to the Antarctic Ocean and from the surface waters down into the deep sea (Jereb and Roper 2010). These exclusively marine animals are characterized by bilateral body symmetry, a prominent head, and a set of arms or tentacles (muscular hydrostats) modified from the primitive molluscan foot. Cephalopods are molluscs with symmetrical bodies on their sides, specific heads, two large eyes on the side of the head, curved around the mouth and movable arms. Their bodies may be spheroid or flat. There are fins on the sides of the body. On the abdominal side there is a mantle cavity consisting of strong mantle muscles where two gills are located. Cephalopods swim backwards with the help of water sprinkled from the siphon of the abdomen located behind the head (Wikipedia 2019j; Atay 1997). Cephalopods occupy most of the depth of the ocean, from the abyssal plain to the sea surface. Their diversity is greatest near the equator and decreases towards the poles (Wikipedia 2019k). The class contains two, only distantly related, extant subclasses: Coleoidea, which includes octopuses, squid, and cuttlefish; and Nautiloidea, represented by Nautilus and Allonautilus. Members of the Cephalopoda class belong to the families Sepiidae, Sepiolidae, Loliginidae, Ommastrephidae and Octopodidae (Atay 1997).

They have separate sexes and direct development (without a truly larval stage); they are carnivores or (more rarely) scavengers; and some cephalopods constantly swim or hover in mid-water, whereas others stay close to the bottom, and some bury in sandy substrata to remain immobile for large parts of the day (thus hiding from predators along with minimizing energy expenditures) (Boletzky and Villanueva 2014). Cephalopods are widely regarded as the most intelligent of the invertebrates and have well developed senses and large brains (larger than those of gastropods). The nervous system of cephalopods is the most complex of the invertebrates and their brain-to-body-mass ratio falls between that of endothermic and ectothermic vertebrates. The brain is protected in a cartilaginous cranium. Some cephalopods can fly for distances of up to 50 m. While cephalopods are not particularly aerodynamic, they achieve these impressive ranges by jet-propulsion; water continues to be expelled from the funnel while the organism is in the air. One species, *Todarodes pacificus*, has been observed spreading tentacles in a flat fan shape with a mucus film between the individual tentacles while another, *Sepioteuthis sepioidea*, has been observed putting the tentacles in a circular arrangement. Cephalopods have advanced vision, can detect gravity with statocysts, and have a variety of chemical sense organs. Octopuses use their arms to explore their environment and can use them for depth perception. Most cephalopods rely on vision to detect predators and prey, and to communicate with one another. Consequently, cephalopod vision is acute: training experiments have shown that the common octopus can distinguish the brightness, size, shape, and horizontal or vertical orientation of objects. The morphological construction gives cephalopod eyes the same performance as sharks; however, their construction differs, as cephalopods lack a cornea, and has an everted retina. The gills of cephalopods are supported by a skeleton of robust fibrous proteins. The lack of mucopolysaccharides distinguishes this matrix from cartilage (McKenzie and Parker 2019).

Although there is a long fossil record of many different groups, all living cephalopods belong to two 'subclasses' the Coleoidea, which includes the major groups known as squids, cuttlefishes sensulato, octopods and vampires, and the Nautiloidea, containing two genera, Nautilus and Allonautilus, the only surviving cephalopods with an external shell. FAO's fishing catalogue published in 1984 states that the total number of living species of cephalopods is less than 1000 distributed in 43 families (Roper et al. 1984). In the FAO's 2010 fisheries catalogue, the number of commercial cephalopod species has been reported to have grown significantly since 1984 as a result of increasing demand in fisheries, market demand, and the spread of fishing activities to new fishing areas and deeper waters (Jereb and Roper 2010).

Cephalopods (Class Cephalopoda) are represented by two extant subclasses, Nautiloidea (Nautilus and Allonautilus) and Coleoidea and one extinct subclass, Ammonoidea. Members of the subclass Coleoidea includes two subdivisions, the Belemnoidea, which is the primitive form of cephalopod possessing ink sac and ten equally sized arms, became extinct during the cretaceous period and Neocoleoidea (cuttlefish, squid and octopus) where the shell has been internalized and reduced, completely lost. The major division of Coleoidea is based upon the number of arms or tentacles and their structure. The living coleoids can be segregated into two superorders, Decapodiformes and Octopodiformes. The Decapodiformes has fourth arm pair modified into long tentacles. The Decapodiformes contains two orders; the order Teuthoidea, which includes two suborders [yopsida (closed-eye squids) and Oeopsida (open-eye squids)] and the order Sepioidea which includes families like Idiosepiidae (pygmy squid), Sepiidae (cuttlefish), Sepiolidae (bobtail squids), Spirulidae (ram's horn squid), and Sepiadariidae (bottletail squids). The Octopodiformes includes the orders Octopoda (pelagic and benthic octopuses) and Vampyromorpha (vamphire squid). Octopodiformes has modifications to second arm pair; it is drastically reduced as a sensory filament in the Vampyromorphida, while Octopoda species have totally lost that arm pair. The Octopoda contains two suborders: Cirrata (deep-sea finned octopuses) and Incirrata (pelagic and benthic octopuses including the argonautiods and blanket octopuses) (Venkatesan and Mohamed 2015).

In the past 2 years, China and Morocco were the largest exporters of octopus, while China, Peru and India were the top three exporters of squid and cuttlefish. Japan, the United States, and larger southern European countries such as Spain and Italy are the most important consumer markets. China and Thailand are also large importers, although much of this volume is raw material for processing and re-export. The growing worldwide popularity of Japanese cuisine, as well as Hawaiian poke (fish salad) and Spanish tapas, has helped to boost demand for cephalopods, particularly squid and octopus. In 2017, and traded prices rose strongly (FAO 2018). In 2016, the share of cephalopods (squid, octopus, and cuttlefish) in the world trade was 3.8%. However, poor catches meant tightened supplies in 2016.

3.2.1 Squids

3.2.1.1 General Information About Squids

Biology of Squids

Squid are cephalopods in the superorder Decapodiformes with elongated bodies, large eyes, eight arms and two tentacles. Like all other cephalopods, squid have a distinct head, bilateral symmetry, and a mantle. They are mainly soft-bodied, like octopuses, but have a small internal skeleton in the form of a rod-like gladius or pen, made of chitin. Squids are soft-bodied molluscs whose forms evolved to adopt an active predatory lifestyle. The head and foot of the squid are at one end of a long body, and this end is functionally anterior, leading the animal as it moves through the water. A set of eight arms and two distinctive tentacles surround the mouth; each appendage takes the form of a muscular hydrostat and is flexible and prehensile, usually bearing disc-like suckers (Wikipedia 2019k).

The body of the squid is covered with a thick mantle of strong muscles. The shell is embedded in the upper part of the mantle and is pen shaped. It has a pair of eyes on the sides of the head. There are eight arms of the same size around the mouth and two longer arms, also called catchers. There are two rows of suction cups in the short arms. The forward or backward movement of the squid occurs when the swimming siphon changes direction. Movement is provided by the strong discharge of water taken from the edge of the head through the siphon. It can change its colour in less than a second by the contraction of black, red, and yellow colour cells (pigment) under the skin. It catches its prey with tentacles and arms. Digestion begins in the mouth. The nutrient, which is broken down by a pair of jaws and radula, is fed through the cylindrical esophagus into the stomach. The intestine where the stomach is opened is thrown out through the anus behind the siphon. A pair of gills located behind the mantle cavity is connected to the heart and kidney (nephridium), which consists of a pair of atria and a ventricle (Anonymous 2011).

Squids are major predators in the world's oceans. Generally thought of as predators of micronekton, they can also act as top predators (Young et al. 2013).

Distribution of Squids

Squids occur in almost all marine habitats of the world. Salinity is a limiting factor in squid's distribution. They are generally restricted to salinity concentrations between 27 and 37‰. However, *Lolliguncula brevis*, which lives and reproduces in waters of 17‰, demonstrates a capacity for a higher degree of salinity tolerance. Some species in habit the Red Sea and the southern coasts of the Iberian Peninsula, where the salinity is higher than 37‰ and other species have been found in waters where salinity ranges between 25 and 18‰. The habitat depth range extends from the intertidal to over 5000 m. Many species of oceanic squids undergo diel vertical migrations: they occur at depths of about 200–700 m during the day, and then at the onset of twilight and increasing darkness, they ascend into the uppermost 200 m for the night. A deeper-living layer of diel migrators occurs from about 1000 to 600 m during the daytime. The abundance of squids varies, depending on genera, habitat and season, from isolated individuals, small schools with a few dozen individuals, to huge schools of neritic and oceanic species with millions of specimens (Jereb and Roper 2010). The squid, characterized by short life spans and rapid growth rates, can more easily react to changes in the environment and tropical structure than other mid-trophic organisms in the open ocean (Young et al. 2013).

Production and Trade of Squids

After 5 years of continuous growth that started in 2010, catches of cephalopods were stable in 2015 but dropped in 2016 when catches of the three major squid species showed a combined loss of 1.2 million tonnes. The three major squid species (jumbo flying squid (*Dosidicus gigas*), Argentine shortfin squid (*Illex argentinus*) and Japanese flying squid (*Todarodes pacificus*)) decreased by 26%, 86% and 34%, respectively, for a combined loss of 1.2 million tonnes between 2015 and 2016 (FAO 2018).

It is reported that squid fishing has increased significantly in the world and has almost doubled compared to 30 years ago. Historically, Japan has always been at the forefront of squid production and consumption, but it is reported that in 2016, China became a leader in squid fishing, but Japan maintained its leadership position as a consumer (Arkhipkin et al. 2015). There was a significant increase in Japanese imports of squid and cuttlefish, from 111,000 tonnes during the first three quarters of 2016, to 133,300 tonnes during the same period in 2017 (FAO-Globefish 2018b).

Squid catches have increased substantially worldwide, and this has highlighted the fact that their populations are highly variable. The *Illex illecebrosus* fishery in the northwest Atlantic which was developed very rapidly by the East Asian squid jigging fleet in the late 1970s and early 1980s, following a decline of *Todarodes pacificus* in the northwest Pacific, collapsed suddenly and led to a rapid switching of effort to the southwest Atlantic in the early 1980s to target *Illex argentinus*. Subsequently catches of *Todarodes pacificus* have continued to fluctuate but the fishery for *Illex illecebrosus* has never returned to Canadian waters although consistent, but lower, catches are taken off the eastern USA further south (Rodhouse 2005).

The giant squid (*Dosidicus gigas*) stocks off Peru and Chile appear to be suffering. Since June 2017, these fisheries have collapsed, and supplies have been scarce. The reason is thought to be the effect of El Niño, which led to a complete fishery stop also in 2016. The collapse of the giant squid fishery has brought on an initiative to assess the giant squid biomass. Giant squid is the second largest fishery in Peru, and consequently of great economic importance to the country. After recent years of scarce squid landings in California due to the El Niño effect, fishing is now improving. The arrival of La Nina with colder waters reaching southern California has brought the squid back. Since October, fishing has improved somewhat. In the

Falkland Islands (Malvinas), Loligo squid landings were good in 2017. At the end of the first season of 2017 total catches amounted to 39,400 tonnes. The second season in 2017 brought larger sized squid, which fetch higher prices. In China, prices for Argentine *Illex squid*, Peruvian giant squid and Japanese flying squid rose from an already high level, while Loligo prices marginally decreased (FAO-Globefish 2018b).

Fishery of Squids

Squid jigging accounts for nearly 40% of the world cephalopod catches followed by trawling, which contributes 25% of the catch. Gillnets are also used for catching the squids, which accounts for nearly 10% of the catch. Gears like shore seines, boat seines, hooks and lines and spearing are the popular methods to catch cephalopods (Pravin et al. 2013).

Modern squid fishing began to develop with the emergence of motor fishing boats in the early twentieth century and the development of specific trawling and jigging equipment. Only after World War II, with the development of the fishing boats in the ocean, the cephalopods in general and the squid began to reach hundreds of thousands of tons per year and then millions of tons. Various types of fishing gear based on nets have been used for catching squids since the early days of exploitation. These include the various trap nets, set nets, and purse seines that have mainly been used in artisanal fisheries. The advent of motorized vessels in the early twentieth century created opportunities for targeting large schools of pelagic and near bottom squids as well as fish (Arkhipkin et al. 2015).

Jigging for squid is less damaging to the marine environment and produces a more valuable product. This technology exploits the natural behaviour of the squid which moves up in the water column toward the surface at night where they can then be attracted using lights toward the fishing vessel and the jigs. Many large-scale fisheries for both ommastrephid and loliginid squids employ jigging with lights. This method results in a higher value product where the squid can be sold whole because the process causes little or no damage to the skin (Arkhipkin et al. 2015). The ommastrephid squid are almost exclusively caught using jigs armed with barbless hooks which are fished in series on lines using automatic machines. The squid are attracted towards the jigs at night with incandescent, metal halide lamps suspended on cables above the deck of the vessel. The lamps mostly emit white light, but small numbers of green lamps are often interspersed in the arrays. Some industrial vessels will also operate one or two underwater lamps which are raised through the water column, and dimmed, as the squid are attracted towards the vessel. Jiggers typically deploy a large parachute drogue to prevent drifting downwind while fishing, thus enabling the jig lines to operate close to the vertical (Rodhouse 2005). Small inshore jiggers may use only a single lamp whilst >150 (2 kW) lamps may be operated on a large industrial, distant-water vessel. Mostly white light is emitted, but the arrays are often interspersed with small numbers of green lamps to create optimal light conditions for jigging. Jigs are also used successfully in some of the smaller long-fin (loliginid) squid fisheries (Hastie et al. 2009).

Trawlers use various types of the trawling gear (pelagic, semi-pelagic, and bottom) which are deployed during daytime to exploit the natural behaviour of squids over the continental shelf as they aggregate near the seabed during daylight. The trawling gear used is essentially the same as that used for finfish. Trawlers use acoustic target-finding technology to locate aggregations of squids. However, squids provide weak acoustic targets because they lack a swim bladder, so the technology has limited use where squid targets are mixed with fish possessing swim bladders. Squid targets can be also confused with aggregations of similar sized fish that do not have a swim bladder, such as the rock cod *Patagonotothen ramsayi* (Arkhipkin et al. 2015). Long-fin (loliginid) squid are typically caught by demersal trawling. The trawls are used during hours of daylight when the squid are concentrated near the seabed. Several trawl gear types are used for squid fishing. These include: conventional otter trawls, which are used to fish on the bottom over rough ground; pelagic trawls, which can be used to fish just off the seabed; and specially designed squid trawls, which tend to have small mesh cod ends and higher head ropes than those normally used to catch fish (Hastie et al. 2009).

Some Important Species of Squids

There are more than 300 known species of squid out there that have been identified. About 30–40 species have substantial commercial importance. The bulk of the global squid catch comprises species from two families, the Ommastrephidae and Loliginidae. The members of the family Ommastrephidae dominate in terms of biomass with five main commercial species.

***Todarodes pacificus* (Japanese Flying Squid)**

Todarodes pacificus is one of the most abundant of the commercially valuable squid in the world. An oceanic and neritic species occurring within abroad temperature range from about 5 to 27 °C, usually in surface waters to 100 m depth and, to a minor extent, down to 500 m depth. During its lifespan of about 1 year a northward migration occurs first, followed by another one in southward direction, usually in close correlation with changes of the main surface currents. Large aggregations occur in small gyres and along oceanic fronts. Three independently breeding subpopulations can be distinguished in Japanese waters. The main group spawns in winter in the East China Sea, the second in autumn, west of Kyushu, and the third, minor group in spring/summer in the Sea of Japan as well as off north eastern Japan. Post-spawning mortality is very high. The males of all three subpopulations mature before the females and transfer their spermatophores on the still immature females (in water temperatures of 13–18 °C). Growth rates are directly related with temperature and inversely with size. Main food items are myctophids, anchovies, crustaceans, gastropod larvae, and chaetognaths. Total coach in 2017 was reported to be 152,839 tons (FAO 2019c). *T. pacificus* is distributed mainly in the northwest Pacific along the Japanese Island including the East China Sea, the Sea of Japan, and southern Okhotsk Sea. This species is fished by using several fishing methods. It is mostly fished by jiggers. It is mostly fished buy coastal jiggers and offshore jiggers in the Sea of Japan. Along the Pacific side of Japanese waters *T. pacificus* is mainly caught by coastal jiggers, but more than a half of total catch along the Pacific side is caught by other fishing methods such as trawlers, purse seines and set nets (Sakurai et al. 2013).

Todarodes sagittatus **(European Flying Squid)**
The European flying squid *Todarodes sagittatus* is one of the most abundant oceanic cephalopods in the Atlantic Ocean and the adjacent Mediterranean Sea. It is found from the polar seas in the north to the equator in the south, living in the open ocean and near the coasts, at the surface and near the bottom at depths down to 2500 m. Although its spawning areas are not known precisely, *T. sagittatus* is thought to make extensive spawning and feeding migrations within the whole northern Atlantic. In winter, *T. sagittatus* migrates into deeper offshore waters; spawning is thought to take place on the continental slope (Piatkowski et al. 1998). Most individuals probably live 12–14 months, but the lifespan of the largest individuals may approach 2 years. Spawning is seasonal, its timing varying with geographic location. *Todarodes sagittatus* may be found both in the open ocean and near the coast. It is known to migrate vertically between the surface at night and near-bottom waters by day. It can be found in surface waters above depths of 4595 m *Todarodes sagittatus* undertakes pronounced migrations, probably mainly related to feeding and spawning. Common fishing methods include jigging off Norway and artisanal hand-jigging in parts of the Mediterranean and Canary Islands where there are directed artisanal fisheries. Over much of its distribution, it is taken mainly as by-catch in trawl fisheries (Piatkowski et al. 2015) (Fig. 3.5).

Loligo forbesii
Loligo forbesii is one of the two loliginid species of significant commercial importance in the Northeast Atlantic. *Loligo forbesii* has prominent longitudinal flame-like stripes of purplish dark chromatophores on the anterior and ventrolateral surfaces of the mantle. This feature is also sometimes seen in large mature males of *L. vulgaris*, although the stripes are usually much smaller and less numerous. In general, the mantle colour is more orange in *L. forbesii* and more violet or purple in

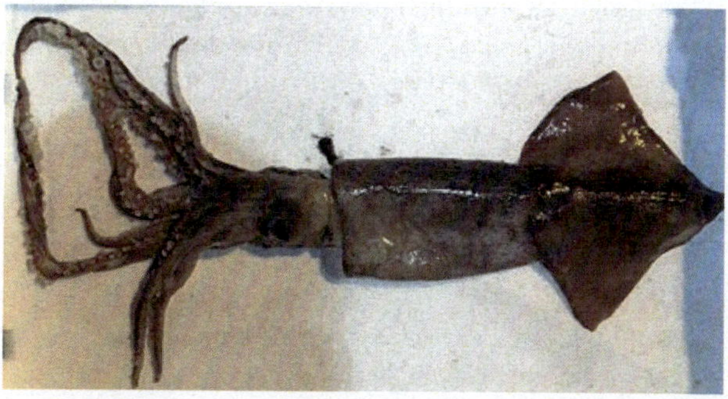

Fig. 3.5 *Todarodes sagittatus*

L. vulgaris. *Loligo forbesii* is one of the largest members of the family Loliginidae. Male *L.* forbesii can grow considerably larger and heavier than females and have faster growth rates. Typically, adult body size reaches 100–650 mm ML (Mantle length) in males (weight range 155–3700 g) and 175–350 mm ML in females (weight range 200–1150 g) throughout the species' range. The veined squid, *Loligo forbesii* Steenstrup, 1856, is found in the Northeast Atlantic, from ca. 60° N to ca. 20° N, and throughout the Mediterranean. A neritic and mainly near-bottom species, it lives in coastal waters and continental shelf seas of the Northeast Atlantic, from the Faroe Islands and the northern North Sea to the southwest coast of Norway. The life cycle of *L. forbesii* is annual, and maximum lifespan is ca. 16 months. It usually spawns in winter, but summer breeders have also been described in some areas (Jereb et al. 2015).

Loligo vulgaris **(European Squid)**
L. vulgaris as a pelagic species inhabits continental shelf up to a depth of 80 m. It feeds on shrimp, fish, and wolves. At night, he comes to the light of fishermen and catches organisms gathered there. Although it may be 45–50 cm in length, it is usually between 20 and 30 cm. The head section around the arms is easily distinguishable from the body. The large eyes are covered with a transparent corneal layer. The body is long and tapered towards the end. The back is slightly rounded compared to the abdominal side. The fins, two-thirds long in the body, are large and round in the middle, thin and pointed towards the ends. The suction cups in the eight of the eight arms, which are shorter than the ten arms arranged around the head, have two rows, whereas the long arms have four rows of suction cups in the hand-widened end portions (Anonymous 2011). The European squid, *Loligo vulgaris* Lamarck, 1798, is found in the Northeast Atlantic from ca. 55°N to ca. 20°S and throughout the Mediterranean. It is one of the most common squids in the coastal waters of the Northeast. *Loligo vulgaris* is an annual species with a maximum lifespan of ca.15 months. Spawning is usually in winter in the northern and eastern portions of its geographic range and year-round with seasonal peaks elsewhere, although there is high spatiotemporal variability in reproductive and growth parameters. Para larvae are planktonic for 2–3 months. *Loligo vulgaris* is neither pelagic nor fully benthic; it is restricted to the sea bottom during the spawning season, but displays pelagic behaviour at other times, e.g. when hunting (Jereb et al. 2015) (Fig. 3.6).

Illex coindetii **(Broadtail Shortfin Squid)**
The broadtail shortfin squid, *Illex coindetii* (Vérany, 1839) is a medium-sized squid, commonly reaching 200–250 mm mantle length throughout its distributional range. The maximum mantle lengths recorded for females and males are 379 and 279 mm, respectively. Females are larger than males, and maximum size varies depending on the population. *Illex coindetii* is found on both sides of the Atlantic and throughout the Mediterranean Sea. In the Northwest Atlantic, it is found from off the northeast coast of the United States to south. Age at maturation varies between 120 and 271 days in males and between 120 and 285 days in females, depending on the geographic area and season considered. *Illex coindetii* females spawn several times during the spawning period, which may last for several weeks. Spawning is

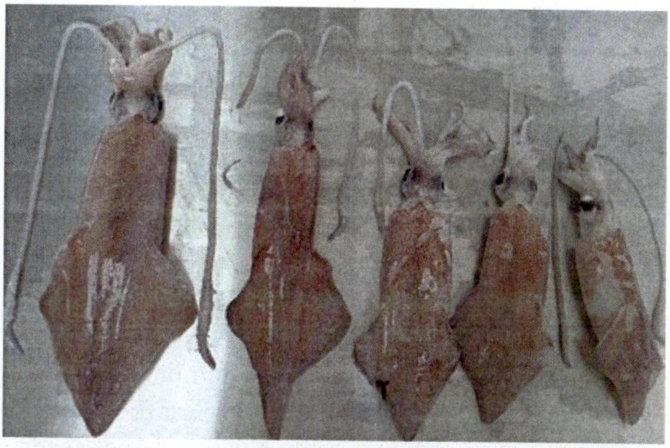

Fig. 3.6 *Loligo vulgaris*

Fig. 3.7 *Illex coindetii*

year-round, but seasonal peaks exist and vary with area throughout the Mediterranean and Atlantic. It lives close to muddy, sandy, and debris-rich bottoms, which are often, covered by *Funiculina* spp. in the middle and lower sub-littoral and upper bathyal domains. *Illex coindetii* is taken throughout the year in the Mediterranean, off West Africa, and in the Northeast Atlantic as bycatch in bottom and pelagic trawls, and, to a lesser extent, with gill- and trammel nets, in depths of 100–400 m. It is of increasing fisheries value and represents a valuable resource in some areas of its distribution range because of the size of the catches (Jereb et al. 2015) (Fig. 3.7).

I. illecebrosus **(Northern Shortfin Squid)**

Northern shortfin squid, *I. illecebrosus*, are distributed across a broad latitudinal range in the Northwest Atlantic Ocean, in continental shelf, slope, and oceanic waters located off the east coast of Florida. Distribution is highly influenced by water temperatures and water masses, and on the eastern USA shelf, temperature preferences during the fall are size-specific. The timing of migrations into the fishing areas varies inter annually and begins earliest in the southern portion of the species' range. Shortfin squid utilize continental shelf, slope, and oceanic habitats during their lifecycle and adults undergo long-distance migrations among boreal, temperate, and subtropical waters. A portion of the landings by the USA fleet are sold domestically as bait, but a majority is exported as food.

Doryteuthis pealeii **(Longfin Inshore Squid)**

Longfin inshore squid inhabits the continental shelf and upper slope waters between southern Newfoundland and the Gulf of Venezuela, including the Gulf of Mexico and the Caribbean Sea. Water temperatures have a major influence on longfin squid growth rates and the effect of an increase in water temperature on growth rate is most pronounced during the first 3 months of life.

Dosidicus gigas
D. gigas is an oceanic and neritic species occurring from the surface to 500 m depth. It is most abundant and largest off South America, where adults are found in water temperatures of between 26 and 28 °C to much colder. In nearshore waters, it occurs near to the surface day and night. Their seasonal migration is similar to those of other ommastrephids. The cohorts grow at different rates depending on the environmental conditions at the time of hatching, but all recruit into the fishery around May each year. Longevity is about 1 year for the population in the northern hemisphere. Mortality after spawning is high. This species feeds on larvae of pelagic fishes such as lantern fishes, sardines, mackerels, and sauries, and on crustaceans. Cannibalism is common. It is in turn preyed upon by swordfish, sharks, porpoises and other mammals. *D. gigas* has relatively recently expanded its distribution northwards, probably due to a combination of favourable environmental conditions and fishery impacts interacting with physiological mechanisms (FAO 2019e).

Berryteuthis magister
B. magister is a widespread boreal North Pacific species of the oegopsid family gonatidae. This species is highly abundant and plays an important role in various ecosystems of the North Pacific Ocean. It is the only commercially harvested species of the family Gonatidae. It is a bentopelagic species. Most of commercial catch is taken using bottom trawls and lesser amounts of squid are taken as by-catch using mesopelagic trawl nets towed near the bottom (Katugin et al. 2013).

3.2.2 *Cuttlefishes*

3.2.2.1 General Information About Cuttlefishes

Biology of Cuttlefishes

Cuttlefishes are primarily demersal inhabitants of near shore and continental shelf zones in warm and temperate waters. Although excellent swimmers, they generally are bottom dwellers. Habitats range from rocky, sandy, and muddy bottoms to grass flats, seaweed beds, coral reefs, etc. Many (but not all) species are known to migrate seasonally in response to temperate changes. Spawning usually takes place with an increase in water temperature, sometimes twice a year in areas where the hydrographical regime is strongly seasonal. During mating, males use their modified arm(s) (hectocotylus) to transfer spermatophores to the females. Cuttlefishes spawn relatively few and large eggs which the female attaches in grape like clusters to plants, debris, gravel and other substrates. Sexual maturity often is attained as early as a few months after hatching and it is not uncommon to find juveniles from the spring brood participating in autumn spawning. Post spawning mortality is usually high, particularly in females. Longevity ranges between 1 and 3 years (Roper et al. 1984).

The body, consisting of the head and body, is oval and flattened from the top. The mouth is between the legs and beak shaped. The eyes are on both sides of the head and are large. There are 10 ft on the head, two of which are longer than the others. The long legs have four lined suction cups. The colour is dull whitish dark stained, and the underside is white. They migrate to deep waters in winter to reach sexual maturity. They come to mossy shores in the spring to mate and lay eggs, and usually die after laying in late summer. Eggs adhere to algae or hard soils in a capsule, hatch at the end of summer, fry migrate to the deep waters in autumn. Maximum length: male is 30 cm, female is 25 cm. The average length is male 15–20 cm female10–18 cm. Reproduction time is from March to May. Cuttlefishes have been used as medicines in the treatment of various diseases as well as nutrients since ancient times. For this reason, "common cuttlefish" is called officinalis (medicine) (Anonymous 2011).

The multifunctional mantle cavity is important for cuttlefish locomotion, giving the animal its characteristic jet propulsion ability. To move away from a predator, the cuttlefish sucks water into the mantle cavity and then uses its strong mantle muscles to expel water with great force, forcing the cuttlefish in the opposite direction. While the cuttlefish uses its mantle cavity for jet propulsion, it relies on its specialized fins for basic mobility and maintaining consistent speeds. The muscular fin can manoeuvre the cuttlefish in nearly any direction: backward, forward, even in circles, with such movement being more energetically efficient than jetting. The cuttlefish has three hearts, with two hearts pumping blood to its large gills and one circulating the oxygenated blood to the rest of its body. The blood itself is blue green in colour because of hemocyanin, a copper-containing protein that transports oxygen throughout their bodies. Reproductive systems are highly complex structures with ducts, glands and storage organs. In males, the sperms are produced in the testis located in the posterior end of the mantle. Female reproductive system consists of a single ovary, the single oviduct having thin walled as well as glandular portions, the paired nidamental glands and the paired accessory nidamental glands. All cuttlefish are active carnivores feeding on live prey during their entire life cycle. They are opportunistic feeders, switching easily from one prey to another. Preferred diet of cuttlefish is crabs and fish; they feed on small shrimp soon after hatching (Sasikumar et al. 2015).

Production and Trade of Cuttlefishes

There was a significant increase in Japanese imports of squid and cuttlefish, from 111,000 tonnes during the first three quarters of 2016, to 133,300 tonnes during the same period in 2017. All the major suppliers increased their shipments: China increased by 16.4% to 70,400 tonnes, Chile by 14.95% to 13,300 tonnes, and Peru by 53.8% to 12,900 tonnes. Spanish imports of squid and cuttlefish also increased to 221,200 tonnes during this period, albeit by a much more modest factor. The main suppliers Peru and the Falkland Islands (Malvinas) registered impressive increases in their shipments to Spain, whereas China shipments declined by 13% compared to the same period in 2016. The squid and cuttlefish imports into the Republic of Korea increased by 32.3% during the first 9 months of 2017, to 73,700 tonnes. Supplies from Chile, Peru and China accounted for 93.5% of the imports. US squid and cuttlefish imports rose by 5% during the first three quarters of 2017 to 49,200 tonnes. China dominates this trade, accounting for 72.5% of total US imports. India and Peru are the second and third largest suppliers, accounting for 8.5% and 5.0% of the total, respectively (FAO-Globefish 2018b).

Fishery of Cuttlefishes

In the industrial fisheries, cuttlefishes usually are taken only as by-catch to other target species in bottom trawls, even in cases where they make up a sizeable portion of the catch. In the artisanal fisheries, on the other hand, they are actively sought with highly selective gear and fishing techniques based on knowledge of the biology and behaviour of the species. Such techniques include the use of aggregation devices (e.g. light, substrates for egg deposition), live or artificial lures (for example, live sexually mature females are used to attract males), and a variety of fishing gear, such as harpoons, spears, trammel nets, pound nets, hoop nets, lines, jigs, baited pots, etc. (Reid et al. 2005).

Basket traps has been the most popular cuttlefish fishing method since olden times. Basket traps were used for capture of cuttlefish around Inland Sea in Japan, Atlantic coast in Europe and in countries around the Mediterranean Sea. Benthic trapping and potting is carried out mostly in reef areas where fish and other animals are concentrated by the sheltered nature of the bottom, either for protection or feeding purpose. Japanese fishermen have been using cuttlefish trap for *Sepia esculenta* as early as 1660s. Full scale trap fishery began in 1920s hatching (Sasikumar et al. 2015).

Some Important Species of Cuttlefishes

Sepia officinalis

It is a demersal, neritic species occurring predominantly on sandy to muddy bottoms from the coastline to about 200 m depth, but most abundant in the upper 100 m. In the western Mediterranean, in early spring, large individuals leave the deeper water, where they spend the winter, to migrate into shallower water. Spawning occurs in shallow waters, throughout the year, with peaks at water temperatures from 13 to 15 °C. Growth rate varies directly with temperature and inversely with size (Pascual 1978). Food consists of small molluscs, crabs, shrimps, other cuttlefishes, and juvenile demersal fishes. Cannibalism is common and has been interpreted as "strategy" to overcome temporary shortage of adequately sized prey (FAO 2019d).

Cuttlefish are among the most important commercial cephalopod resource in European waters and the highest yielding cephalopod group harvested in the north east Atlantic. Several cuttlefish species are present in the area but landings of the common cuttlefish (*Sepia officinalis*) dominate. The English Channel supports the main fishery for this species. On both sides of channel landings are dominated by offshore trawlers with inshore traps consistently contributing only a minor proportion (Bloor et al. 2013). The species is usually caught with trawls as a target species and as by-catch in demersal fisheries. The artisanal fisheries utilise a larger variety of highly selective gear types, such as spears, pots, and traps, often combined with the use of light. Several species of cuttlefish are fished, but the species most frequently landed is the common cuttlefish (*Sepia officinalis*). Common cuttlefish is frequently marketed as fresh and frozen and is a highly attractive food item in Japan, South Korea, Italy, and Spain (EUMOFA 2018b) (Fig. 3.8).

Sepia aculeate

This species is the third most important commercial cuttlefish around Hong Kong where it is caught with set nets and seines during the spawning season. It is also one of the most important commercial cuttlefish in southwest India, where is mainly caught by trawl, with peak landings in October and November. The species is fished commercially in southern China, Taiwan Province of China, Sri Lanka, and Thailand, where it is caught by otter trawl, pair trawl, squid light-lures, traps and push nets. Captures by traps in Thai waters are most abundant in January and February, when most animals are fully mature, and females are predominantly caught (Reid et al. 2005).

Sepia hierredda

Sepia hierredda is the most commercially important cuttlefish in the east central Atlantic waters (from Cape Blanc, 21°N, to Cape Bojador, 26°N). It represents the dominant cuttlefish caught off Western Sahara and in Mauritania waters. It is distributed from South-eastern Atlantic to Africa, Cape Blanc, Mauritania, Tigres Bay, and Angola (Reid et al. 2005).

Sepia latimanus

Sepia latimanus is an important fisheries species. It is taken by trawl, set net, jig, hand line and spear. It supports local fisheries in western Japan and the Philippines where it is caught with jigs, hand lines, set nets and spears and it is common as by-catch in south-east Asian trawl fisheries. *Sepia latimanus* is fished in small quantities in the Ryukyus, China, near Taiwan Province of China and in the waters of Indochina (Reid et al. 2005).

Sepia lycidas

Sepia lycidas is an important commercial species in Japan, China, South Korea, Vietnam, and Thailand. Most cuttlefishes are caught off Thailand using otter trawl, with smaller catches made using pair trawl and to a lesser extent, squid light-lures, traps and push nets; bottom otter and pair trawls are used offshore, while push nets and lift nets are used in inshore and coastal waters. This species is the second most important commercial cuttlefish in Hong Kong and Japan, caught as by-catch in

Fig. 3.8 *Sepia officinalis*

trawls and with set nets and jigs, or by using live cuttlefishes as lures during the spawning season, and by hook, baited with live prawns or crabs, in other seasons. The mantle flesh is thick and tasty and is, therefore, highly esteemed. The species has been reared successfully in aquaculture experiments under the name *Sepia subaculeata*. Research indicates that waste materials obtained during the processing of this species have potential in supplementing the skin of land vertebrates as a source of collagen (Reid et al. 2005).

Sepia orbignyana
Sepia orbignyana is one of the most abundant cephalopod species in some areas of its distributional range (e.g. within the Mediterranean: Aegean, southern Adriatic, and Tyrrhenian Seas and in the Sicilian Channel). It is taken mainly as a by-catch throughout the Mediterranean and in the West African trawl fisheries. Separate statistics are not reported, but *S. orbignyana* represents a very significant percentage of the catches in some areas. In the Mediterranean Sea it is marketed along with *S. elegans* and small *S. officinalis* and constitutes a valuable resource locally. In the Sicilian Channel, research studies showed an exploitation rate of 0.60 for this species, which suggests an intense fishing pressure on this resource. It is marketed fresh and frozen (Reid et al. 2005).

Sepia pharaonis
This species supports industrial or artisanal fisheries throughout its range. With *S. esculenta* Hoyle, 1885, it is the most abundant cuttlefish species caught in the Philippines and the Samar and Visayan Seas, with the highest catches reported in the Lingayen Gulf and Carigara Bay. In Iran, the fishing activity occurs during the spawning season, when adults migrate from deeper waters to shallower waters in the littoral zone. *Sepia pharaonis* is caught by bottom trawlers in the Oman Sea and by traps in the Persian Gulf and is one of the most important cuttlefish species fished in both areas. The species is important to the commercial cephalopod fishery of Thailand, being highly abundant in the Gulf and the Andaman Sea, where it is the most common species of cuttlefish caught. The species contributes about 90% of the cuttlefishes caught off Australia by Chinese pair trawlers (Reid et al. 2005).

3.2.3 *Octopuses*

3.2.3.1 General Information About Octopuses

Octopuses are members of the Cephalopoda class of molluscs. They are in the Octopodidae family and most of the species are placed in the genus Octopus. The life span of octopuses varies with species from about 6 months to 3–5 years. Octopuses are found in many diverse habitats, including rocks, kelp holdfasts, and soft bottom substrates such as sand and mud. Most octopuses live in rocky areas where small caves, crevices, and rocks serve as dens or homes. Some species live in sandy areas and apparently do not have permanent homes. Octopuses are thought to be territorial because some individuals are known to defend their territory against other octopuses entering their area. They feed on crustaceans, other molluscs, and small fishes. Some species feed exclusively at night, while others are only active during the day or at dawn or dusk (Price 2000).

As the name suggests, the species in this group have eight legs or arms lined up around the head. On the inner surface of these arms, there are suckers providing them adhering to prey animals or hard objects. Octopuses do not have fins, and some species have thin membrane skin connections on the bottom of the arms. The body is spherical. His arms are longer than his body. They are bottom-living molluscs with moderate swimming ability.

Like most other cephalopods, octopuses have a relatively short life span (usually 6 months to 4 years, depending on the species), though there is strong evidence that deep-sea species live longer. Most species are semelparous, having one reproductive event near the end of life. Mating occurs opportunistically throughout life but is reported to be more frequent as the end of the lifespan approaches. Octopuses are typically crepuscular hunters, leaving their dens at dawn and dusk to survey their feeding territories and speculatively pounce on likely prey locations. Octopuses largely prefer crustacean prey, but they are highly opportunistic and also known to consume fish, polychaetes, chaetognaths, echinoderms and other cephalopods. Octopuses contain no hard parts in their bodies except for a beak and a radula. Similar to other cephalopods, but to a more extreme degree, the octopus also possesses a cartilaginous "skull" surrounding and enveloping the great majority of the central nervous system. Since there is no rigid skeleton, bodily support and movement are accomplished via muscular hydrostat. This mechanism is especially important to the movement of octopus arms. In the arms, four different bundles of muscle are recognized: longitudinal, transverse, circular, and oblique. These provide support and contract against each other to allow a large variety of movements. The arms of the octopus are incredibly flexible and strong, with unlimited degrees of freedom and the ability to bend at any point. They originate from the area around the mouth (the brachial crown) and serve a variety of purposes for the animal, including searching in crevices for prey, capturing prey, transferring food to the mouth, exploring their surroundings, certain forms of locomotion, self-cleaning, intra- and inter-specific signalling, mating and defence against predators. Another astonishing feature of the octopus and it cephalopod relatives is its skin's ability to change colour, iridescence and texture. Near-instantaneous changes in colour and luster are achieved via the contraction and relaxation of thousands of specialized pigment-containing cells, the chromatophores, along with an underlying layer of two other cell types, leucophores and iridiphores, which reflect light. Changes in skin texture are achieved via expansion and contraction of specialized patches of tissue (papillae) that can alternatively form bumps and "horns" or become almost completely flat. Because these structures are directly innervated by neural centres and controlled by the brain, changes in body appearance can occur in less than 30 ms. These abilities, combined with changes in behaviour and posture, enable the octopus to camouflage against the natural background, or even to mimic objects and animals in the environment (e.g., coconut, sea snake) (O'Brien et al. 2019).

Trade and Market of Octopuses

It has been reported that octopus consumption is limited to Japan, Republic of Korea and Northern Mediterranean countries and national authorities have taken strict steps to protect species due to the excessive exploitation of octopus resources, especially in the Eastern Mid-Atlantic. In 2004, Moroccan closed octopus fishing for several months as a protection measure. China is by far the main octopus fishing nation. This country started octopus catching rather recently. China is catching octopus in the Eastern Central Atlantic, under fishing agreements with several African countries. Japanese catches are stable. China became the main octopus exporter in 2004, when the Moroccan production was hit by a major crisis due to an imposed ban on octopus fishing. Despite the Moroccan recovery in 2005, China continued to be the main exporter also during this year. In value terms, however, Morocco is still the main octopus exporter, followed by Spain. China is only number three of octopus exporters in value terms (Josupeit 2008).

The Moroccan octopus season ended in September with lower catches than in 2016 and consequently higher prices. Cold storage holdings in Morocco are practically empty, as stock is being sold as soon as it comes in. The winter season had a slow start in December, with mostly smaller sized octopus catches, while demand for larger sizes has been high, especially in the USA. Morocco decided to delay opening the winter season for octopus by a few days until 5 December 2017 in the northern and southern areas. The TAC for the first month of the season was set at 815 tonnes. At the same time, Mauritania opened its winter season. Despite the relatively high quota for the first month, prices for Moroccan octopus continued to rise. Octopus landings in northern Spain dropped 47% from the beginning of the season in July to the end of September 2017. Portugal stopped fishing for 1 month in the last quarter of 2017 to let the stocks recover. The ArmAlgarve Octopus Fishing Association suggested a longer fishing ban, but the authorities have not reacted to this suggestion. Octopus prices in Indonesia, India and Pakistan have increased recently, and landings were very scarce. Most of the octopus from India and Pakistan was exported to China, where it was processed and re-exported to other markets such as Europe. Venezuela catches of octopus were reportedly good in 2017 and sold at high prices in the European market. International trade of octopus was mixed for the first 9 months of 2017. Japanese octopus imports dropped by almost 5% compared to the same period in 2016, from 35,300 tonnes to 33,600 tonnes. The main suppliers were Morocco, Mauritania, and China, which together accounted for 92% of the imports during this period. Spanish octopus imports increased 8% during this period, to 43,500 tonnes, compared to 2016. The main suppliers to Spain were Morocco, Mauritania, and Portugal (FAO-Globefish 2018b).

Fishery of Octopuses

The techniques for the exploitation of Octopus are very diverse and the harvesting strategies are related to the biological and ecological characteristics of the species. When the density of an octopus population is high and the bottom is flat, then bawling is possible (Guerra 1997). Octopus is caught using various methods, including trawling, clay pots and traps, long lines using hooks, by hand during SCUBA diving and spearing. Large fisheries for *Octopus vulgaris* are concentrated on the northwest African coast, in the Mediterranean Sea and Japan, while small-scale fisheries exist in countries such as Mexico, Canary Islands, Chile and Hawaii. While octopus fishing is mostly done with trawlers in the northwest coasts of Africa, there is also the use of pot besides trawl fishing in the Mediterranean (Oosthuizen 2003). The largest octopus fishing in the world has been reported to be Sahara fishing on the northwest coast of Africa (Guerra 1997; Balguerias et al. 2000; Tsangridis et al. 2002).

While trawling is done in deep waters, pot is used in coastal waters. Large octopuses are caught with a pot and peak values reach during the winter months. Trawler fishing is more intensely in the summer months. Trawling, pots, long lines and angling are used for octopus fishing in Japanese waters (Oosthuizen 2003). The use of pots is very old and perhaps the most widespread of all fishing methods for *Octopus vulgaris* and many other species. Pots are un-baited open structures that afford a safe, dark hiding place either for resting between catching periods or as a safe place in which a female may lay and brood her eggs. The fishery of octopus with pots is also very common and traditional in regions along the Spanish Atlantic and Mediterranean coasts (Guerra 1997).

Traps are essentially pots with some type of lid or trap door to prevent the octopus from leaving the container. Several different types have been used in various countries. The traps are baited in contrast to the pots (Guerra 1997). Traps and pots are traditionally used along the Iberian Peninsula (Banon et al. 2017). In Portuguese waters, especially in the southern coast, the main fishing methods for octopus are reported as pots and traps (Sonderblohm et al. 2017).

Drift fishing for octopus is an artisanal method used in many countries. This is fastened to a handline that the fisherman drags along or over the bottom. It is baited with a dead portunid crab. An octopus attracted to the crab seizes it and is caught by the hook as the line tightens, alerting the fisherman who pulls it in (Guerra 1997).

In some areas, octopuses are fished from the shore by hand, with spears or hooks. In Northern Spain, a traditional fishing method combines a baited stick and a hook mounted at the end of a stick and is used in the intertidal zone at low tide. Usually, these sticks are baited with fish and/or squid flesh. The baited stick is placed inside the octopus den and the fisherman waits until it grasps the lure. Then, the octopus is slowly taken outside its den with the baited stick and extracted with the hook (Sauer et al. 2019).

Spearing is accomplished either by wading or from small skiffs. Octopuses are located by using boxes with a glass in the bottom, SCUBA masks, or by clearing the surface with a few drops of oil. When an octopus is found in the open, it is speared. If an octopus den is found, a short pole with a hook on the end is thrust into the hole (Guerra 1997).

Some Important Species of Octopuses

Octopus maya
This species is common in shallow grassy bottom. These animals are mainly fished from a small boat with numerous baited lines drawn slowly across the bottom.

Octopus macrocus
This species is distributed worldwide in warm waters. It lives on rocky and sandy to grassy bottom. Since it lives in somewhat deeper water it will probably best be caught with a trawl.

Octopus aegina
It lives in muddy coastal water found subtidally on soft substrastes to depths of at least 40 m. This octopus is a major fisheries species throughout coastal Mainland Asia, important in commercial trawl fisheries, particularly from the Gulf of Thailand and South China Sea. It is distributed in coastal waters of continental Asia, from China south to Malaysia and west to at least Madras, India (Dunning et al. 1998).

Octopus marginatus
This species is found in tropical continental waters of the Indian Ocean, from Red Sea and East Africa and Southeast Asia and eastern Australia. It lives in coastal muddy waters on mud and sand substrates; from subtidal to depths at least190 m. It is important fisheries species collected by trawlers, pots, and lines (Dunning et al. 1998) (Fig. 3.9).

Nutritive Composition of Cephalopods

The cephalopods were reported to be rich source of long chain n-3 polyunsaturated fatty acids, essential amino acids, antioxidants and minerals, such as, selenium (Zlatanos et al. 2006). Proximate composition of cephalopods varies with habitat, maturation, feed and feeding, species and season. Composition also varies among species due to geographical differences of fishing grounds. On the other hand, it can vary with body part or organ (Lawal-Are et al. 2018). Many researchers have revealed these differences between the organs (Ballantyne et al. 1981; Salman et al. 2007; Vairamani 2010; Ramasamy et al. 2010; Nurjanah et al. 2012). The highest protein content of octopus (*Octopus aegina*) was reported in gonad followed by muscle, gill, and liver (Kavitha and Ponni 2018). Feeding and feed type is also effective on biochemical composition of cephalopods. Effects of diet on biochemical composition of octopus have been reported by Biandolino et al. (2010). In their study, octopuses were fed with three experimental diets (Diet group I: 80% crab, 15% bogue fish and 5% mussels; diet group II: fish; diet group III: mussels). The wild octopuses and the bogue-fed group had significantly higher lipid content than the mixed and mussel dietary groups. The proportion of the fatty acids changed significantly between the feeding treatments. In another study, octopuses were fed using five different experimental diets (white crab, blue crab, bogue fish, white crab + discarded bogue, blue crab + discarded bogue) and it has been reported that the highest n-3 HUFA (highly unsaturated fatty acid) content was observed in the bogue-fed group and in the mixed diet group (Estefanell et al. 2011). Various researchers have reported that wild cephalopods contain in high levels of DHA and DHA/EPA ratio and DHA levels are correlated with the dietary input in cultured cephalopods (Sinanoglou and Miniadis-Meimaroglou 1998; Navarro and Villanueva 2000, 2003). It has also been determined that temperature and body weight affect the fatty acid composition of octopus (Miliou et al. 2006).

Many researchers have reported high protein and low-fat content in cephalopods. This feature makes cephalopods very suitable for human consumption and especially for the elderly population (Bano et al. 1992; Thanonkaew et al. 2006; Remyakumari et al. 2018). It is reported that when compared to fish, cephalopods have 20% more protein, 80% less ash, 50–100% less lipid and 50–100% less carbohydrate. Moreover, it was reported that cephalopod mantle does not store lipid, or its storage is below 1 g of its wet weight (Lee 1994).

Cephalopods contain small amounts of fat, which is rich in n-3 fatty acids. They can be used as a good source of the n-3. In a study, most abundant fatty acid in squid and cuttlefish was found as docosahexaenoic acid (DHA) followed by palmitic acid and eicosapentaenoic acid (EPA). On the other hand, in octopus, palmitic acid was found slightly higher than DHA and the third most abundant fatty acid was again EPA (Zlatanos et al. 2006). Greater PUFA quantities than those SFA and MUFA were reported in different cephalopod species (*Amphioctopus neglectus*, *Cistopus indicus*, *Uroteuthis duvauceli*, *Sepia pharaonis* and *Sepiella inermis*). Moreover, the mean total of n-3 PUFA content was reported to be greater as compared with n-6. DHA (22:6n-3) was found to be the predominant n-3 PUFA in these species, which represented greater than half of the total PUFA content followed by EPA (Chakraborty et al. 2016b).

Cholesterol is an important lipid component in cell membranes, and the body uses it to create a range of hormones and vitamin D. In many marine species, cholesterol is the main sterol, making up more than 90% of all sterols. It can be found in some shellfish creatures that may be as low as 25%. Cephalopods usually contain

Fig. 3.9 *Octopus* spp.

higher levels of cholesterol, for example, about 140 mg of cholesterol/100 g in European squid (Nunes et al. 2011) (Tables 3.8, 3.9, 3.10 and 3.11).

3.2.3.2 Quality Changes in Cephalopods

The cephalopods deteriorate very quickly because of endogenous and bacterial proteases. Protease plays an important role in protein breakdown, and the protease activity of cephalopods has been reported to be higher than in various fish species (Stanley and Hultin 1984; Hurtado et al. 1999). After death, cephalopods undergo high protein degradation by both endogenous and bacterial enzymes. Such rapid protein degradation results in high levels of nitrogen release from the muscle, which promotes bacterial growth that leads to rapid degradation (Gómez-Guillén et al. 1996; Hurtado et al. 1999). In the mantle muscle of Squid (*Ommastrephes loani pacificus*), autolytic activity was observed in the acidic pH range (Sakai and Matsumoto 1981). It has been reported that for *Illex illecebrosus* and *Loligo pealei*, proteolytic activity is mostly observed in the acidic pH range and minor activity in the alkaline pH range (Lebanc and Gill 1982). Hurtado et al. (1999) observed proteolytic activity in octopus (*O. vulgaris*) muscle in an acid range at pH 2.5 and temperature optimum at 40 °C.

As a result of enzymatic and bacterial activity, changes occur in the cephalopod muscles, which reduce the sensory acceptability of the product. To protect these products from deterioration, refrigeration or ice storage is applied. They are conserved until they maintain their sensory acceptability. Sensory rejection for octopus was reported to be 6–7 days at 2.5 °C (Hurtado et al. 1999), 7 days at 1–2 °C (Civera et al. 1999) and 8 days in ice (Barbosa and Vaz-Pires 2004) it was reported to be 9 days at 2 °C (Unal-Sengor et al. 2018) and 12 days in ice (Prafulla et al. 2000) for cuttlefish (*Sepia pharaonis*) 9 days for *Sepia officinalis* (Vaz-Pires et al. 2008). Panellists observed a loss of sensory quality with a decrease in the brightness of the cuttlefish skin, an increase in eye opacity, intensification of the yellowish colour of the mantle, unacceptable odour development, softening of flesh and loss of mucus (Unal-Sengor et al. 2018). As storage progressed, instead of fresh sea weedy odour, sheen and glossy appearance, characteristic white colour, sweet flavour, firm and juicy texture; loss of bloom, flabby texture and discolouration were observed in the cuttlefish (*Sepia pharaonis*) (Prafulla et al. 2000). According to the sensory quality based on quality index method (QIM), while the skin of octopus (*Cistopus indicus*) stored in ice which was initially very bright, with well-marked colour, elastic and white, later changed to slightly pink with loss of brightness on day 5 and finally became dull and pinkish on day 8 (Shalini et al. 2015). Manimaran et al. (2016a) reported sensory shelf life to be 5–7 days for whole octopus (*Cistopus indicus*) and 9 days for gutted octopus. Gullian-Klanian et al. (2016) determined gradual changes in the texture, skin colour, and muscle odour of octopus (*Octopus maya*) during the storage at 4 °C. They found a sensorial shelf life of 5 days in this condition. Paarup et al. (2002a) reported to be 12 days in ice for whole and gutted squid (*Todaropsis eblanae*). Quality index Method (QIM) has been developed for some cephalopod species based on the analysis of some freshness quality parameters, as follows: appearance/colour, odour and mucus of the skin; texture of the flesh; cornea and pupil of the eyes; colour, odour and mucus of the mouth region and finally presence of material in the sucker of arms (Barbosa and Vaz-Pires 2004; Vaz-Pires and Seixas 2006).

For some time after capture, the dorsal surface of the squid shows a dark brown colour due to certain pigment cells. With subsequent decrease in freshness, these cells contract and the meat turn white. When the meat shows an alkaline reaction, the cells decompose, and the pigment particles redden the meat. Thus, the freshness of the squid can be roughly estimated by the change in skin colour with progressive deterioration (Takashi 1965).

Table 3.8 Proximate compositions of some squid, octopus, and cuttlefish species (g/100 g)

Species	Moisture	Protein	Fat	Ash	Reference
Unknown	78.3	18.0	0.9	1.5	Zlatanos et al. (2006)
U. duvauceli	80.40	17.50	0.52	1.31	Remyakumari et al. (2018)
	84.60	14.20	0.70	0.90	Mehta and Nayak (2017)
Loligo vulgaris	78.5	18.5	1.4	1.5	Atayeter and Ercoskun (2011)
Loligo sp.	75	19.9	0.5	0.9	Gopakumar (1997)
Loligo plei	74.2	14.4	2.0	1.7	Lapa-Guimaraes et al. (2005)
Unknown	81.2	15.5	1.6	2.1	Zlatanos et al. (2006)
Enteroctopus zealandicus	78.1–82.0	12.2–15.3	0.8–2.4		Meynier et al. (2008)
Cistopus indicus	84.15	13.33	0.64	0.31	Shalini et al. (2015)
Octopus vulgaris	79.90	16.30	0.56	1.86	Vaz-Pires and Barbosa (2004)
Unknown	80.4	15.8	1.2	1.3	Zlatanos et al. (2006)
Sepia officinalis	73.33	20.94	4.33	0.89	Lawal-Are et al. (2018)
	79.55	16.60	0.09	1.39	Sykes et al. (2009)
Sepia recurvirostra	83.65	13.51	0.79	0.69	Nurjanah et al. (2012)
Sepia pharaonis	83.02	13.9	0.73	0.94	Brita Nicy et al. (2016)
Sepia prabahari	85.08	12.59	0.75	0.96	Brita Nicy et al. (2016)
Sepia ramani	81.41	16.47	0.58	1.20	Brita Nicy et al. (2016)
Sepia sp.	83.65	13.51	0.79	0.69	Santi et al. (2019)
	78.8	18.1	0.2	1.4	Gopakumar (1997)

Table 3.9 Macro element compositions of some squid, octopus, and cuttlefish species (mg/100 g)

Species	Macro elements					References
	Na	K	Ca	Mg	P	
Squid						
Uroteuthis duvauceli	171.98	176.13	72.82	127.36	85.63	Chakraborty et al. (2016b)
Loligo vulgaris	15.7	26.1	13.6	43.5	–	Lourenco et al. (2009)
Octopus						
Amphioctopus neglectus	172.45	144.56	73.72	92.83	86.07	Chakraborty et al. (2016b)
Cistopus indicus	170.14	184.83	54.04	104.78	66.33	Chakraborty et al. (2016b)
Octopus vulgaris	57.2	22.3	21.3	93.8	14.7	Lourenco et al. (2009)
Cuttlefish						
Sepia recurvirostra	153.27	27.75	18.62	6.48	56.96	Nurjanah et al. (2012)
Sepia pharaonis	170.18	201.2	108.41	127.28	80.4	Chakraborty et al. (2016b)
Sepiella inermis	172.77	134.67	79.73	109.76	72.37	Chakraborty et al. (2016b)
Sepia officinalis	26.6	28.9	13.4	56.7	24.9	Lourenco et al. (2009)

Table 3.10 Micro element compositions of some squid, octopus, and cuttlefish species (mg/100 g)

Species	Micro elements						References
	Mn	Cu	Zn	Fe	Cr	Se	
Squid							
Uroteuthis duvauceli	6.57	1.33	7.21	10.42	9.54		Chakraborty et al. (2016b)
Loligo vulgaris	0.01	0.15	1.26	0.17		<QL	Lourenco et al. (2009)
Octopus							
Amphioctopus neglectus	6.39	3.55	13.08	7.44		9.31	Chakraborty et al. (2016b)
Cistopus indicus	6.51	7.54	14.17	6.60		9.45	Chakraborty et al. (2016b)
Octopus vulgaris	0.03	0.38	1.77	0.42		<QL	Lourenco et al. (2009)
Cuttlefish							
Sepia recurvirostra	–	0.57	1.96	0.4	–	0.002	Nurjanah et al. (2012)
Sepia pharaonis	6.92	5.57	15.18	7.79	–	8.95	Chakraborty et al. (2016b)
Sepiella inermis	6.43	5.33	8.34	8.45	–	9.34	Chakraborty et al. (2016b)
Sepia officinalis	0.01	0.45	1.08	0.14	–	<QL	Lourenco et al. (2009)

QL Quantification Limit

Trimethylamine (TMA) and Total Volatile Basic Nitrogen (TVB-N) have been used as chemical indices to test freshness of cephalopods. Acceptable limit of TVB-N for squid (*Illex illecebrosus*) has been reported as 45 mg/100 g. TVB-N values have been classified as A quality with less than 30 mg/100 g, B quality 30–45 mg/100 g and unacceptable with more than 45 mg/100 g (Ke et al. 1984). On the other hand, some researchers (Ohashi et al. 1991) reported that TVB-N is not a suitable indicator for cephalopods, while others reported that limit values are variable depending on the type and storage conditions (Civera et al. 1999; Paarup et al. 2002a, b). TMA-N, caused by the bacterial reduction of osmoregulatory agents such as TMA-O, is a volatile amine, often associated with the typical 'fishy' odour of seafood degradation. Both TVB-N and TMA-N are relatively constant in the mantle

Table 3.11 Fatty acid composition of some squid, octopus and cuttlefish species (% in total fatty acids)

Species	SFA	MUFA	PUFA	n-3	EPA + DHA	Reference
Squid						
U. duvauceli	30.12	1.98	67.9	60.49	60.49	Remyakumari et al. (2018)
	34.9	9.9	55.2			Torrinha et al. (2014)
Loligo vulgaris	37.2	10.2	59	47.7	47.1	Salman et al. (2007)
	34.3	9.3	56.4			Torrinha et al. (2014)
Loligo gahi	32	10.7	57.3			Torrinha et al. (2014)
Loligo opalescens	30.9	9.6	59.5			Torrinha et al. (2014)
Loligo reynaudii	33.5	10.1	56.4			Torrinha et al. (2014)
Sepioteuthis australis	38.6	9.6	51.5	46.8	45.7	Phillips et al. (2002)
Gonatus antarcticus	19.9	21.8	58.1	52.9	51.9	Phillips et al. (2002)
Moroteuthis robsoni	26.5	20.3	53.0	47.1	45.8	Phillips et al. (2002)
Todarodes spp.	29.7	13.6	56.5	54.3	53.8	Phillips et al. (2002)
Octopus						
Octopus vulgaris	30.05–34.07	7.10–10.71	51.25–53.59	43.45–45.17	43.27–44.61	Ayas (2012)
Eledone moschata	29.69–31.09	14.35–15.06	46.30–47.09	37.51–38.92	37.03–38.39	Ayas (2012)
Enteroctopus zealandicus	47.41	32.60	19.9	15.89	14.95	Meynier et al. (2008)
Cuttlefish						
Sepia officinalis	29.5–36.8	7.81–9.84	43.7–49.6	37.9–44.5	37.8–44.4	Ozyurt et al. (2006)
	42.72–54.20	10.29–12.81	35.20–54.11	31.40–48.19	25.9–45.7	Tir et al. (2015)

of fresh caught cephalopods (Ruiz-Capillas et al. 2002a; Sykes et al. 2009). TVB-N content increases in seafood during the storage depending on storage conditions. Increases in TVB-N reported for cuttlefish (Paarup et al. 2002a; Joseph and Sherief 2003; Ganesan et al. 2005; Boumpalos and Lougovois 2005; Unal-Sengor et al. 2018) squid (Moral et al. 1983; Jeyasekaran et al. 2010) and octopus (Dhananjaya and Venkatappa 2006; Gullian-Klanian et al. 2016; Manimaran et al. 2016a, b; Yuan et al. 2017). TVB-N values increased above 30 mg N $(100\ g)^{-1}$ after 9 days of storage of squid (*Ommastrephes bartrami*) in flake ice (Yuan et al. 2017). TVB-N level of octopus (*Cistopus indicus*) stored in ice reached 144.9 mg/100 g in 5 days (Shalini et al. 2015).

There are many differences between fish and cephalopods. There are no clear Z-lines and no protein band of the characteristic in the cuttlefish and squid mantle. The degradation pathway of ATP is different from vertebrates in invertebrates. Squid doesn't accumulate inosine-5′-phosphate (IMP) and therefore, textural changes of squid during cold storage differ from those of vertebrates (Kagawa et al. 2002).

Yamanaka et al. (1987) argued that agmatine is a useful potential index for evaluating the freshness of common squid (*Todarodes pacificus*). They also stated that the amino acid arginine is extremely abundant in free state in invertebrates and Agmatine can be easily formed from arginine decarboxylation. In their study, these researchers determined that agmatine detected small amounts even in fresh squid muscle and increased during storage. They reported that they detected agmatine at 40 mg/100 g level in the advanced stage of the deterioration. Paarup et al. (2002b) found that the level of agmatine biogenic amine determined in trace amount in fresh squid (*Todaropsis eblanae*) muscle reached 120 mg/100 g in 9 days of iced storage. The agmatine produced could be, subsequently, converted in putrescine, spermidine and spermine (Vaz-Pires et al. 2008). On the 13th day of storage, other biogenic amines, cadaverine, tyramine and putrescine were detected. It has been determined that the content of biogenic amines, especially putresine, cadaverine and tyramine in the octopus purchased at supermarket exceeded 100 mg/kg after 5 days of storage at 4 °C (Hu et al. 2012). However, histamine level ranged 1.3–9 mg/kg. Similarly, histamine level has been reported as a maximum of 42.5 mg/kg after 100 h of storage at 4 °C for of octopus (O. *maya*) (Gullian-Klanian et al. 2016). The low histamine concentration in octopus samples may be related to the low levels of the decarboxylase enzyme of histidine. While initial putrecine content of musky octopus (*Eledone moschata*) was found 5 mg/kg, other biogenic amines were not detected. However, after 12 and 24 h at 22 °C putrecine increased to 33 mg/kg and 56.7 mg/kg, respectively. The increase in other biogenic amines was very low (Prester et al. 2010).

Microorganisms play an important role in the quality change and deterioration of cephalopods. Studies have reported increases in microbial load during storage. The increase from an initial level of 4–7 log cfu/g in aerobic plate count of squid (*Todaropsis eblanae*) mantles stored at 4 °C, when it was sensorially rejected has been reported (Paarup et al. 2002b). Initial aerobic plate count (APC) of 4.023 cfu/g remained almost constant during the storage of the octopus (*Cistopus indicus*) in ice (Shalini et al. 2015). In *Octopus vulgaris* a bacterial load of 5–6 log cfu/cm^2 was detected, when was rejected (Vaz-Pires and Barbosa 2004). This indicates that the deterioration in the octopus is mainly enzymatic rather than bacterial (Hurtado et al. 1999; Shalini et al. 2015). It was reported that the initial load of 6 log cfu/g^{-1} of squid (*Ommastrephes bartrami*) stored in flake ice maintained after 12 days of storage (Yuan et al. 2017).

Although there are studies showing that gutting process is effective to extend the shelf life of cephalopods immediately after caught (Paarup et al. 2002a, b; Manimaran et al. 2016a), there are also studies reporting that it is little effective (Boumpalos and Lougovois 2005) or not effective (Park and Hur 1990). As the gut and skin of octopus contain most of bacteria and relevant proteolytic enzymes responsible for spoilage, it has been suggested to perform gutting prior to storage in with ice for extending their shelf life (Manimaran et al. 2016a).

3.2.3.3 Processing of Cephalopods

Chilling/Refrigerating

With chilling, a limited shelf life is provided to seafood. However, it is an important preservation method that ensures the quality of the products, from catching to the landing and to processing on land. Icing and refrigeration are the common chilling methods applied to cephalopods. Icing and refrigerating are mostly used together. While octopus (*Cistopus indicus*), packed individually in polythene bags and kept in an insulated container over the layers of ice, had a shelf life of 7 days (Shalini et al. 2015), *Octopus vulgaris* had a 8 day shelf life (Vaz-Pires and Barbosa 2004). Shelf life of squid (*Todaropsis eblanae*) (Paarup et al. 2002a) and (*Illex coindetii*) (Vaz-Pires et al. 2008) stored in ice was 10 days. Whole cuttlefish (*Sepia officinalis*), iced in self-draining polystyrene boxes and hold in a refrigerator at 2 ± 2 °C, had a 10-day shelf life (Boumpalos and Lougovois 2005; Vaz-Pires et al. 2008). In one study, it was determined that squid (*Loligo vulgaris*) at room temperature (20 °C) was stored for 1 day and at refrigerator (4 °C) for 4 days and consequently refrigeration was effective on the quality of squid (Gokoglu et al. 1999).

Cephalopods that suffer a rapid loss of quality immediately after capture should be cooled quickly. Icing is an effective cooling method applied to seafood. Cephalopods can be cooled by various icing methods. In a comparative study with flake ice and slurry ice, the quality changes of squid (*Ommastrephes bartrami*) were examined. It has been reported that more effective results are obtained with slurry ice and it is recommended to use slurry ice to preserve the quality during the storage and transportation of squid (Yuan et al. 2017). Slurry ice is defined as a binary mixture of small spherical ice crystals surrounded by seawater at sub-zero temperature. For this reason, it is a promising technique for the preservation of aquatic food products. It is also called to be fluid ice, slush ice, liquid ice, or flow ice. Compared to flake, ice slurry ice has two main features, such as faster cooling rate due to higher heat exchange capacity and reduced physical damage to seafood due to microscopic spherical particles (Rodriguez et al. 2006). Besides protecting the product from the effects of oxygen via completely covering the product surface, additives such as ozone and melanosis inhibitors can be added to slurry ice. It has also an easily pumpability feature (Huidobro et al. 2002).

It was determined that the use of 100% dry ice in squid (*Loligo duvauceli*) preservation provides a longer shelf life compared to 100% water ice use. However, the combination of dry ice (20%) and water ice (50%) has been tried since dry ice is expensive to use alone and sometimes causes freezing. Despite the fact that this combination has a similar effect with 100% water ice on the shelf life of squid, the combined use of dry ice and wet ice in the short-term transport of squid was recommended to the seafood industry (Jeyasekaran et al. 2010). In a similar study, both dry ice (at the ratio of 1:1) and a combination of dry ice and wet ice (in the ratio of 1:0.2:0.5) improved the quality and increased the shelf life of cuttlefish (*Sepia pharaonis*) fillets by about 33%, when compared to wet ice alone (in the ratio of 1:1) (Jeyasekaran et al. 2011).

In some studies, icing has been applied by direct and indirect contact method to protect them from the melted ice and to avoid leaching. Melting ice in contact with the fish permits water to penetrate the skin, thereby reducing muscle translucency (Lougovois et al. 2008). Lapa-Guimaraes et al. (2002) reported that non-contact ice storage method presented no advantages when compared with the contact ice storage method with respect to the quality preservation of squid (*Loligo plei*) under laboratory conditions on the other hand. Whereas Prafulla et al. (2000) indicated that indirect icing preserves most of the nutrients in squid and cuttlefish, but with shorter shelf life. On the other hand, it has been reported that contact icing is not suitable because it causes colour, texture, and aroma changes in squid (*Illex illecebrosus*) (Ke et al. 1991). Manimaran et al. (2016b) did not advice the storage in non-contact with ice due to the rapid multiplication of spoilage bacteria as reflected by the reduction in NPN as well as accumulation of water soluble ammoniacal nitrogen, TVB-N and FFA.

Different packaging methods have been used to increase the shelf life of cephalopods in addition to chilling. In a study, cuttlefish were packed in three different atmospheres (MAP1: 20% CO_2-80% N_2; MAP2: 50% CO_2-50% N_2; MAP3: 70% CO_2-30% N_2) and under vacuum and stored at 2 °C. As a result of the study, it was determined that packaging in a modified atmosphere is effective in increasing the shelf life. Especially MAP2 and MAP3 samples with high CO_2 concentration had a longer shelf life (8 days) compared to MAP1 and vacuum-packed samples. Vacuum packed samples did not differ from control samples (Bouletis et al. 2015). Longer shelf lives have been reported for squid (*Illex coindetii*) packed with the same gas ratios. Again, in this study, the best results were obtained with A2 and A3 atmospheres, and the shelf life of 8 and 10 days was determined respectively (Bouletis et al. 2014). While a shelf life of 9 days was reported for cuttlefish packed under modified atmospheres (A1: 50% CO_2-50% N_2; A2: 80 CO_2%-20% N_2; A3: 65% CO_2-35%), control samples were rejected on day 5 (Caglak et al. 2014). In a study examining the effect of modified atmosphere and active packaging on the shelf life of Cuttlefish (*Sepia officinalis*), a group of cuttlefish was packed in a modified atmosphere (20% CO_2/80% N_2), while the other group was packaged in the same modified atmosphere but with the addition of an adsorbent with high water retention in the package. The total number of psychrotrophic microorganisms and TVB-N and TMA-N values of the samples in the adsorbent-added packages were lower than that of the MAP and control samples (Albanese et al. 2005).

It was also tried to use of protective additives for octopus (*Cistopus indicus*) in chilled storage. Two commercial additives which one of them consisted of a mixture of sodium citrate and hydrogen peroxide, and other one is a mixture of sodium citrate and sodium bicarbonate, were applied to octopus and stored with flake ice at 5 °C. However, it has been determined that these protective additives do affect neither the shelf life nor the sensory quality of the octopus (Manimaran et al. 2016b).

Freezing of Cephalopods

Since squid catches have increased substantially worldwide, it is an important export product. Freezing is the most important and large scale preservation method that facilitates exports for squid (Sukumar et al. 2014). As squid is a soft-bodied species, it requires special handling during harvesting, processing and preservation to maintain its quality as it can suffer physical and mechanical damage such as breakage and tearing. Otherwise, market value and quality decrease. Squid also contain multi-coloured chromatophores on the skin surface. These alter the colour of the squid to match the ambient colour when swimming in the ocean. However, after death, these chromatophores can expand or contract depending upon processing and storage temperatures (Learson 2000). In long-term storage, pink discoloration is seen on the skin. This is the breakdown of the chromophores in the skin. It is generally reported that large squid are more sensitive to pink coloration than small ones (Sungsri-in 2010). As the squid grading is mostly based on colour in the market, its transport and storage techniques should be designed to preserve the surface colour and texture. Especially the colour of the surface is important for the squid to be marketed as a whole, and less important for the squid which to be skinned or subjected to further processing. Freezing at sea is the recommended processing method for the whole, frozen squid market. Immediately after capture the squid should be placed in refrigerated seawater (RSW) at 0°C to 4 °C. The cold seawater prevents the squid from drying out and eliminates discoloration of the skin surface. The squid should then be quickly sorted and packed, and either blast- or plate-frozen at −30°C to −40 °C. Rapid freezing and constant storage at low temperatures are essential to avoid changes in colour and adverse textural changes. Fluctuating storage temperatures can also result in reddening of the product surface which reduces market value (Learson 2000).

The quality of cephalopods changes during frozen storage. Besides protein denaturation, oil oxidation and desiccation, odour and discoloration can be seen in frozen storage. Changes in squid during frozen storage have been studied by several researchers. It has been reported that changes in seafood meat during freezing and frozen storage are mainly related to the modification of myofibrilar proteins and the potential impact of lipids on protein denaturation in relation to free fatty acids (Sikorski et al. 1976). In frozen stored squid (*Illex argentinus*), a significant decrease in protein solubility was detected in 5-month storage, and then remained unchanged during storage (Paredi et al. 2006). Similar decreases have been reported for squid (*Loligo vulgaris*), octopus (*Octopus vulgaris*) and cuttlefish (*Sepia officinalis*)

(Gokoglu et al. 2018), cuttlefish (*Sepia aculeata*) (Joseph and Perigreen 1988). During frozen storage, protein solubility increased in pota (*Todaropsis eblanae*) and white octopus (*Eledone cirrhosa*) after 6 and 2 months respectively, and then gradually decreased (Moral et al. 2002). However, lower protein solubility was found in squid muscle than those of octopus and cuttlefish. The authors explained this difference by reporting that squid contains a different class of myofibrillar proteins and is therefore less susceptible to freezing and more prone to thermal denaturation (Gokoglu et al. 2018). Contrary to these reports, there are also researchers who reported that there was no significant change in protein solubility during frozen storage of squid (*Illex argentinus*) (Mignino et al. 2008) and (*Loligo vulgaris*) (Gomez-Guillen et al. 2003). Besides the protein solubility, Gokoglu et al. (2018) also reported changes indicating protein denaturation, such as decreases in water retention capacity and increases in loss of cooking with total free amino acid content during frozen storage in squid, octopus and cuttlefish muscles. Increases in free fatty acids (FFA) content were reported during frozen storage of squid (*Illex argentinus*). Ruiz-Capillas et al. (2002b) observed decreases in the viscosity and emulsifying capacity of protein extracts from frozen-stored squid (*Illex coindetti*) (Paredi et al. 2006) (*Loligo vulgaris*) (Atayeter and Ercoskun 2011).

Freezing and frozen storage affect the texture of the cephalopod muscles. The frozen mantles of *Illex argentinus*, *Loligo edulis*, *Sepia pharaonis* were tougher than the unfrozen ones. Toughness increased with increasing frozen storage time. Injury of muscle fibres associated with the formation and development of ice crystals caused protein aggregation and hence toughening (Ueng and Chow 1998). Similarly, frozen North Atlantic squids, *Loligo pealei* and *Illex illecebrosus*, were found tougher compared to non-frozen (Stanley and Hultin 1982). The cuttlefish (*Sepia aculeata*) texture, initially defined as "firm" and "chewy", changed to "rubbery" in frozen storage for 10 months, and 16 months later it was defined as "hard to chew" (Joseph and Perigreen 1988). The increase in shear force values of squid (*Loligo formosana*) during frozen storage of 10 months has been reported and stated that this increase caused by protein aggregation due to injured muscle fibres (Benjakul et al. 2012).

Sensory changes are also observed in cephalopod meats in frozen storage. Squid (*Sepioteuthis lessonian*), which was kept on ice using flake ice for 20 h and then frozen with a horizontal contact plate freezer and stored at −20 °C, was reported to be sensorially fair and acceptable until the end of the sixth month (Sukumar et al. 2014). Cuttlefish (*Sepia aculeata*) fillet inside colour turned pale yellow in 7 months storage and its density increased with storage. The surface colour began to fade after 10 months of storage (Joseph and Perigreen 1988).

Thermal Processing of Cephalopods

The use of cephalopod molluscs as canned raw materials is low. One of the reasons for this is the low yield in the final products due to the high losses of the nutrients during processing, making them unprofitable as a canned food under current conditions. Another reason is that the sensory properties of the product worsen because of the reaction between reducing sugars and amino acids, called the "maillard reaction" during thermal sterilization in the canning process of cephalopods. The melanoidin formed during this reaction causes the change of colour of the sterilized meat to brown, forming caramelization in appearance and odour and to the loss of flavour properties (Valencia-Perez et al. 2008; Shul'gina et al. 2013). The necessity of removal of cephalopod skin which has nutritive tissue containing amino acids such as proline and oxyproline by the thermal enzymatic and or mechanical methods in conventional canning technologies has been shown as another reason of loss of raw material during canning process by Shul'gina et al. (2013). The same researchers reported that in a study of cephalopods (squid and octopus) used as a raw material, the nutrient parts of cephalopods were approximately two-thirds of their total body masses before the skin was removed, and the nutritional mass losses after removal of the skin were from 18.8 to 33.6%. In this study, where they tried to bring a new approach to canning technology, they found that removing the skin of cephalopod from canning technology can significantly increase output and reduce the cost of the finished product. They also determined that the use of spicy oily extracts as an oil filling in cephalopod canning technology improves product quality by reducing the required thermal effect level and the degree of heat damage in the food during sterilization.

When fresh squid meat overheats, it contracts and hardens, making it almost unacceptable as food. If Squid is heated without removing the skin, it shows a significant contraction. It is shortened by 6% of the initial length. Longitudinal contraction can be reduced to some degree when boiled after peeling the skin. It is thought that the longitudinal hydrothermal shrinkage occurring in unpeeled meat is caused by collagen in the skin. When very fresh meat is cooked at a temperature of about 100 °C, transverse contractions occur, but few; More stale meat stored in the cold for a day or two after death shows hydrothermal shrinkage of about 7.5%. As the cooking temperature rises to 50–60 °C, squid meat begins to lose weight as well as water content; at 100 °C the meat has about one half its initial weight and the water content has been reduced about 15% (Takashi 1965).

To reduce the negative effects of overheating such as muscle contraction, nutrient and water loss, colour and texture changes in squid, it was cooked using sous-vide cooking technique in different temperature combinations (Cui et al. 2019). Sous-vide cooking means cooking the food in vacuumed plastic bags under controlled temperatures. With the increase of cooking heating temperature, the pH, hardness, elasticity, and chewiness of squid increased first and then decreased. The squid under 60 °C heating condition had better texture characteristics. With 60 °C sous-vide cooking, texture and cooking losses were reduced.

Although little literature knowledge is available on canning of cephalopods, canned squid is produced in Japan, China and India and is known to be a popular product.

Marinating of Cephalopods

The pieces of cephalopod mantles and arms are marinated in salt, acid (vinegar or citrus-fruit juice), or sauces like soy sauce. Other interesting marinades can be made of miso or sake lees (sake kasu) that serve both to tenderize the meat and to impart umami taste. When marinating in acidic media one should know acids make the muscle proteins contract and can lead to a firmer texture in the short run and only tenderizing over longer times. Squid and cuttlefish mantles cut in fine strips along the long direction of the mantle are particularly suited for marinating. These strips, called ika-somen in Japan, look like a kind of fettuccine pasta and are often served on top of a bowl of rice (chirashi-zushi) or in a soup broth (Mouritsen and Styrbæk 2018).

Drying of Cephalopods

Cephalopods contain high amounts of water, resulting in their short shelf life. Drying is an effective method to increase the shelf life of such products. Dried and seasoned products are most popular products in Asian countries, especially in China and Japan. Dried and seasoned squid products are consumed as popular snack foods worldwide because of their delicious taste. The demand for dried squid products is growing steadily (Deng et al. 2014). Dried-seasoned squid is produced with addition of amino acids, saccharides, and organic acids, and smoke flavours in China, Japan, and other Asian countries. Main squid species in dried-seasoned squid production are neon flying squid (*Ommastrephes bartrami*), Argentina squid (*Illex argentines*) and jumbo squid (*Dosidicus gigas*) (Chiou et al. 2010; Zhu et al. 2016). These products are popular in these countries due to their pleasant flavour. Colour and appearance are important features in these products. The mantles are skinned, cooked at 65–80 °C for 3–5 min, cooled, graded, and seasoned at a temperature below 20 °C for more than 4 h with sugar, salt, sorbitol, phosphate, sweetener, and organic acid. They are then dried at 40–45 °C for 12–20 h, until the moisture content drops to 40% and depending on the size of the squid. The semi-finished product is kept in a cold environment for 2 weeks or more. Then it is dried again at 110–120 °C for 3–5 min. It is then mechanically crushed and seasoned a second time and dried to 25–27% moisture. The product colour is slightly yellowish or brownish (Yean et al. 1998).

Drying causes different changes on the composition, physicochemical and structural properties of foods. The conformational stability of muscle proteins changes and muscle proteins degrade. This results in changes in the microstructure, such as pore formation and compactness (Deng et al. 2014). Squid turns into a hard structure when dried. This hard structure makes the consumption of dried squid especially difficult for the elderly and children (Zhao et al. 2017). It is reported that the hardening of dried squid is usually caused by coagulation of sarcoplasmic proteins, breakdown of the sarcoplasm, fibre shrinkage and dehydration (Benjakul et al. 2000). For this reason, some methods are used to soften the dried squid, such as immersion in hot water or alkaline solution. It is applied to soften dried squid before cooking (Zhao et al. 2017). The softening process of dried squid can be carried out by various dehydration methods. Immersion in alkaline solution is considered the most effective and common method to promote softening. Alkaline application destroys various bonds that stabilize the fibre structure, causing structural change and enhancing water absorption. Appropriate alkaline application is required to soften the dried squid structure. Alkaline selection and concentration should be appropriate. Strong alkaline solution damages fibre structure, while weak alkaline solution increases inter-fibre space and water absorption of muscle fibres (Benjakul et al. 2000). In a previous study, softening of dried squid in different conditions was tried to compare raw squid with dried squid. In this study, the necessity of softening in three stages, namely pre-soaking, alkaline-soaking, and post-soaking, has been determined to swell dried squid to its original form. The same researchers have reported that soaking in alkaline solution may cause leakage of low molecular weight compounds; especially free amino acids, trimethylamine oxide, sarcoplasmic proteins, and myofibrillar proteins (Kugino et al. 1993). In another study, Benjakul et al. (2000) examined the effect of alkaline application on the textural properties of squid, and for this purpose they treated the dried squid with different proportions of sodium hydroxide and sodium carbonate. It has been reported that the most effective application in terms of appearance and textural properties is 0.15 mol/kg sodium carbonate with a squid/alkaline solution ratio of 1:10 (w/v) for 20 h. In china dried squid are usually softened in water and then the same water is used for making soup. The remaining softened squid are soaked in alkaline solution and then soaked in water again. The softened squid become swollen and their mantles become thicker than those of fresh squid (Konishi et al. 2003) (Fig. 3.10).

Brown discolouration change caused by sugar-amino reactions called "maillard reaction" during processing and subsequent storage in dried squid products is an important quality problem. It is reported that the sensitivity of squid species to browning varies according to the species and that neon flying squid and Atlantic short finned squid species are more sensitive than others. High content of amino acids, which are the precursors of the maillard reaction, have been shown as the reason why these squid species are more sensitive to browning (Haard and Arcilla 1985). It has been reported that during drying at 35 °C, the maillard reaction is observed related to the decrease in proline and taurine amino acids in the squid (*Illex argentinus*) mantle and the browning is increased (Tsai et al. 1991). O_2, light, heat and water activity are other factors that promote browning. Drying temperature is an important factor and drying in hot air causes more browning than drying in cold air (Fu et al. 2007). The water activity of seasoned squid products is usually between 0.70 and 0.75. In this Aw range, lysine, histidine, and glycine in dried squid constitute 75% of the total free amino acids and show high browning rates compared to other amino acids. Browning accelerates at 23 °C storage temperature and increases with the increase in temperature. Since browning increases during prolonged storage of dried squid products, colour is used as an indicator of storage history. Excessive browning is considered an undesirable situation for consumers (Yean et al. 1998). It has been reported that air drying temperature affects the chemical, physical and nutritional properties of jumbo squid (*Dosidicus gigas*), colour change at higher temperatures is more pronounced, and non-enzymatic browning and protein denaturation are effective in this (Vega-Gálvez et al. 2011).

There are many methods for drying of cephalopods. Traditionally, the most widely used method is sun drying. Since the sun drying method does not require special equipment and energy consumption costs, it is an economical method. However, this method has many disadvantages such as being time consuming, weather dependent, requiring intense labour, loss of nutritional value, degradation of physical characteristics, exposure to environmental contaminants such as insect damage and contamination and bird excrement and feathers (Zhang and Mao 2004; Deng et al. 2011a, b; Chen et al. 2013). Sun drying is currently main drying method used in many developing countries in Asia, Africa, and the Pacific region due to suitability of these regions in terms of plenty of sunshine and dry air.

Drying with hot air is a drying method that uses heated air as a carrier of heat and moisture, where its parameters such as temperature, speed, relative humidity and various sources of pollution can be well controlled, so relatively high quality products are obtained (Zhang and Mao 2004; Chen et al. 2013). Low investment costs and ease of operation is another advantage of this method. However, hot air drying requires high energy consumption and long drying time (Wang et al. 2011).

New techniques have been developed to avoid the disadvantages of traditional drying methods and to improve the quality of the dried product. Heat pump drying technique is one of these new methods, it is a drying technique that can increase the energy efficiency and control the operating parameters independently. Besides, this method has a wide range of drying conditions and controls drying conditions excellently for high value products and reduces energy consumption for low value products (Chou and Chua 2001). Since drying takes place at low temperatures, it is a suitable technology for temperature sensitive products. It has been reported that when combined with far infrared radiation (FIR) drying techniques, it can shorten the drying time and improve the nutritional, sensory and functional properties of the products. It is especially suitable for temperature sensitive products as it allows drying at low and medium temperatures. Squid (*Illex illecebrosus*) fillets dried using a heat pump (HP) dryer combined with far infrared radiation (FIR) at 500, 1000 and 2000 W showed changes in hardness and colour increased with an increase in the power supplied to FIR (Deng et al. 2011a). Squid fillets dried in a heat pump (HP)

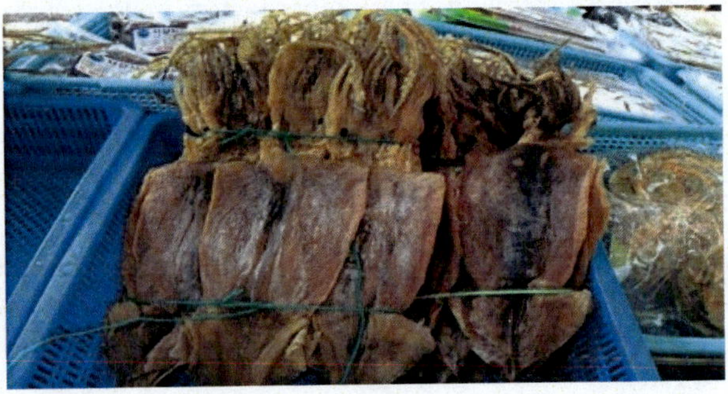

Fig. 3.10 Dried cephalopods

dryer alone or combining with far-infrared radiation (FIR) showed that FIR in combination with HP decreased TMAO'ase, TMA, TVBN, polyunsaturated fatty acids (PUFA), and total aerobic bacterial counts, but increased both saturated (SFA) and Monounsaturated (MUFA) fatty acid content, compared with HP alone (Deng et al. 2011b).

Although the drying process is mostly applied to the squid, the octopus is also dried. Octopuses are usually sun/air or machine dried. Dried octopus is processed in some districts as a traditional food. The head of a small octopus (1–3 kg) is split open and viscera and eyeballs removed. After washing, the base of the arm is cut open. A U-shaped piece of split bamboo is inserted into the trunk to expand it into a spherical shape. Thus shaped, it is sun-dried for several days in an airy place. Dried octopus is eaten with soy sauce after being torn to bits and roasted or immersed in boiling water for a short time (Josupeit 2008). In Mediterranean countries such as Greece and Spain, air drying is traditionally applied in the processing of the octopus and then grilled. Thus, while the dried octopus loses half its weight, it gains a crispy and chewy texture. In addition to the Mediterranean countries, this drying and grilling technique is used for both octopus and squid in Japan, and the traditional Japanese breakfast consists of these products (Mouritsen and Styrbæk 2018).

3.2.3.4 Tenderizing of Cephalopods to Improve Texture

Cephalopods have a high rate of connective tissue in the muscular system. Cephalopods are known to be difficult to prepare because they are famous for being chewy and rubberlike. This feature of the cephalopod muscle discourages for consumers (Hurtado et al. 2001; Mouritsen and Styrbæk 2018). Toughness of cephalopod muscle is related to collagen structure, content, and aggregation of collagen in the cephalopod muscles (Morales et al. 2000). There are various applications to tender the cephalopods and make them consumable. Tenderization has been achieved through applications such as freezing, heating, enzyme application, marination with organic acids addition, sodium chloride addition, phosphate salts addition mechanical massaging (tumbling) and pressurizing (Sikorski and Kolodziejska 1986; Kugino and Kugino 1995; Collignan and Montet 1998; Katsanidis 2004; Pietrasik et al. 2010; Ketnawa and Rawdkuen 2011; Ha et al. 2012; Gokoglu et al. 2017a, b, 2018; Mouritsen and Styrbæk 2018).

Cephalopods are generally recommended to be frozen before all procedures as freezing will eliminate possible parasites. It is also reported that the freezing process to be tenderize the meat. In the freezing process, which is defined as a physical process involving the conversion of water molecules from an amorphous state to highly structured ice crystals, the phase change of water can cause protein denaturation. Ice-crystal formation will break-up the muscle fibres (Xiong 1997; Mouritsen and Styrbæk 2018). During freezing, the muscle fibres undergo cellular degradation and the meat is tenderized in relation to the rupture and stretching of the muscle fibres. The amount of ice crystals formed between the cells is affected by factors such as freezing rate, freezing time and temperature (Shanks et al. 2002).

Tenderization has been achieved for some cephalopod species (*Loligo vulgaris*, *Octopus vulgaris*, *Sepia officinalis*) frozen at −45 °C and stored at −18 °C for 30 days (Gokoglu et al. 2018). The decreases in hardness and chewiness values measured by the tissue analyser device were determined after freezing, and the sensory texture scores supported these results. On the other hand, there are researchers who report otherwise. These researchers reported that the water holding capacity decreased with the freezing and as a result, a harder and drier structure appeared (Stanley and Hultin 1982; Mackie 1993; Ueng and Chow 1998). It was thought that this hardening after freezing may have caused partial dehydration, inorganic salts, and the oxidation of lipids and the production of formaldehyde.

Heating is another method used to tender cephalopods. Heat treatment is applied by various cooking methods such as boiling, frying, baking, sous-vide cooking. It is reported that cephalopod muscles soften after cooking (Otwell and Haman 1979; Stanley and Hultin 1982; Kolodziejska et al. 1987; Kugino and Kugino 1995; Ando 1996, 1997). Hardness connective tissue protein is associated with collagen. During the cooking process, the heat turns collagen into gelatin, which provides the muscle to soften. It is shown as the main cause of softening as the destruction of muscle cells during heat treatment (Kugino and Kugino 1995). In a study, a significant softening was found in the mantle muscle after 30% gelatinization of collagen by cooking in five squid species (Ando 1996).

As in fish and land animals, the meats of the cephalopods can also be tenderized with certain enzymes. Enzymes such as papain, bromelain and ficin from plants have been used as meat tenderizers (Ha et al. 2012; Ketnawa and Rawdkuen 2011; Koak et al. 2011; Navaeena et al. 2004). Plant proteases are superior to bacterial enzymes because there are safety problems, such as pathogenicity, to bacterial enzymes. These enzymes break down muscle proteins when mixed with meat. These enzymes also hydrolyze collagen and elastin proteins. Decreased hardness and shear force values of some cephalopod species (*L. vulgaris, O. vulgarsi, S. officinalis*) treated with papain and bromelain were detected. It has been determined that enzyme applications cause a decrease in gumminess and chewiness values besides hardness (Gokoglu et al. 2017b). Seven different enzymes (Alcalase, Neutrase, Flavourzyme, Protamex, Collupulin, Alphalase, and Bromelain) have been injected to squid (*Dosidicus gigas*) body. Six other enzymes, except flavourzyme, were found to be effective in softening jumbo squid. In addition to the enzyme type, it has been determined that enzyme concentration is also effective, and high concentrations cause tissue destruction and result in the loss of the shape of the squid. Therefore, enzyme concentrations should be optimized (Eom et al. 2015).

Cephalopod muscles can be softened by mechanical means. The traditional way to overcome the toughness of octopus has been the repeated "beating" of the freshly caught octopus on the rocks of the sea. This procedure has been adopted by the industry, and the mechanical beating of the octopus has been carried out in special tumbling equipment where the octopus is tumbled in water (Gokoglu et al. 2017a). The tumbling process is reported to have a positive effect on softening (Krause et al. 1978a, b; Lachowicz et al. 2003). Another positive effect of tumbling is that it increases cooking efficiency by keeping the added water in the muscle (Pietrasik and Shand 2004, 2005). Tumbling is a process applied with rollers called drums (tumblers), which gives kinetic energy to the meat with free fall and impact, affects various features of meat such as juiciness, brittleness, and colour. In the tumbling system, there is an impact force caused by the fall of the meat pieces coming to the top of the rotating drum with the effect of gravity. The machine used is tumbler, like concrete mixers and there are metal sills inside the drum. The drum can be adjusted from the adjacent handle and operated at the desired angle (Oztan 2003). In the tumbling process, kinetic energy is given to the muscle tissues and with this energy; muscle cells break down, muscle structure changes, brittleness increases, cell membrane permeability increases, muscle proteins swell and strengthen (Dolata et al. 2004). The main purpose of the physical strains applied is the solubility and extraction of some salt-soluble proteins such as actin, myosin, and tropomyosin. The more important advantage of the tumbling process for the producers is that the cooking efficiency increases and the final product shrinks. Pietrasik and Shand (2004), in their study on cooked beef roast, revealed that tumbling affects hydration properties and thermal stability, resulting in low cooking loss and high water-retention capacity. In another study, samples of squid, octopus and cuttlefish have been tumbled in three different times. It has been found a decrease in the water holding capacity of the samples and an increase in cooking losses. Total protein and total free amino acid contents have increased in squid and octopus and had no effect in cuttlefish. While the tumbling process improved the texture of squid and octopus, it was not seen in the cuttlefish. It has been reported that the tumbling process is generally suitable for squid and octopus and especially 6 h of application is more effective (Gokoglu et al. 2017a).

It is known that the acid marination process improves the texture of the meat. Organic acids are widely used in marinating of meats. Common marinating ingredients are vinegar, lemon, grapefruit, and orange juices. Various sauces and wines are also used for marination. The two most used acids in softening are citric and lactic acid. It is reported that softening is related to muscle pH (Ke et al. 2009). When the pH decreases and the muscle pH reach the isoelectric point, the plus and minus loads will be equalized. When pH is more acidic, balance will be disturbed as a shot at positive loads.

References

Aaraas R, Hernar IJ, Vorre A, Bergslien H, Lunestad BT, Skeie S, Slinde E, Mortensen S (2006) Sensory, histological, and bacteriological changes in Flat oysters, *Ostrea edulis* L., during different storage conditions. J Food Sci 69:205–210

Abad M, Ruiz C, Martinez D, Mosquera G, Sanchez JL (1995) Seasonal variations of lipid classes and fatty acids in flat oyster, *Ostrea edulis*, from San Cibran. Comp Biochem Physiol 110(2):109–118

Ablett RF, Gould SP, Sherwood DA (1986) Frozen storage performance of cooked cultivated mussels (*Mytilus edulis* L.)-Influence of ascorbic acid and chelating agents. J Food Sci 51:1118–1121

Akinjogunla VF, Lawal-Are AO, Soyinka OO (2017) Proximate compositiona and mineral content of mangrove oyster (*Crassostrea gasar*) from Lagos lagoon, Lagos, Nigeria. Niger J Fish Aquacult 5(2):36–49

Albanese D, Cınquanta L, Lanorte MT, Matteo MD (2005) Squid (*Sepia officinalis*) stored in active packaging: some chemical and microbiological changes. Ital J Food Sci 3(17):325–332

Almonacid S, Bustamante J, Simpson R, Urtubia A, Pinto M, Teixeira A (2012) Commercially sterilized mussel meats (*Mytilus chilensis*): a study on process yield. J Food Sci 77:127–135

Almonacid S, Bustamante J, Simpson R, Pinto M (2015) Shellfish (mussel) processing and components. In: Preedy V (ed) Processing and impact on active components in food. Academic, pp 447–454

Alpbaz A (1993) Crustaceans and arthropodsculture. Ege University Fisheries Faculty Publications No:26, İzmir, pp 26–82

Ando M (1996) Textural change of frozen-thawed several squid mantles by heating. Memories Narabunka Women's Junior College 27:121–127

Ando M (1997) Correrspondence of collagen on heat induced softening of frozen-thawed cuttlefish Sepia officinalis mantle. J Home Econ Jpn 48:315–321

Andrews LS, Park DL, Chen YP (2000) Low temperature pasteurization to reduce the risk of vibrio infections from raw shell-stock oysters. Food Addit Contam 17(9):787–791

Andrews LS, DeBlanc S, Veal CD, Park DL (2003) Response of *Vibrio parahaemolyticus* 03: K6 to a hot water/cold shock pasteurization process. Food Addit Contam 20:331–334

Anibal J, Esteves E, Rocha C (2011) Seasonal variations in gross biochemical composition, percent edibility, and condition index of the clam *Ruditapes decus*satus cultivated in the Ria Formosa (South Portugal). J Shellfish Res 30:17–23

Anonymous (2011) MEGEP (Strengthening the Vocational Education and Training System Project. Maritime.Mollusc and crustacea.624B00002, Ankara (In Turkish)

Arason S, Nguyen MV, Thorarinsdottir KA, Thorkelsson G (2014) Preservation of fish by curing. In: Boziaris IS (ed) Seafood processing technology, quality and safety. Wiley Blackwell, Chichester, pp 129–160

Arcales JAA, Nacional LM (2018) Effect of lactic and citric acid pretreatment on quality changes of green mussel (*Perna Viridis*) during chilled storage. Curr Res Nutr Food Sci J 6(3):862–872

Arellano-Martínez M, Racotta IS, Ceballos-Vázquez BP, Elorduy-Garay JF (2004) Biochemical composition, reproductive activity, and food availability of the lion's paw scallop Nodipecten subnodosus in the laguna Ojo de Liebre. Baja California Sur, Mexico. J Shellfish Res 23:15–23

Arias-Pérez A, Insua A, Frere R, Méndez J, Fernández-Tajes J (2013) Genetic studies on commercially important species of veneridae. In: da Costa F (ed) Clam fisheries and aquaculture. Nova Science Publishers, Inc., pp 73–105

Arik-Colakoglu F, Ormanci HB, Berik N, Kunili IE, Colakoglu S (2011) Proximate and elemental composition of *Chamelea gallina* from the Southern Coast of the Marmara Sea (Turkey). Biol Trace Elem Res 143:983–991

Arik-Colakoglu F, Çardak M, Colakoglu S, Kunili IE (2014) Depuration times of Donax trunculus and *Tapes decussatus*. Braz J Microbiol 45(3):1017–1022

Arkhipkin AI, Rodhouse PGK, Pierce GJ, Sauer W, Sakai M, Allcock L, Arguelles J, Bower JR, Castillo G, Ceriola L, Chen CS, Chen X, Diaz-Santana M, Downey N, González AF, Amores JG, Green CP, Guerra A, Hendrickson LC, Ibáñez C, Ito K, Jereb P, Kato Y, Katugin ON, Kawano M, Kidokoro H, Kulik VV, Laptikhovsky VV, Lipinski MR, Liu B, Mariátegui L, Marin W, Medina A, Miki K, Miyahara K, Moltschaniwskyj N, Moustahfid H, Nabhitabhata J, Nanjo N, Nigmatullin CM, Ohtani T, Pecl G, Perez JAA, Piatkowski U, Saikliang P, Salinas-Zavala CA, Steer M, Tian Y, Ueta Y, Vijai D, Wakabayashi T, Yamaguchi T, Yamashiro C, Yamashita N, Zeidberg LD (2015) World Squid Fisheries. Rev Fish Sci Aquacult 23(2):92–252

Asha KK, Anandan R, Mathew S, Lakshmanan PT (2014) Biochemical profile of oyster *Crassostrea madrasensis* and its nutritional attributes. Egypt J Aquat Res 40:35–41

Atay D (1997) Shellfish and production technique. University of Ankara Publications. No: 1478. Text book: 441, Ankara

Atayeter S, Ercoskun H (2011) Chemical composition of European squid and effects of different frozen storage temperatures on oxidative stability and fatty acid composition. J Food Sci Technol 48(1):83–89

Aveiro M, Pellizzaro QC, Amboni RDMC, Batista CRV, Beirao HL, Barreto PLM (2007) Chemical, microbiological and sensory changes of marinade mussel (*Perna perna* linné 1758) storage at 4°C. Alimentos e Nutrição Araraquara 18:121–126

Ayas D (2012) Seasonal variations of fat and fatty acid composition in muscle tissues of Mediterranean octopuses. Iran J Fish Sci 11(4):724–731

Azanza MRV, Ventura SR (2005) Heat shocking of Philippine green mussels, Perna viridis. Int J Food Sci Technol 40:689–694

Azpeitia K, Ríos Y, Garcia I, Pagaldai J, Mendiola D (2017) A sensory and nutritional validation of open ocean mussels (*Mytilus galloprovincialis* Lmk.) cultured in SE Bay of Biscay (Basque Country) compared to their commercial counterparts from Galician Rías (Spain). Int Aquat Res 9:89–106

Baker SM, Hornbach DJ (2001) Seasonal metabolism and biochemical composition of two unionid mussels *Actinonaias ligementina* and *Amblema plicata*. J Molluscan Stud 67:407–416

Balasundari S, Abraham J, Shanmugam SA, Jasmin IG (1997) Frozen storage performance of edible oyster *Crassostrea madrasensis* (Preston). J Food Sci Technol 34(5):434–436

Balguerias E, Quintero ME, Hernandez-Gonzalez C (2000) The origin of the Saharan Bank cephalopod fishery. ICES J Mar Sci 57:15–23

Ballantyne JS, Hochachka PW, Mommsen TP (1981) Studies on the metabolism of the migratory squid, *Loligo opalescens*: enzymes of tissues and heart mitochondria. Mar Biol Lett 2:75–85

Banks A, Dassow JA, Feiger EA, Novak AF, Peters JA, Lavin JWS, Waterman J (1977) Freezing of shellfish. In: Desrosier NW, Tressler DK (eds) Fundamentals of food freezing. Avi Publishing Company Inc., Westport, pp 318–356

Bano A, Sadioa S, Askari B, Quadri RB (1992) Amino acids content of sea squids. J Chem Soc Pak 14:184–184

Banon R, Otero J, Campelos-Álvarez JM, Garazod A, Alonso-Fernández A (2017) The traditional small-scale octopus trap fishery off the Galician coast (Northeastern Atlantic): historical notes and current fishery dynamics

Barbosa A, Vaz-Pires P (2004) Quality index method (QIM) development of a sensorial scheme for common octopus (*Octopus vulgaris*). Food Control 15:161–168

Barile NB, Nerone E, Scopa M, Recchi S (2018) Depuration of striped venus clams following artificial vibriones contamination. Ital J Food Sci 30:602–613

Barrento S, Powell A (2016) The effect of transportation and re-watering strategies on the survival, physiology and batch weight of the blue mussel, *Mytilus edulis*. Aquaculture 450:194–198

Barrento S, Lupatsch I, Keay I, Shields S (2013) Protocol on best practice handling and transportation of live mussels. Technical Report supported by Project No. 243452, FP7-SME MusselsAlive, p 59

Basu S, Gupta SS (1984) Studies on the ice storage characteristics of blood clam *Anadara granosa* meat. Fish Technol 21(1):6–10

Bindu J, Mallick AK, Gopal TKS (2014) Thermal processing of fishery products in flexible and rigid containers. Fish Technol 51:137–148

Beltran-Lugo AI, Maeda-Martinez AN, Pacheco-Aguilar R, Nolasco-Soria HG (2006) Seasonal variations in chemical, physical, textural, and microstructural properties of adductor muscles of Pacific lions-paw scallop (*Nodipecten subn*odosus). Aquaculture 258:619–632

Beninger PG, Pennec ML (2006) Structure and function in scallops. In: Shumway SE, Parsons GJ (eds) Scallops: biology, ecology and aquaculture. Developments in aquaculture and fisheries science, vol 35, 2nd edn. Elsevier, San Diego, pp 123–211

Benjakul S, Visessanguan W, Tanaka M, Ishizaki S, Taluengphol A, Chichanan U (2000) Physicochemical and textural properties of dried squid as affected by alkaline treatments. J Sci Food Agric 80:2142–2148

Benjakul S, Sungsri-in R, Kijroongrojana K (2012) Effect of treating of squid with sodium chloride in combination with oxidising agent on bleaching, physical and chemical changes during frozen storage. Food Bioprocess Technol 5:2077–2084

Berik N, Çankiriligil C (2013) Determination of proximate composition and sensory attributes of scallop (*Flexopecten glaber*) gonads. Mar Sci Technol Bull 2(2):5–8

Berik N, Çankiriligil C, Gül G (2017) Mineral content of smooth scallop (*Flexopecten glaber*) caught Canakkale, Turkey and evaluation in terms of food safety. J Trace Elem Med Biol 42:97–102

Berlin DL, Herson DS, Hicks DT, Hoover DG (1999) Response of pathogenic Vibrio species to high hydrostatic pressure. Appl Environ Microbiol 65:2776–2780

Berthelin C, Kellner K, Mathieu M (2000) Storage metabolism in the Pacific oyster (*Crassostrea gigas*) in relation to summer mortalities and reproductive cycle (West Coast of France). Comp Biochem Physiol 125B:359–369

Biandolino F, Portacci G, Prato E (2010) Influence of natural diet on growth and biochemical composition of *Octopus vulgaris* Cuvier, 1797. Aquac Int 18:1163–1175

Biji KB, Shamseer RM, Mohan CO, Ravishankar CN, Mathew S, Gopal TKS (2015) Effect of thermal processing on the biochemical constituents of green mussel (*Perna viridis*) in Tin-free steel cans. J Food Sci Technol 52(10):6804–6809

Bindu J, Gopal TKS, Unnikrishnan Nair TSU (2004) Ready to-eat mussel meat processed in retort pouches for the retail and export market. Packag Technol Sci 17(3):113–117

Bindu J, Ginson J, Kamalakanth CK, Gopal TKS (2015) High pressure treatment of green mussel *Perna viridis* Linnaeus, 1758: effect on shucking and quality changes in meat during chill storage. Indian J Fish 62(2):70–76

Bloor ISM, Attrill MJ, Jackson EL (2013) A review of the factors influencing spawning, early life stage survival and recruitment variability in the common cuttlefish (*Sepia officinalis*). In: Lesser M (ed) Advances in marine biology. Elsevier, Amsterdam, pp 2–54

Boletzky SV, Villanueva R (2014) Cephalopod biology. In: Iglesias J, Fuentes L, Villanueva L (eds) Cephalopod culture. Springer Science + Business Media, Dordrecht, pp 3–16

Bongiorno T, Tulli F, Comi G, Sensidoni A, Andyanto D (2018) Sous vide cook-chill mussel (*Mytilus galloprovincialis*): evaluation of chemical, microbiological and sensory quality during chilled storage (3 °C). LWT-Food Sci Technol 91:117–124

Bouletis DA, Arvanitoyanis SI, Hadjichristodoulou C, Neofitou C, Sakkomitrou M, Kolokythopoulou F (2014) The effect of modified atmosphere packaging on the microbiological, physical, chemical and sensory characteristics of broadtail squid (Illex coindetii). Int J Food Sci Technol 49:329–336

Bouletis AD, Arvanitoyannis IS, Hadjichristodoulou C, Neofitou C, Parlapania FF, Gkagtzis DC (2015) Quality changes of cuttlefish stored under various atmospheremodifications and vacuum packaging. J Sci Food Agric 96:2882–2888

Boumpalos AP, Lougovois VP (2005) Shelf life assessment of whole and eviscerated cuttlefish (Sepia officinalis) stored in melting ice. In: Lazos ES (ed) Assessing useful technologies—optimizing food safety and nutrition. Proceedings of the 4th International Congress On Food Technology, Athens, 18–19 Feb 2005, pp 168–174, vol II. Hellenic Association of Food Technologists, Athens

Brand AR (2006) Scallop ecology: distributions and behaviour. In: Shumway SE, Parsons GJ (eds) Scallops: biology, ecology and aquaculture. Developments in aquaculture and fisheries science, vol 35, 2nd edn. Elsevier, San Diego, pp 651–713

Bremner HA, Statham JA (1983) Effect of potassium sorbate on refrigerated storage of vacuum packed scallops. J Food Sci 48(4):1042–1047

Brenton C, Flick GJ, Microbiological Quality and Safety of Quahog Clams, Pierson MD, Croonenberghs RE, Peirson M (2001) *Mercenaria mercenaria*, during Refrigeration and at Elevated Storage Temperatures. J Food Prot 64(3):343–347

Brita Nicy A, Velayutham P, Sukumar D, Shanmugam SA, Kanaga V (2016) Comparative study on proximate composition of cuttlefish species from Thoothukudi coast. Int J Fish Aquat Stud 4(5):96–97

Britton JC, MORTON B (1979) Corbicula in North America: The evidence reviewed and evaluated, In: J. C. Britton (ed.). Proc. First International Corbicula Sym posium. Texas Christian Univ. Research Foundation, Fort Worth, p. 247–287

Brown CT (1967) Apparatus for removing scallops from their shells. United States Patent No. 3,320,631

Cabi (2019a) Mytilus galloprovincialis (Mediterranean mussel). Invasive Species Compendium. https://www.cabi.org/isc/datasheet/73756. Accessed 4 Oct 2019

Cabi (2019b) *Corbicula fluminea* (Asian clam). Invasive Species Compendium. https://www.cabi.org/isc/datasheet/88200. Accessed 4 Nov 2019

Cabi (2019c) *Crassostrea gigas* (Pacific oyster). Invasive Species Compendium. https://www.cabi.org/isc/datasheet/87296. Accessed 12 Nov 2019

Cabi (2019d) *Crasostrea virginica* (Eastern oyster). https://www.cabi.org/isc/datasheet/87298. Accessed 12 Nov 2019

Cabi (2019e) *Patinopecten yessoensis* (Japanese scallop). https://www.cabi.org/isc/datasheet/70927. Accessed 26 Nov 2019

Caers M, Coutteau KC, Cure K, Morales V, Gajardo G, Sorgeloos P (1999) The Chilien Scallop *Argopecten Pupuratus* (Lamarck, 1819): I. Fatty acid composition and lipid content of six organs. Comp Biochem Physiol 123(B):89–96

Caglak E, Cakli S, Kilinc B (2008) Microbiological, chemical and sensory assessment of mussels (*Mytilus galloprovincialis*) stored under modified atmosphere packaging. Eur Food Res Technol 226:1293–1299

Caglak E, Cakli S, Kilinc B (2014) Comparison chemical, sensory, microbiological and textural changes of cuttlefish (Sepia officinalis) stored under different packaging. Bulg J Agric Sci 2:1046–1053

Calik H, Morrissey MT, Reno PW, An H (2002) Effect of high-pressure processing on *Vibrio parahaemolyticus* strains in pure culture and Pacific oysters. J Food Sci 67:1507–1510

Cao R, Xue C-H, Liu Q (2009a) Changes in microbial flora of Pacific oysters (*Crassostrea gigas*) during refrigerated storage and its shelf-life extension by chitosan. Int J Food Microbiol 131:272–276

Cao R, Xue C-H, Liu Q, Xue Y (2009b) Microbiological, chemical, and sensory assessment of Pacific Oysters (*Crassostrea gigas*) stored at different temperatures. Czech J Food Sci 27:102–108

Carboni S, Kaur G, Pryce A, McKee K, Desbois AP, Dick JR, Galloway SDR, Hamilton DL (2019) Mussel consumption as a "Food First" approach to improve omega-3 status. Nutrients 11:1381

Castagna M (1990) Shellfish-mollusks. In: Martin RE, Flick GJ (eds) The seafood industry. Springer Science + Business Media, New York, pp 77–87

Chae M (2007) Low temperature post harvest processing at reducing *Vibrio parahaemolyticus* and Vibrio vulnificus in raw oysters. Master's Thesis, Oregon State University

Chai T, Liang KT, Pace J, Schlimme DV (1991) Effect of heat processing on quality of pasteurized oyster. J Food Sci 56:1292–1294

Chakraborty K, Chakkalakal SJ, Joseph D, Joy M (2016a) Nutritional composition of edible oysters (*Crassostrea madrasensis* L.) from the Southwest Coast of India. J Aquat Food Prod Technol 25(8):1172–1189

Chakraborty K, Joy M, Vijayagopal P (2016b) Nutritional qualities of common edible cephalopods at the Arabian Sea. Int Food Res J 23(5):1926–1938

Chandrapavan A, Kangas MI, Sporer EC (2012) Seasonal, spatial, and postharvest variability in the survival of repeatedly discarded saucer scallops in Shark Bay, Western *Australia*. J Shellfish Res 31:1161–1171

Changhu X, Yisheng W, Jamil K, Jie LZ, Hong L (1998) Changes in lipids of musssel during storage. Pak J Mar Sci 7(1):27–34

Chen L (2011) Sensory and chemical characteristics of eastern oysters (*Crassostrea virginica*). Master thesis submitted to the Faculty of the Virginia Polytechnic Institute and State University, Blacksburg, 106 p

Chen H, Zhang M, Fang Z, Wang Y (2013) Effects of Different Drying Methods On The Quality Of Squid Cubes. Dry Technol 31:911–1918

Cherifi H, Sadok S (2016) Effects of marinating process on mussels physicochemical and mıcrobiologıcal quality attributes during refrigerated storage. In: Proceedings of the closing seminar: "quality and innovation for the promotion of aquaculture products: the Tunisian-Sicilian cross-border contribution", Hammamet, Tunisia on 18–20 Apr 2016

Chinnamma G (1973) Technological aspects of preservation and processing of edible shell fishes V. Cold storage changes in mussels (*Mytilus edulis*) and clams (*Villortia* [i.e. Villorita] sp.). Fish Technol 11(2):22–27

Chinnamma PL, Chaudhuri DR, Pillai VK (1970) Technological aspects of processing of edible mussels, clams and crabs. Pt. 1. Spoilage during ice storage. Fish Technol 7(2):137–142

Chiou TK, Chang HK, Lo L, Lan HL, Shiau CY (2010) Changes in chemical constituents and physical indices during processing of dried-seasoned squid. Fish Sci 66:708–715

Chou SK, Chua KJ (2001) New hybrid drying technologies for heat sensitive foodstuffs. Trends Food Sci Technol 12:359–369

Civera T, Grassi MA, Pattono D (1999) Chemical and microbiological characteristics of cephalopods during storage. Ind Aliment 38:933–937

Coan EV, Valentich-Scot P (2006) Marine bivalves. In: Sturm CF, Pearce TA, Valdés A (eds) The mollusks: a guide to their study, collection, and preservation. American Malacological Society/ Universal Publisher, Pittsburgh, pp 339–347

Colakoglu S, Colakoğlu FA (2013) Economic status of bivalve molluscs. J Food 73–78 (in Turkish)

Colakoğlu S, Tokac A (2017) The fishing gear used in commercial bivalve fishery in Turkey. Ege J Fish Aquat Sci 34(4):415–421

Collignan A, Montet D (1998) Tenderizing squid mantle by marination at different pH and temperature Levels. LWT-Food Sci Technol 31:673–679

Cook DW, Ruple AD (1992) Cold storage and mild heat treatment as processing aids to reduce the numbers of *Vibrio vulnifcus* in raw oysters. J Food Prot 55(12):985–989

Correa AA, Rigotto C, Moresco V, Kleemann CR, Teixeira AL, Poli CR, Simoes CMO, Barardi CRM (2012) The depuration dynamics of oysters (*Crassostrea gigas*) artificially contaminated with hepatitis A virus and human adenovirus. Mem Inst Oswaldo Cruz 107(1):11–17

Cragg SM (2006) Development, physiology, behaviour and ecology of scallop larvae. In: Shumway SE, Parsons GJ (eds) Scallops: biology, ecology and aquaculture. Developments in aquaculture and fisheries science, vol 35, 2nd edn. Elsevier, San Diego, pp 45–105

Croci L, Suffredini E, Cozzi L, Toti L (2002) Effects of depuration of molluscs experimentally contaminated with Escherichia coli, *Vibrio cholera* O1 and *Vibrio parahaemolyticus*. J Appl Microbiol 92:460–465

Cruz-Romero M, Kelly AL, Kerry JP (2007) Effects of high-pressure and heat treatments on physical and biochemical characteristics of oysters (*Crassostrea gigas*). Innov Food Sci Emerg Technol 8:30–38

Cruz-Romero M, Kerry JP, Kelly AL (2008) Changes in the microbiological and physicochemical quality of high-pressure treated oysters (*Crassostrea gigas*) during chilled storage. Food Control 19:1139–1147

Cui Z, Dubova H, Mo H (2019) Effects of sous vide cooking on physicochemical properties of squid. J Hyg Eng Des 29:35–40

Dare PJ, Edwards DB (1975) Seasonal changes in flesh weight and biochemical composition of mussels (*Mytilus edulis* L.) in the Conway Estuary, North Wales. J Exp Mar Biol Ecol 18:89–97

De Moreno JE, Pollero RJ, Moreno VJ, Brenner RR (1980) Lipids and fatty acids of the mussel (*Mytilus platenses*, d'Orbigny) from South Atlantic waters. Elsevier/North-Holland Biomedical Press. Contribution INIDEP.377, pp 263–277

De Vidode Mattio N, Paredi ME, Crupkin M (2001) Post mortem changes in the adductor muscle of scallop (*Chlamys tehuelchus*) in chilled and frozen storage. J Aquat Food Prod Technol 10(3):49–60

Dean LM (2000) Cookery. In: Martin RE, Carter EP, Flicj GJ Jr, Davis LM (eds) Marine and freshwater products handbook. Technomic Publishing Co., Inc., Lanchaster, pp 795–796

Deng Y, Qian B, Wu J, Su S, Feng X (2011a) Characteristics of squid (Illex illecebrosus LeSueur) fillets dried using a combination of heat pump drying and far infrared radiation. Phillippine Agric Sci 94(3):270–277

Deng Y, Liu Y, Qian B, Su S, Wu J, Song X, Yang H (2011b) Impact of far-infrared radiation-assisted heat pump drying on chemical compositions and physical properties of squid (Illex illecebrosus) fillets. Eur Food Res Technol 232:761–768

Deng Y, Luo Y, Wanga Y, Yue J, Liu Z, Zhong Y, Zhao Y, Yang H (2014) Drying-induced protein and microstructure damages of squid fillets affected moisture distribution and rehydration ability during rehydration. J Food Eng 123:23–31

Dernekbası S, Öksüz A, Çelik YM, Karayücel I, Karayücel S (2015) The fatty acid composition of cultured mussels (*Mytilus galloprovincialis* Lamarck 1819) in offshore longline system in the Black Sea. J Aquacult Mar Biol 2(6):46–53

Dhananjaya S, Venkatappa (2006) Studies on the ice storage characteristics of octopus, *Octopus embranaceus* (Quoy and Gaimard). Fish Technol 43:154–161

Dimitriadou D, Zotos A, Petridis D, Taylor AKD (2008) Improvement in the production of smoked trout fillets (*Salmo Gairdnerii*) steamed with liquid smoke. Food Sci Technol Int 14(1):67–77

Doiron FA (1949) Scallop Shucking Machine. United States Patent No. 2,476,962

Dolata W, Piotrowska E, Wajdzik J, Tritt-Goc J (2004) The use of the MRI technique in the evaluation of water distribution in tumbled porcine muscle. Meat Sci 67:25–31

Dong FM (2001) The nutritional value of shellfish. School of Aquatic and Fishery Sciences, University of Washington Seattle, Washington, DC, p 8

Dore WJ, Lees DN (1995) Behaviour of *Escherichia coli* and male specific bacteriophage in environmentally contaminated bivalve molluscs before and after depuration. Appl Environ Microbiol 61(8):2830–2834

Dridi S, Romdhane MS, Elcafsi M (2007) Seasonal variation in weight and biochemical composition of the Pacific oyster, *Crassostrea gigas* in relation to the gametogenic cycle and environmental conditions of the Bizert lagoon, Tunisia. Aquaculture 263:238–248

Dunning MC, Norman MD, Reid AL (1998) Cephalopods. In: Carpenter KE, Niem VH (eds) The living marine resources of the Western Central Pacific, vol 2. Cephalopods, crustaceans, holothurians and sharks. FAO species identification guide for fishery purposes. FAO, Rome, pp 687–800

Ekanem EO, Achinewhu SC (2000) Effects of shucking method on opening, meat yield and selected quality parameters of west African clam, *galatea paradoxa* (born). J Food Process Preserv 24:365–377

Ekanem EO, Achinewhu SC (2006) Mortality and quality indices of live west african hard-shell clams (*Galatea paradoxa* born) during wet and dry postharvest storage. J Food Process Preserv 30:247–257

Ekin I, Bashan M, Sesen R (2011) A comparison of the fatty acid composition of the phospholipid and neutral lipid of *Unio elongatulus* (Bourguignat, 1860) (Bivalvia: Unionidae) mussels from 4 diff erent localities in southeastern Anatolia, Turkey. Turk J Zool 35(6):837–849

Eldredge LG, Smith CM (2001) A guidebook of introduced marine species in Hawaii. Bihosp Museum Technical Report 21

El-Shenawy NS (2004) Heavy-metal and microbial depuration of the clam *Ruditapes decussatus* and its effect on bivalve behavior and physiology. Environ Toxicol 19:143–153

Eom VS, Lee H, Chun YG, Park CU, Park DJ (2015) Softening of jumbo squid dosidicus gigas via enzyme injection. Fish Aquat Sci 18(2):229–233

Erkan N (2005) Changes in quality characteristics during cold storage of shucked mussels (*Mytilus galloprovincialis*) and selected chemical decomposition indicators. J Sci Food Agric 85:2626–2630

Ersoy B, Sereflisan H (2010) The proximate composition and fatty acid profiles of edible parts of two freshwater mussels. Turk J Fish Aquat Sci 10(1):71–74

Espinoza LA (2013) Development of safe and ready to eat frozen oyster products using microwave steam-venting technology. A Doctoral Dissertations. Submitted to the Graduate Faculty of the Louisiana State University and Agricultural and Mechanical College1339

Estefanell J, Socorro J, Tuya F, Izquierdo M, Roo J (2011) Growth, protein retention and biochemical composition in *Octopus vulgaris* fed on different diets based on crustaceans and aquaculture by-products. Aquaculture 322–323:91–98

EUMOFA (European Market Observatory for Fisheries and Aquaculture) (2018a) Species profile: oysters. https://www.eumofa.eu/documents/20178/137160/Oyster_31-1.pdf. Accessed 12 Nov 2019

EUMOFA (European Market Observatory Fisheries and Aquaculture Products) (2018b) Monthly highlights. No: 8

EUMOFA (2019) Case study: fresh mussel in the EU- Price structure in the supply chain. Focus on Denmark, Germany and Italy. http://www.eumofa.eu/documents/20178/151118/PTAT+Fresh+Mussel_EN.pdf. Accessed 4 Oct 2019

FAO (2008) Bivalve depuration: fundamental and practical aspects. FAO Fisheries Technical Paper 511, p 139

FAO (2009) *Ostrea edulis* (Linnaeus, 1758). Cultured Aquatic Species Information Programme Food and Agriculture Organisation of the United Nations: Fisheries and Aquaculture Department. http://www.fao.org/tempref/FI/DOCUMENT/aquaculture/CulturedSpecies/file/en/en_europeanflatoyster.htm. Accessed 12 Nov 2019

FAO (2016) Yearbook of fishery statistics. Summary tables Fisheries and Aquaculture Department, Rome, 200 p. http://www.fao.org/3/i9942t/I9942T.pdf

FAO (2018) The State of World Fisheries and Aquaculture 2018—Meeting the sustainable development goals. Rome

FAO (2019a) Cultured aquatic species information programme. Food and Agriculture Organization of the United Nations. Fisheries and Aquaculture Department, Rome. http://www.fao.org/fishery/culturedspecies/mytilusedulis/en. Accessed 1 Oct 2019

FAO (2019b) Fisheries and Aquaculture Department. Species Fact Sheets. Pecten maximus. http://www.fao.org/fishery/species/3516/en. Accessed Nov 2019

FAO (2019c) Species fact sheets. *Dosidicus gigas* (Orbigny, 1935). Food and Agriculture Organization of the United Nations. Fisheries and Aquaculture Department, Rome. http://www.fao.org/fishery/species/2721/en. Accessed 9 Jan 2020

FAO (2019d) Species fact sheets. *Sepia officinalis*. Food and Agriculture Organization of the United Nations. Fisheries and Aquaculture Department, Rome. http://www.fao.org/fishery/species/2711/en

FAO/WHO (2000) Joint FAO/WHO food standards programme codex committee on fish and fishery products, Twenty-fourth session, Ålesund, Norway, 5–9 June 2000, 157 p

FAO-Globefish (2018a) Globefish highlights. A quarterly update on world seafood markets. FAO, Rome, 73 p

FAO-Globefish (2018b) Globefish highlights. A quarterly update on world seafood markets. FAO, Rome, p 31

FAO-Globefish (2019) Globefish highlights. A quarterly update on world seafood markets. FAO, Rome, 69 p

Fernandez-Piquer J, Bowman JP, Ross T, Tamplin ML (2012) Molecular analysis of the bacterial communities in the live Pacific oyster (*Crassostrea gigas*) and the influence of postharvest temperature on its structure. J Appl Microbiol 112:1134–1143

Fisher RA (2000) Biology of certain commercial Mollusk Species: Scallops. In: Martin RR, Carte EP, Flick GJ, Davis LM (eds) Marine and Freshwater handbook. Technomic Publishing Co., Inc., Lancaster, pp 49–125

Fiz Da Costa G (2013) Introduction to the biology of clams. In: da Costa F (ed) Clam fisheries and aquaculture. Nova Science Publishers, Inc., pp 1–26

Freites L, Labarta U, Fernandez-Reiriz MJ (2002) Evolution of fatty acid profiles of subtidal and rocky shore mussel seed (*Mytilus galloprovincialis*, Link.). Influence of environmental parameters. J Exp Mar Biol Ecol 268:185–204

Fu XY, Xue CH, Miao BC, Li ZJ, Zhang YQ, Wang Q (2007) Effect of processing steps on the physico-chemical properties of dried-seasoned squid. Food Chem 103:287–294

Ganesan P, Jeyasekaran G, Maheswari K, Shakila RJ, Sukumar D (2005) Quality evaluation in chilled cuttlefish fillets. J Aquat Food Prod Technol 14:37–49

Garrido-Handong L (1990) Oyster culture. In: Selected papers on mollusk culture. Fisheries and Aquaculture Department. Food and Agriculture Organization. http://www.fao.org/3/ab737e/AB737E03.htm

Gaspar MB, Barracha I, Carvalho S, Vasconcelos P (2013) Clam fisheries worldwide: main species, harvesting methods and fishing impacts clam fisheries worldwide: main species, harvesting. In: da Costa F (ed) Clam fisheries and aquaculture. Nova Science Publishers, Inc., pp 291–328

Global Invasive Species Database (2019) Species profile: *Mytilus galloprovincialis*. http://www.iucngisd.org/gisd/species.php?sc=102. Accessed 4 Oct 2019

Gokoglu N (2002) Seafood processing technology. Su Vakfı Publications, Book, 157 p

Gokoglu N, Metin S, Baygar T, Ozden O, Erkan N (1999) Determination of the quality changes of squid (*Loligo vulgaris*, Lamarck) stored at the different temperatures. Turk J Vet Anim Sci 23:511–514

Gokoglu N, Erkan N, Ozden O (2000) The effect of frozen storage on proximate composition and quality of mussels (*Mitylus galloprovin*cialis). J Aquat Food Prod Technol 9(2):83–88

Gokoglu N, Gokoglu M, Yerlikaya P (2006) Seasonal variations in proximate and elemental composition of pearl oyster *Pinctada Radiata* from the Gulf of Antalya, Turkey. J Sci Food Agric 86(13):2161–2165

Gokoglu N, Topuz OK, Gokoglu M, Tokay FG (2017a) Characterisation of protein functionality and texture of tumbled squid, octopus and cuttlefish muscles. J Food Meas Charact 11:1699–1705

Gokoglu N, Yerlikaya P, Ucak I, Yatmaz HA (2017b) Effect of bromelain and papain enzymes addition on physicochemical and textural properties of squid (*Loligo vulgaris*). J Food Meas Charact 11:347–353

Gokoglu N, Topuz OK, Yerlikaya P, Yatmaz HA, Ucak I (2018) Effects of freezing and frozen storage on protein functionality and texture of some cephalopod muscles. J Aquat Food Prod Technol 237(2):211–218

Gómez-Guillén C, Solas T, Borderías J, Montero P (1996) Ultrastructural and rheological changes during the gelation of giant squid (*Dosidicus gigas*) muscle. Z Lebensm Unters Forsch 202:215–220

Gomez-Guillen MC, Martinez-Alvarez O, Montero P (2003) Functional and thermal gelation properties of squid mantle proteins. Affect by chilled and frozen storage. J Food Sci 68:1962–1967

Gopakumar K (1997) Biochemical composition of Indian food fishes. Central Inst. of fisheries technology, India, 44 p

Gosling E (1992) The mussel mytilus: geology, physiology, genetics and culture. Elsevier, Amsterdam, p 590

Gosling E (2015) Marine bivalve molluscs, 2nd edn. Wiley-Blackwell, Hoboken, 520 p

Goulas AE (2008) Combined effect of chill storage and modifi ed atmosphere packaging on mussels (*Mytilus galloprovincialis*) preservation. Packag Technol Sci 21:247–255

Goulas AE, Chouliara J, Nessi E, Kontominas MG, Savvaidis IN (2005) Microbiological, biochemical and sensory assessment of mussels (*Mytilus galloprovincialis*) stored under modified atmosphere packaging. J Appl Microbiol 98:752–760

Grahl-Nielsen O, Jacobsen A, Christophersen G, Magnesen T (2010) Fatty acid composition in adductor muscle of juvenile scallops (*Pecten maximus*) from five Norwegian populations reared in the same environment. Biochem Syst Ecol 38:478–488

Greenberg EP, Duboise M, Palhof B (1982) The survival of marine vibrios in *Mercenaria mercenaria*, the hardshell clam. J Food Saf 4:113–123

Guerra A (1997) Octopus vulgaris: review of the world fishery. In: Lang MA, Ilochbcrg FC (eds) Proceedings of thc workshop of the fishery and market potential of octopus in California. Smithsonian Institution, Washington, DC, pp 95–97

Guldas M, Hecer C (2012) Influences of the selected additives on the weight loss and organoleptic properties of marinated mussels and squids. Acta Vet Brno 81:263–267

Gullian-Klanian M, Sanchez-Solis MJ, Terrats-Preciat M, Delgadillo M, Aranda J (2016) Quality indicators and shelf life of red octopus (*Octopus maya*) in chilling storage. Food Sci Technol Campinas 36(2):304–312

Ha M, Bekhit AEA, Carne A, Hopkins DL (2012) Characterization of commercial papain, bromelain, actinidin, zingibain protease preparations and their activities toward meat proteins. Food Chem 134:95–105

Haard NF, Arcilla R (1985) Precursors of Maillard browning in Atlantic short finned squid. Can Inst Food Sci Technol 5(18):326–331

Haby MG, Coale CW Jr (1990) Merchandising and managing of fresh seafood department. In: Martin RE, Flick GJ (eds) The seafood industry. Springer Science + Business Media, New York, pp 227–254

Hackney CR (1990) Processing mollusks. In: Martin RE, Flick GJ (eds) The seafood industry. Springer Science + Business Media, New York, pp 165–173

Hackney CR, Rippen TE (2000) The molluscan shellfish industry. In: Martin RE, Carter EP Jr, Flicj GJ, Davis LM (eds) Marine and freshwater products handbook. Technomic Publishing Co., Inc., Lanchaster, pp 323–338

Hamm R (1986) Functional properties of the myofibrillar-system and their measurements. In: Bechtel PJ (ed) Muscle as food. Academic, Orlando, pp 135–199

Hanks Jr F, Grieb Jr JW (1971) Method and apparatus for removing meat from the shells of bivalve mollusks. United States Patent. No. 3, 594, 859

Hasanspahić H (2011) The effects of refrigeration on sensory characteristics of oysters (*Ostrea edulis*). MESO XIII, pp 369–372

Hastie L, Pierce G, Pita C, Viana M, Smith J, Wangvoralak S (2009) Squid fishing in UK waters. A Report to SEAFISH Industry Authority. School of Biological Sciences University of Aberdeen, pp 15–17

Hatano M, Lida H, Seki K, Takahashi K (1990) Occurrence of unacceptable taste and coloring of giant pacific oyster. Nippon Suisan Gakkaishi 56(9):1481–1484

He H, Adams RM, Farkas DF, Morrissey MT (2002) Use of high-pressure processing for oyster shucking and shelf-life extension. J Food Sci 67(2):640–645

Heath WA (2001) Handling and shipping of live northeast pacific scallops: larvae to adults. Marketing and shipping live aquatic product. In: Proceedings of the second international conference and exhibition, November 1999, Seattle, WA. University of Alaska Sea Grant, AK-SG-01-03, Fairbanks, pp 111–125

Heinonen K (2013) Seafood watch: farmed scallops. Monterey Bay Aquarium, 44 p

Heinonen K (2014) Seafood watch: farmed oysters. Monterey Bay Aquarium, 54 p

Héral M, Deslous-Paoli JM (1991) Oyster culture in European Countries. In: Menzel W (ed) Estuarine and marine bivalve mollusk culture. CRC Press Inc., pp 154–190

Hiltz DF, Dyer WJ (1970) Principal acid soluble nucleotides in adductor muscle of the scallop *Placopecten magellanicus* and their degradation during postmortem storage in ice. J Fish Res Board Can 27:83–92

Hiltz DF, Dyer WJ (1973) Hexose monophosphate accumulation and related metabolic changes in unfrozen and thawed adductor muscle of the sea scallop (*Placopecten magellanicus*). J Fish Res Board Can 30:45–52

Hirimuthogoda NY, Gamage CS, Marapana RAUJ, Gunarathna S (2016) Growth characteristics and meat quality attributes of clam (*Geloina vexans*) and oysters (*Crassostrea madrasensis*) harvested from two lagoons of Sri Lanka. Indian J Geo-Mar Sci 45(6):772–777

Holdsworth D, Simpson R (2007) Introduction. In: Holdsworth D, Simpson R (eds) Thermal processing of packaged foods, 2nd edn. Springer, New York, pp 1–13

Hu XP, Mallikarjunan PK, Vaughan D (2008) Development of non-destructive methods to evaulate oyster quality by electronic nose Technology. Sens Instrumen Food Qual 2:51–57

Hu Y, Huang Z, Li J, Yang H (2012) Concentrations of biogenic amines in fish, squid and octopus and their changes during storage. Food Chem 135(4):2604–2611

Huidobro A, López-Caballero ME, Mendes R (2002) Onboard processing of deepwater pink shrimp (*Parapenaeus longirostris*) with liquid ice: effect on quality. Eur Food Res Technol 214:469–475

Hurtado JL, Borderías J, Montero P, An H (1999) Characterization of proteolytic activity in octopus (*Octopus vulg*aris) arm muscle. J Food Biochem 23:469–483

Hurtado JL, Montero P, Borderias J (2001) Chilled storage of pressurized octopus (Octopus vulgaris) muscle. J Food Sci 66:400–406

Inglis SD, Kristmundsson A, Freeman M, Levesque M, Stokesbury KDE (2016) Gray meat in the Atlantic sea scallop, *Placopecten magellanicus*, and the identification of a known pathogenic scallop apicomplexan. J Invertebr Pathol 141:66–75

James SJ, James C (2014) Chilling and freezing of foods. In: Clark S, Jung S, Lamsal B (eds) Food processing: principles and applications, 2nd edn. Wiley, pp 79–105

Jay JM (2000) Seafoods. In: Modern food microbiology, 6th edn. Aspen Publisher Inc., USA, pp 101–112

Jayasooriya MCN, Jayamanne VS, Ranathunge MTN (2014) Development of new processed mussel product using local mussel species (*Perna viridis*). Sabaragamuwa Univ J 13(1):57–64

Jeong BY, Ohshima T, Koizumi C, Kanou Y (1990) Lipid Deterioration and Its Inhibition of Japanese Oyster *Crassostrea gigas* during Frozen Storage. Nippon Suisan Gakkaishi 56(12):2083–2091

Jereb P, Roper CFE (2010) Cephalopods of the world. An annotated and illustrated catalogue of cephalopod species known to date, vol 2. Myopsid and oegopsid squids. FAO species catalogue for fishery purposes, No. 4, vol 2. Rome, FAO, 605 p

Jereb P, Allcock AL, Lefkaditou E, Piatkowski U, Hastie LC, Pierce GJ (eds) (2015) Cephalopod biology and fisheries in Europe: II. Species Accounts. ICES Cooperative Research Report, No. 325, 360 p

Jeyasekaran G, Shakila JR, Sukumar D, Ganesan P, Anandaraj R (2010) Quality changes in squid (*Loligo duvaucelli*) tubes chilled with dry ice and water ice. J Food Sci Technol 47(4):401–407

Jeyasekaran G, Shakila RJ, Sukumar D (2011) Microbiological quality of Cuttlefish (*Sepia pharaonis*) fillets stored in dry and wet ice. Food Sci Technol Int 18(5):455–464

Jiménez-Ruiz EI, Ocaño-Higuera VM, Maeda-Martínez AN, Castillo-Yáñez FJ, Varela-Romero A, Graciano-Verdugo AZ, Márquez Ríos E (2012) Quality and shelf life of the adductor muscle of lions paw scallop Nodipecten subnodosus transported and stored whole in refrigeration. Interciencia 37:464–469

Joseph J, Perigreen PA (1988) Studies on frozen storage of Cuttlefish Fillets. Fish Technol 25(1):32–35

Joseph SM, Sherief PM (2003) Effect of treatments onthe iced storage shelf life of Cuttlefish (*Sepia aculeata*) fillets. Fish Technol 40:32–35

Josupeit H (2008) World octopus market. Globefish Research programme, vol 94. FAO, Rome, p 65

Kaba N, Erkoyuncu I (2005) Determination of the sensory and chemical quality of differently processed mussels (*Mytilus galloprovincialis* Lamarck, 1819) during frozen storage. Atatürk Univ J Agric Fac 36(2):193–200. (In Turkish)

Kagawa N, Matsumoto M, Yoneda C, Mitsuhashi T, Htae K (2002) Changes in meat texture of three varieties of squid in the early stage of cold storage. Fish Sci 68:783–792

Karnjanapratum S, Benjakul S, Kishimura H, Yung-Hsiang T (2013) Chemical compositions and nutritional value of Asian hard clam (*Meretrix lusoria*) from the coast of Andaman Sea. Food Chem 141:4138–4145

Katsanidis E (2004) Impact of physical and chemical pretreatments on texture of octopus (Eledone moschata). J Food Sci 69(7):264–267

Katugin ON, Shevtsov GA, Zuev MA, Didenko VD, Kulik VV, Vanin NS (2013) *Berryteuthis magister*, School master Gonate Squid. In: Rosa R, Pierce G, O'dor R (eds) Advances in squid biology, ecology and fisheries. Part II. Nova Science Publisher Inc., pp 1–48

Kavitha S, Ponni AC (2018) Seasonal variations in the protein content from different tissues of *Octopus aegina* in Pazhayar coastal waters, Nagapattinam District, Tamilnadu. Int J Sci Res Biol Sci 5:90–93

Kawashima K, Yamanaka H (1992) Effects of storage temperatures on the Post-mortem biochemical changes in scallop adductor muscle. Nippon Suisan Gakkaishi 58(11):2175–2180

Ke PJ, Burns BG, Woyewoda AD (1984) Recommended procedures and guidelines for quality evaluation of Atlantic short-fin squid (*Illex illecebrosus*). LWT-Food Sci Technol 17:276–281

Ke PJ, Fierheller M, Lemon DW (1991) Studies on processing technology for atlantic short-"n squid (Illex illecebrosus). LWT-Food Sci Technol 24:328–333

Ke S, Huang Y, Decker EA, Hultin HO (2009) Impact of citric acid on the tenderness, microstructure and oxidative stability of beef muscle. Meat Sci 82:113–118

Ketnawa S, Rawdkuen S (2011) Application of bromelain extract for muscle foods tenderization. Food Nutr Sci 2:393–401

Kiin-Kabari DB, Hart AD, Nyeche PT (2017) Nutritional composition of selected shellfish consumed in Rivers State, Nigeria. Am J Food Nutr 5(4):142–146

Kipcak AS (2017) Microwave drying kinetics of mussels (*Mytilus edulis*). Res Chem Intermed 43:1429–1445

Koak JH, Kim HS, Choi YJ, Baik MY, Kim BY (2011) Characterization of a protease from overmatured fruits and development of a tenderiser using an optimisation technique. Food Sci Biotechnol 20:485–490

Koivunen J, Heinonen-Tanski H (2004) Inactivation of enteric microorganisms with chemical disinfectants, UV irradiation and combined chemical/UV treatments. Water Res 39:1519–1526

Kolbe E, Kramer D (2007) Planning for seafood freezing. Alaska Seagrant College Program, Cooper Publishing, Alaska, p 118

Kolodziejska I, Sikorski ZE, Sadowska M (1987) Texture of cooked mantle of squid Illex argentinus as influenced by specimen characteristics and treatments. J Food Sci 52:932–935

Konishi F, Fukunaga Y, Yoneda C, Shimomura M, Kasai M, Hatae K (2003) Changes in physicochemical properties and microstructure of dried squid during softening treatment. J Sci Food Agric 83:1565–1570

Krause RJ, Ockerman HW, Krol B, Moerman PC, Plimpton RF Jr (1978a) Influence of tumbling, tumbling time, trim and sodium tripolyphosphate on quality and yield of cured hams. J Food Sci 43:853–855

Krause RJ, Plimpton RF, Ockerman HW, Cahill VR (1978b) Influence of tumbling and sodium tripolyphosphate on salt and nitrite distribution in porcine muscle. J Food Sci 43:190–192

Kugino M, Kugino K (1995) Changes in the microstructure and rheological properties of squid mantle due to cooking in a NaCl solution and soy sauce. J Home Econ Jpn 46:751–757

Kugino M, Kugino K, Wu ZH (1993) Rheological properties of dried squid mantle change on softening. J Food Sci 58:321–324

Lachowicz K, Sobczak M, Gajowiecki L, Zych A (2003) Effects of massaging time on texture, rheological properties, and structure of three pork ham muscles. Meat Sci 63:225–233

Landman NH, Mikkelsen PM, Bieler R, Bronson B (2001) Pearls: a natural history. In: Abrams HN (ed) Association with the American museum of natural history and the field museum. New York, 232 p

Lapa-Guimaraes J, Eduardo de Felicio P, Segundo E, Guzmán C (2005) Chemical and microbial analyses of squid muscle (*Loligo plei*) during storage in ice. Food Chem 91:477–483

Lapa-Guimaraes J, Azevedo da Silva MA, Eduardo de Felcio P, Guzman EC (2002) Sensory, colour and psychrotrophic bacterial analyses of squids (*Loligo plei*) during storage in ice. LWT-Food Sci Technol 35:21–29

Lawal-Are AO, Moruf OR, Junaid DA, Moruf OO (2018) Chemical bio-compounds and functional properties of raw and processed cuttlefish, *Sepia officinalis* (Mollusca: Cephalopoda). Food Environ Saf 7(3):332–340

Learson RJ (2000) Preservation of squid quality. In: Martin RE, Carter EP, Flick GJ Jr, Davis LM (eds) Marine and freshwater poducts handbook. Technomic Publishing Co. Inc., Lancherster, pp 339–343

Lebanc EL, Gill TA (1982) Comparative study of proteolysis in short-finned (*Illex illecebrosus*) and long-finned (*Loligo peuleileseur*) squid. Comp Biochem Physiol 73B:201–210

Lee PG (1994) Nutrition of cephalopods: fueling the system. Mar Freshw Behav Physiol 25:35–51

Lee R, Lovatelli A, Ababouch L (2008) Bivalve depuration: fundamental and practical aspects. Food and Agriculture Organization of the United Nations, p 139

Lee HH, Jung WY, Lee WK, Min JG (2017) Initial freshness of Pacific oyster (*Crassostea gigas*) affects its quality and self-life during freezing storage. J Food Nutr Res 5(9):629–635

Legaspi AS (1979) Post-harvest handling of mussels and oysters. In: Technical consultation on available aquaculture technology in the Philippines, 8–11 Feb 1979. Tigbauan, Iloilo, Philippines: Aquaculture Department, Southeast Asian Fisheries Development Center, pp 262–265

Lekjing S, Karrila S, Siripongvutikorn S (2017) Thermal Inactivation of Listeria monocytogenes in Whole Oysters (Crassostrea belcheri) and Pasteurization Effects on Meat Quality. J Aquat Food Prod Technol 26(9):1107–1120

Leslie RW, Lipinski MR (2018) Phylum mollusca–class cephalopoda. In: Atkinson LJ, Sink KJ (eds) Field guide to the offshore marine invertebrates of South Africa. Malachite Marketing and Media, Pretoria, pp 321–391

Li D, Zhang Y, Sinclair AJ (2007) Seasonal variations of lipid content and composition in *Perna viridis*. Lipids 42(8):739–747

Lima M, Sıga C, Leitempergher F, Lerin LA, Soares LS, Tosati JV, Rodrigues NB, Monteiro AR (2017) Mussel (*Perna perna*) processing by an alternative method and packaging under modified atmosphere to improve physicochemical and microbiological characteristics. J Food Process Preserv 41:e12923

Lin H, Jiang J, Xue CH, Zhang B, Xu JC (2003) Seasonal changes in phospholipids of mussel (*Mytilus edulis* Linne). J Sci Food Agric 83:133–135

Linehan GL, O'Connor TP, Burnell G (1999) Seasonal variation in the chemical composition and fatty acid profile of Pacific oysters (*Crassostrea gigas*). Food Chem 64:211–214

Lira GM, Pascoal JCM, Torres EAFS, Soares RAM, Mendonça S, Sampaio GR, Correia MS, Cabral CCVQ, Cabral Júnior CR, López AMQ (2013) Influence of seasonality on the chemical composition of oysters (*Crassostrea rhizophorae*). Food Chem 138(2–3):786–790

Liu C, Lu J, Su YC (2009) Effects of flash freezing, followed by frozen storage, on reducing *Vibrio parahaemolyticus* in Pacific raw oysters (*Crassostrea gigas*). J Food Prot 72(1):174–177

Lopes Da Costa VD (2018) Effects of storage on product quality of bivalves (oysters, scallops, and clams). A master thesis in Aquaculture Biology, Faculty of Mathematics and Natural Sciences University of Bergen, Norway

Lopez-Caballero ME, Perez-Mateos M, Montero P, Bonderıas AJ (2000) Oyster preservation by high-pressure treatment. J Food Prot 63:196–201

Lougovois VP, Kolovou MK, Savvaidis IN, Kontominas MG (2008) Spoilage potential of ice-stored whole musky octopus (*Eledone moschata*). Int J Food Sci Technol 43:1286–1294

Lourenco HM, Anacleto P, Afonso C, Ferraria V, Martins MF, Carvalho ML, Lino AR, Nunes ML (2009) Elemental composition of cephalopodsfrom Portuguese continental waters. Food Chem 113:1146–1153

Lovatelli A (1987) Characteristics of the farmed species. Networkof Aquaculture centres in Asia. Status of scallop farming: a review of techniques. National Inland Fisheries Institute Kasetsart University Campus Bangkhen, Bangkok. Thailand NACA-SF/WP/87/1

Lutz G (2012) Oyster profile. Agricultural Marketing Resource Center. https://www.agmrc.org/commodities-products/aquaculture/aquaculture-non-fish-species#Oysters. Accessed 12 Nov 2019

Mackie IM (1993) The effects of freezing on flesh proteins. Food Rev Int 9(4):575–610

Mackie IM, Pirie L, Ritchie AH, Yamanaka H (1997) The formation of non-volatile amines in relation to concentrations of free basic amino acids during postmortem storage of the muscle of scallop (*Pecten maximus*), herring (*Clupea harengus*) and mackerel (*Scomber scombrus*). Food Chem 60(3):291–295

Maffei M, Vernocchi P, Lanciotti R et al (2009) Depuration of striped Venus clam (*Chamelea gallina* L.): effects on microorganisms, sand content and mortality. J Food Sci 74:M1–M7

Malarvanan G, Edward JKP (2002) Bacterial quality and depuration of the green mussel *Perna viridis* from natural beds. J Indian Fish Assoc 29:71–75

Manimaran U, Shakila RJ, Sivaraman B, Shalini R, Jeyasekaran G, Shanmugam SA (2016a) Biochemical quality changes during iced storage of Indian octopus (*Cistopus indicus*). J Food Qual 39:487–495

Manimaran U, Shakila RJ, Shalini R, Sivaraman B, Sumathi G, Selvaganapathi R, Jeyasekaran G (2016b) Effect of additives in the shelflife extension of chilled and frozen stored Indian octopus (*Cistopus indicus*). J Food Sci Technol 53(2):1348–1354

Manley LHW (2008) Use of postharvest processes to reduce *Vibrio parahaemolyticus* in raw oysters: blast freezing and heat shock. A master thesis presented to the Graduate School of the University of Florida

Manthey-Karl M, Lehmann I, Ostermeyer U, Rehbein H, Schröder U (2015) Meat composition and quality assessment of king scallops (*Pecten maximus*) and Frozen Atlantic Sea Scallops (*Placopecten magellanicus*) on a retail level. Foods 4:524–546

Martin DE, Hall SG (2006) Oyster shucking technologies: past and present. Int J Food Sci Technol 41:223–232

Martinez O, Rodriguea-Calleja JM, Santos JA, Otero A, Garcia-Lopez ML (2009) Foodborne and indicator bacteria in farmed molluscan shellfish before and after depuration. J Food Prot 72(7):1443–1449

Martino RC, Cruz GM (2004) Proximate composition and fatty acid content of the mangrove oyster *Crassostrea rhizophorae* along the year seasons. Braz Arch Biol Technol 47(6):955–960

Masniyom P, Benjama O (2007) Effect of lactic, acetic and citric acids on quality changes of refrigerated green mussel, *Perna viridis* (Linnaeus, 1758). Songklanakarin J Sci Technol 29(4):1124–1134

McCarron P, Emteborg H, Hess P (2007) Freeze-drying for the stabilisation of shellfish toxins in mussel tissue (*Mytilus edulis*) reference materials. Anal Bioanal Chem 387:2475–2486

McKenzie W, Parker B (2019) Mollusca. In: Aquatic invertebrate cell culture. Ed-Tech Press, Waltham Abbey Essex, pp 189–251

Mehta NK, Nayak BB (2017) Bio-chemical composition, functional, and rheological properties of fresh meat from fish, squid, and shrimp: a comparative study. Int J Food Prop 20(51):S707–S721

Melody K, Senevirathne R, Janes M, Jaykus LA, Supan J (2008) Effectiveness of icing as a post-harvest treatment control of Vibrio vulnificus and Vibrio parahaemolyticus in the Eastern oyster (*Crassostrea virginica*). J Food Prot 71(7):1475–1480

Mercaldo-Allen R, Goldberg R (2011) Review of the ecological effects of dredging in the cultivation and harvest of molluscan shellfish. NOAA technical memorandum NMFS-NE-220, p 84

Merdzhanova A, Dobreva DA, Georgieva S (2016) Nutritional evaluation of aquaculture mussels (*M. Galloprovincialis*) from the Black Sea, Bulgaria. Ovidius Univ Ann Chem 27(1):1–7

Merdzhanova A, Dobreva DA, Panayotova V (2017) Assessment of proximate and bioactive lipid composition of black sea mussels (*M. galloprovincialis*) from Bulgaria. In: Ray S (ed) Biological resources of water. IntechOpen

Meynier L, Morel PCH, Mackenzie DDS, Macgibbons A, Chilvers BL, Duignan PJ (2008) Proximate composition, energy content, and fatty acid composition of marine species from Campbell Plateau, New Zealand. N Z J Mar Freshw Res 42:425–437

Mignino LA, Crupkin M, Paredi ME (2008) Surface hydrophobicity and functional properties of myofibrillar proteins of mantle from frozen-stored squid (Illex argentinus) caught either jigging machine or trawling. LWT-Food Sci Technol 41:678–685

Mikkelsen PM, Henne R (2011) The teacher-friendly guide to evolution using bivalves as a model organism. Paleontological Research Institution, Ithaca, New York, http://teacherfriendlyguide.org/bivalves. Accessed 26 Sept 2019

Mladineo I, Peharda M, Orhanovic S, Bolotin J, Pavela-Vrancic M, Treursic B (2007) The reproductive cycle, condition index and biochemical composition of the horse-bearded mussel *Modiolus barbatus*. Helgol Mar Res 61:183

Miliou H, Fintikaki M, Tzitzinakis M, Kountouris T, Verriopoulos G (2006) Fatty acid composition of the common octopus, *Octopus vulgaris*, in relation to rearing temperature and body weight. Aquaculture 256:311–322

Montanhini MTM, Neto RM (2015) Changes in the microbiological quality of Mangrove oysters (*Crassostrea brasiliana*) during different storage conditions. J Food Prot 78:164–171

Moral A, Tejada M, Borderias AJ (1983) Frozen storage behaviour of squid (*Loligo vulgaris*). Int J Refrig 6(1):54–57

Moral A, Morales J, Ruiz-Capillas C, Montero P (2002) Muscle protein solubility of some cephalopods (pota and octopus) during frozen storage. J Sci Food Agric 82(6):663–668

Morales J, Montero P, Moral A (2000) Isolation and partial characterization of two types of muscle collagen in some cephalopods. J Agric Food Chem 48:2142–2148

Mortensen LM, Frøst MB, Skibsted LH, Risbo J (2012) Effect of time and temperature on sensory properties in low-temperature long-time sous-vide cooking of beef. J Culinary Sci Technol 10(1):75–90

Mouritsen OG, Styrbæk K (2018) Cephalopod gastronomy—a promise for the future. Front Commun 3:38

Muntada-Garriga JM, Rodriguez-Jerez JJ, Lopez-Sabater EI, Mora-Ventura MT (1995) Effect of chill and freezing temperatures on survival of *Vibrio parahaemolyticus* inoculated in homogenates of oyster meat. Lett Appl Microbiol 20:225–227

Muraleedharan V, Nair TSU, Joseph KG (1979) Smoke curing of mussel. Fish Technol 16:29–31

Murchie L, Cruz-Romero M, Kerry JP, Linton M, Patterson MF, Smiddy M, Kelly AL (2005) High pressure processing of shellfish: a review of microbiological and other quality aspects. Innov Food Sci Emerg Technol 6:257–270

Naidu KS (1987) Efficiency of meat recovery from Iceland scallops (*Chlamys islandica*) and sea scallops (*Placopecten magellanicus*) in the Canadian offshore fishery. J Northwest Atl Fish Sci 7:131–136

Narváez M, Freites L, Guevara M, Mendoza H, Guderley CJ, Lodeiros CJ, Salazar G (2008) Food availability and reproduction affects lipid and fatty acid composition ofthe brown mussel, *Perna perna*, raised in suspension culture. Comp Biochem Physiol B 149:293–302

Navaeena BM, Mendiratta SK, Anjaneyulu ASR (2004) Tenderization of buffalo meat using plant proteases from Cucumis Trigonus Roxb and Zingiber Officinalis Roscoe. Meat Sci 68:363–369

Navarro JC, Villanueva R (2000) Lipid and fatty acid composition of early stages of cephalopods: an approach to their lipid requirements. Aquaculture 183:161–177

Navarro JC, Villanueva R (2003) The fatty acid composition of *Octopus vulgaris* paralarvae reared with live and inert food: deviation from their natural fatty acid profile. Aquaculture 219:613–631

Nguyen MV, Arason S, Eikevik TM (2014) Drying of fish. In: Boziaris IS (ed) Seafood processing technology, quality and safety. Wiley Blackwell, Chichester, pp 161–175

Noel L, Chekri R, Millour S, Vastel C, Kadar A, Sirot V, Leblanc JC, Guérin T (2012) Li, Cr, Mn Co, Ni, Cu, Zn, Se and Mo levels in foodstuffs from the 2nd French TDS. Food Chem 132:1502–1513

NRC (National Research Council) (2004) General biology of oysters. In: Nonnative oysters in the chesapeake bay. The National Academies Press, Washington, DC, 325 p

NSSP (National Shellfish Sanitation Program) (2017) Guide for the control of molluscan shellfish (2017 Revision). US Food and Drug Administration. https://www.fda.gov/food/federalstate-food-programs/national-shellfish-sanitation-program-nssp. Accessed 15 Oct 2019

Nunes ML, Bandarra NM, Batista I (2011) Health benefit associated with seafood consumption. In: Alasalvar C, Shahidi F, Miyashita K, Wanasunddara U (eds) Handbook of seafood quality safety and health applications. Wiley-Blackwell, Oxford, pp 369–380

Nurjanah AM, Jacoeb RN, Suhana SN, Siti K (2012) Proximate, nutrient and mineral composition of cuttlefish (*Sepia recurvirostra*). Adv J Food Sci Technol 4(4):220–224

O'Brien CE, Ponte G, Fiorito G (2019) Octopus. In: Choe J (ed) Encyclopedia of animal behavior, 2nd edn. Academic Press is an imprint of Elsevier, London, pp 142–148

Obodai EA, Nyarko HD, Amponsah SL (2010) Effect of depuration on microbial content of mangrove oyster (*Crassostr Ea Tulipa*) from Benya Lagoon, Ghana. Ethiop J Environ Stud Manag 3(2):47–53

Ocano-Higuera VM, Maeda-Martínez AN, Lugo-Sanchez ME, Pacheco-Aguılar R (2006) Postmortem biochemical and textural changes in the adductor muscle of Catarina scallop stored at 0°C. J Food Biochem 30:373–389

Odeyemi OA, Burke CM, Bolch CJS, Stanley R (2018) Evaluation of spoilage potential and volatile metabolites production by *Shewanella baltica* isolated from modified atmosphere packaged live mussels. Food Res Int 103:415–425

Ohashi F, Okamoto M, Ozawa A, Fujita T (1991) Characterization of common squid using several freshness indicators. J Food Sci 56:161–174

Okazaki RK, Panietz MH (1981) Depuration of twelve trace metals in tissues of the oysters *Crassostrea gigas* and *C. virginica*. Mar Biol 63:113–120

Olmez M, Atar HH, Seçer S (2003) A study on meat yield and nutrition contents of striped venus (*Chamelea gallina* L. 1758). Gıda 28(2):169–173. (In Turkish)

Oosthuizen A (2003) A development and management framework for a new *Octopus vulgaris* fishery in South Africa.PhD thesis. Rhodes University

Orban E, Di Lena G, Nevigato T, Casini I, Marzetti A, Caproni R (2002) Seasonal changes in meat content, condition index and chemical composition of mussels (*Mytilus galloprovincialis*) cultured in two different Italian sites. Food Chem 77(1):57–65

Orban E, Di Lena G, Nevigato T, Casini I, Caproni R, Santaroni G, Giulini G (2006) Nutritional and commercial quality of the striped venus clam, *Chamelea gallina*, from the Adriatic Sea. Food Chem 101:1063–1070

Othman ASB, Goh GHS, Todd PA (2010) The distribution and status of giant clams (family tridacnidae)—a short review. Raffles Bull Zool 58(1):103–111

Otwell WS, Haman DD (1979) Textural characterization of squid (*Oligo pealei* L.): instrumental and panel evaulation. J Food Sci 44:1636–1643

Overaa T (2001) Live transport of the great scallop (*Pecten maximus*) (2001). Marketing and shipping live aquatic product. In: Proceedings of the second international conference and exhibition, November 1999, Seattle, WA. University of Alaska Sea Grant, AK-SG-01-03, Fairbanks, pp 105–109

Ovissipour M, Rasco B, Tang J, Sablani SS (2013) Kinetics of quality changes in whole blue mussel (*Mytilus edulis*) during pasteurization. Food Res Int 53:141–148

Ozden O, Erkan N, Ulusoy S (2009) Seasonal variations in the macronutrient mineral and proximate composition of two clams (*Chamelea gallina* and Donax trunculus). Int J Food Sci Nutr 60(5):402–412

Oztan A (2003) Meat science and technology, 4th edn, TMMOB Chamber of Food Engineers Publications Books Series Publication No: 1, Ankara (in Turkish)

Ozyurt O, Duysak O, Akamca E, Tureli C (2006) Seasonal changes of fatty acids of cuttlefish *Sepia officinalis* L. (Mollusca: Cephalopoda) in the north eastern Mediterranean Sea. Food Chem 95:382–385

Paarup T, Sanchez JA, Moral A, Christensen H, Bisgaard M, Gram L (2002a) Sensory, chemical and bacteriological changes during storage of iced squid (*Todaropsis eblanae*). J Appl Microbiol 92:941–950

Paarup T, Sanchez JA, Pelaez C, Moral A (2002b) Sensory, chemical and bacteriological changes in vacuum-packed pressurized squid (*Todaropsis eblanae*) stored at 4°C. Int J Food Microbiol 74:1–12

Pacheco-Aguilar R, Marquez-Rios E, Lugo-Sanchez ME, Garcia-Sanchez G, Maeda-Martinez AN, Ocano-Higuera VN (2008) Postmortem changes in the adductor muscle of Pacific lions-paw scallop (*Nodipecten subnodosus*) during ice storage. Food Chem 106:253–259

Pan L, Lin C, Zhang GC, Mu G, Yang XT (2018) The effect of pre-process and transport strategies on survival, microbiologic, and physiologic of *Patinopecten yessoensis*. Food Sci Nutr 6:1422–1430

Paredi ME, De Vido de Mattio NA, Crupkin M (1996) Biochemical properties of actomyosin and expressible moisture of frozen stored striated adductor muscles of *Aulacomya ater ater* (Molina) effects of polyphosphates. J Agric Food Chem 44(10):3108–3112

Paredi ME, Roldan HA, Crupkin M (2006) Changes ın myofıbrıllar proteıns and lıpıds of squıd (*Illex argentınus*) durıng frozen storage. J Food Biochem 30:604–621

Park H-Y, Hur J-W (1990) A study on the processing aptitude and storage of common-European squid (*Loligo vulgaris*). 1. Changes of freshness during storage. J Korean Soc Food Nutr 19:168–174

Pascual E (1978) Growth and feeding of laboratory reared Sepia officinalis. Investigación Pesquera 42(2):421–442

Patterson J (2001) Canning of smoked brown mussel, Perna indica. Phuket Mar Biol Center Spec Publ 25(1):219–220

Patterson J (2004) Smoking—an ideal method to preserve mollusc meat. SPC Trochus Information Bulletin #11–September 2004, pp 8–10

Pennarun AL, Prost C, Haure J, Demaimay M (2003) Comparison of two microalgal diets. 1. Influence on the biochemical and fatty acid composition of raw oysters (*Crassostrea gigas*). J Agric Food Chem 51:2006–2010

Perdue J, Erickson G (1984) A comparison of the gametogenetic cycle between the Pacific oyster *Crassostrea gigas* and the suminoe oyster *Crassostrea rivularis* in Washington state. Aquaculture 37:231–237

Pereira MA, Nunes MM, Nuernberg L, Schulz D, Batista CRV (2006) Microbiological quality of oysters (*Crassostrea gigas*) produced and commercialized in the coastal region of Florianopolis-Brazil. Braz J Microbiol 37:159–163

Petridis D, Zotos A, Kampouris T, Roumelioti Z (2012) Optimization of a steaming with liquid smoke smoking process of Mediterranean mussel (*Mytilus galloprovincialis*). Food Sci Technol Int 19(1):59–68

Phillips KL, Nichols PD, Jackson GD (2002) Lipid and fatty acid composition of the mantle and digestive gland of four Southern Ocean squid species: implications for food-web studies. Antarct Sci 14(3):212–220

Piatkowski U, Hernández-García V, Clarke MR (1998) On the biology of the European flying squid *Todarodes sagittatus* (Lamarck, 1798) (Cephalopoda, Ommastrephidae) in the Central Eastern Atlantic. S Afr J Mar Sci 20(1):375–383

Piatkowski U, Zumholz K, Lefkaditou E, Oesterwind D, Jereb P, Pierce GJ, Allcock AL (2015) *Todarodes sagittatus* (Lamarck, 1798). In: Jereb P, Allcock AL, Lefkaditou E, Piatkowski U, Hastie LC, Pierce GJ (eds) Cephalopod biology and fisheries in Europe: II: Species Accounts. (ICES Cooperative Research Report, vol 325). International Council for the Exploration of the Sea, Denmark, pp 194–205

Pietrasik Z, Shand PJ (2004) Effect of blade tenderization and tumbling time on the processing characteristics and tenderness of injected cooked roast beef. Meat Sci 66:871–879

Pietrasik Z, Shand PJ (2005) Effects of mechanical treatments and moisture enhancement on the processing characteristics and tenderness of beef semimembranosus. Meat Sci 71:498–505

Pietrasik Z, Aalhus JL, Gibson LL, Shand PJ (2010) Influence of blade tenderization, moisture enhancement and pancreatin enzyme treatment on the processing characteristics and tenderness of beef semitendinosus muscle. Meat Sci 84(3):512–517

Poole SE, Wilson P, Mitchell GE, Wills PA (1990) Storage life of chilled scallops treated with low dose irradiation. J Food Prot 53(9):763–766

Prafulla V, Francis L, Lakshmanan PT (2000) Effect of different methods of icing on the quality of squid and cuttlefish during storage. Fish Technol 37:81–88

Prato E, Danieli A, Maffia M, Biandolino F (2010) Lipid and fatty acid compositions of *Mytilus galloprovincialis* cultured in the mar Grande of Taranto (southern Italy): feeding strategies and trophic relationships. Zoologica 49(2):211–219

Prato E, Biandolino F, Parlapiano I, Papa L, Kelly M, Fanelli G (2018) Bioactive fatty acids of three commercial scallop species. Int J Food Prop 21(1):519–532

Prato E, Biandolino F, Parlapiano I, Papa L, Denti G, Fanelli G (2019) Seasonal changes of commercial traits, proximate and fatty acid compositions of the scallop *Flexopecten glaber* from the Mediterranean Sea (Southern Italy). PeerJ 7:e5810

Pravin P, Aneesh Kumar KV, Meenakumari B (2013) Squid jigging operations in the Southern Ocean. In: Ramesh R, Sudhakar M, Chattopadhyay S (eds), Scientific and geopolitical interests in arctic and antarctic. Proceedings of international conference on science and geopolitics of Arctic and Antarctic, (iSaGAA), March 2013, LIGHTS, Research Foundation, pp 203–211

Prester L, Orct T, Macan J, Vukusic J, Kipcic D (2010) Determınatıon of biogenic amines and endotoxin ın squid, musky octopus, Norway lobster, and mussel stored at room temperature. Arch Ind Hyg Toxicol 61:389–397

Price RJ (2000) Octopus. In: Martin RE, Carter EP Jr, Flicj GJ, Davis LM (eds) Marine and freshwater products handbook. Technomic Publishing Co., Inc., Lanchaster, pp 121–125

Ramasamy P, Subhapradha N, Sudharsan S, Seedevi P, Shanmugam V, Shanmugam A (2010) Nutritional evaluation of the different body parts of cuttlefish *Sepia kobiensis* Hoyle, 1885. Afr J Food Sci 6(22):535–538

Rato A, Pereira LF, Joaquim S, Gomes R, Afonso C, Cardoso C, Machado J, Gonçalves JFM, Vaz-Pires P, Magnoni LJ, Matias AM, Matias D, Bandarra NM, Ozório ROA (2019) Fatty acid profile of pacific oyster, *Crassostrea gigas*, fed different ratios of dietary seaweed and microalgae during broodstock conditioning. Lipids 54:531–542

Reid A, Jereb P, Roper CFE (2005) Cuttlefishes. In: Jereb P, Roper CFE (eds), Cephalopods of the worldan annotated and illustrated catalogue of cephalopod species known to date. FAO species catalogue for fishery purposes, No. 4, vol 1, pp 56–62

Remyakumari KR, Ginson J, Ajeeshkumar KK, Vishnu KV, Asha KK, Suseela M (2018) Biochemical profile and nutritional quality of Indian squid, *Uroteuthis duvauceli*. Int J Fish Aquat Stud 6(3):187–192

Ren T, Su Y-C (2006) Effects of electrolyzed oxidixing water treatment on reducing *Vibrio parahaemolyticus* and *Vibrio vulnificus* in raw oysters. J Food Prot 69:1829–1834

Rodhouse PG (2005) World squid resources. In: Review of the state of world marine fishery resources. FAO fisheries technical paper, Rome, pp 175–187

Rodriguez Ó, Barros-Velazquez J, Pineiro C, Gallardo JM, Aubourg SP (2006) Effects of storage in slurry ice on the microbial, chemical and sensory quality and on the shelf life of farmed turbot (*Psetta maxima*). Food Chem 95:270–278

Roper CFE, Sweeney MJ, Nauen CE (1984) Cephalopods of The World an Annotated and Illustrated Catalogue of Species of Interest to Fisheries. FAO Fisheries Synopsis No. 125, vol 3. Rome, 228 p

Ruiz-Capillas C, Horner WFA, Gillyon CM (2001) Effect of packaging on the spoilage of king scallop (*Pecten maximus*) during chilled storage. Eur Food Res Technol 213:95–98

Ruiz-Capillas C, Moral A, Morales J, Montero P (2002a) Preservation of shelf life of pota and octopus in chilled storage under controlled atmospheres. J Food Prot 65(1):140–145

Ruiz-Capillas C, Moral A, Morales J, Montero P (2002b) The effect of frozen storage on the functional properties of the muscle valador (*Illex coindetti*). Food Chem 78:148–156

Saeedi H, Costello MJ (2013) Aspects of global distribution of six marine bivalve mollusc families. In: Costa F (ed) Clam fisheries and aquaculture. Nova Science Publishers Inc., New York, 17 p

Sakai J, Matsumoto J (1981) Proteolytic enzymes of squid mantle muscle. Comp Biochem Physiol 68B:389–395

Sakurai Y, Kidokoro H, Yamashita N, Yamamoto J, Uchikawa K, Hideo T (2013) Todarodes pacificus, Japanese common squid. In: Rosa R, O'Dor R, Pierce GJ (eds) Advances in squid biology, ecology and fisheries. Part II–oegopsid squids. Nova Publishers, New York, pp 249–272

Salaskar GM, Nayak VN (2011) Nutritional quality of bivalves, *Crassostrea madrasensis* and *Perna viridis* in the Kali estuary, Karnataka, India. Recent Res Sci Technol 3(4):6–11

Salman Y, Salman A, Ozkizilcik S (2007) The fatty acid profile of the marine cephalopod *Loligo vulgaris*. Isr J Aquacult—Bamidgeh 59(3):133–136

Santi A, Metasulach, Genisha D, Mahendradatta M (2019) Proximate and mineral composition of cuttlefish (*Sepia* sp). Int J Sci Res Sci Technol 6(4):130–137

Sasikumar G, Mohamed KS, Rohit P, Sampathkumar G (2015) Policy guidance on cuttlefish fishery using fish aggregating devices. CMFRI Marine Fisheries Policy Series No. 1. 56 p

Sauer WHH, Gleadall LG, Downey-Breedt N, Doubleday Z, Gillespie G, Haimovici M, Ibanez CM, Katugin ON, Leporati S, Lipinski MR, Markaida U, Ramos JE, Rosa, Villanueva R, Arguelles J, Briceno FA, Carrasco SA, Che LJ, Chen C-S, Cisneros R, Conners E, Crespi-Abril AC, Kulik VV, Drobyazin EN, Emery T, Fernández-Álvarez FA, Furuya H, González LW, Gough C, Krishnan P, Kumar B, Leite T, Lu C-C, Mohamed KS, Nabhitabhata J, Noro K, Petchkamnerd J, Putra D, Rocliffe S, Sajikumar KK, Sakaguchi H, Samuel D, Sasikumar G, Wada T, Zheng X, Tian Y, Pang Y, Yamrungrueng A, Pecl G (2019) World octopus fisheries. Rev Fish Sci Aquacult. https://doi.org/10.1080/23308249.2019.1680603

Scarcella G, Cabanelas AM (2016) Research for Pech Committee—The Clam Fisheries Sector in the EU—The Adriatic Sea Case. http://www.europarl.europa.eu/RegData/etudes/STUD/2016/573412/IPOL_STU(2016)573412_EN.pdf. Accessed 5 Nov 2019

Schneider KR, Steslow FS, Sierra FS, Rodrick GE, Noss CI (1991) Ozone depuration of Vibrio vulnificus from the southern quahog clam, *Mercenaria campechiensis*. J Invertebr Pathol 57:184–190

Seaman MNL (1991) Survival and aspects of metabolism in oysters, *Crassostrea gigas*, during and after prolonged air storage. Aquaculture 93:389–395

Sengor FG, Gun H, Kalafatoglu H (2003) The determination of vitamin content and sensorial quality in liquid and traditionally smoked black mussels (*Mytilus galloprovincialis*, l.) canned in different sauce content. Turk J Vet Anim Sci 27:515–520

Sengor GF, Kalafatoglu H, Gun H (2004) The determination of microbial flora, water activity and chemical analyses in smoked, canned mussels (*Mytilus galloprovincialis*, L.). Turk J Vet Anim Sci 28:793–797

Shalini R, Shakila RJ, Jeyasekaran G, Jeevithan E (2015) Sensory, biochemical and bacteriological properties of octopus (*Cistopus indicus*) stored in ice. J Food Sci Technol 52(10):6763–6769

Shanks BC, Wulf DM, Maddock RJ (2002) Technical note: the effect of freezing on warner-bratzler shear force values of beef longissimus steaks across several postmortem aging periods. J Anim Sci 80:2122–2125

Shen X, Su YC, Liu C, Oscar T, DePaola A (2019) Efficacy of *Vibrio parahaemolyticus* depuration in oysters (*Crassostrea gigas*). Food Microbiol 79:35–40

Shi M, Lin Y, Xu G, Xie L, Hu X, Bao Z, Zhang R (2013) Characterization of the zhikong scallop (*Chlamys farreri*) mantle transcriptome and ıdentification of biomineralization-related genes. Mar Biotechnol 15:706–715

Shul'gina VN, Dolbnina NV, Lazhentseva YL, Yu P (2013) New technological approaches to canning cephalopod mollusks. Foods Raw Mater 1(1):40–47

Sikorski ZE, Kolodziejska I (1986) The composition and properties of squid meat. Food Chem 20:213–224
Sikorski ZE, Olley J, Kostuch S (1976) Proteins changes in frozen in frozen fish. Crit Rev Food Sci Nutr 8:97–129
Sikorski ZE (1978) Protein change in muscle food due to freezing and frozen storage. Int J Refrig 1:173–180
Silva JL, Chamul RS (2000) Composition of marine and freshwater finfish and shellfish species and their products. In: Martin RR, Carte EP, Flick GJ, Davis LM (eds) Marine and freshwater handbook. Technomic Publishing, Lancaster, pp 31–47
Sinanoglou VJ, Miniadis-Meimaroglou S (1998) Fatty acid of neutral and polar lipids of (edible) Mediterranean cephalopods. Food Res Int 3:467–473
Singh AS, Nagalakshimi K (2013) Depuration of bivalves for food. Infofish Int 2:46–49
Sirbu R, Bechir A, Negreanu-Pîrjol T, Sava C, Negreanu-Pîrjol B, Zaharia T, Ursache C, Stoicescu R (2011) Studies on the nutrient content od species *Mytilus galloprovincialis* of the Black Sea. Scientific study & research: chemistry & chemical engineering, biotechnology. Food Ind 12(3):221–228
Sirin C (2012) Integration and effects of culture of marine fish-mussel (*Mytilus galloprovincialis* Lamark, 1819) in Vona Bay (Persembe-Ordu). Master Thesis Fisheries Technology Engineering Department, Ordu University Institute of Science and Technology, 58 p
Skipnes D (2014) Heat processing of fish. In: Boziaris IS (ed) Seafood processing technology, quality and safety. Wiley Blackwell, Chichester, pp 61–81
Smaal AC, Ferreira JG, Grant J, Petersen JK, Strand Q (Eds) (2019) Preface. Goods and Services of Marine Bivalves. E-book, Springer Nature Switzerland AG
Sonderblohm CP, Guimaraes MH, Pita C, Rangel M, Pereire J, Gonçalves JMS, Erzini K (2017) Participatory assessment of management measures for *Octopus vulgaris* pot and trap fishery from Southern Portugal. Mar Policy 75:133–142
Songsaeng S (2010) Change in qualities and volatile compounds of oyster (Crassostrea belcheri) during storage and processing. PhD. Thesis submitted to Food Technology Department, Prince of Songkla UniversitSongkhla, Thailand
Songsaeng S, Sophanodora P, Kaewsrithong J, Ohshima T (2010) Effect of different storage conditions on quality of White-Scar oyster (*Crassostrea belcheri*). Int Food Res J 17:491–500
Sorio JC, Peralta JP (2018) Microbiological quality of oyster (*Crassostrea iredalei*) in selected production areas in Dumangas, Iloilo, Philippines. AACL Bioflux 11(2):319–326
Sreedevi KH, James JP, Bindu J, Sreejith S, Gopal N (2018) Chilled storage studies of depurated cooked clam meat in two different packaging materials. Fish Technol 55:114–119
Srilatha G, Chamundeeswari K, Ramamoorthy K, Sankar G, Varadharajan D (2013) Proximate, amino acid, fatty acid and mineral analysis of clam, *Meretrix casta* (Chemnitz) from Cuddalore and Parangipettai Coast, South East Coast of India. J Mar Biol Oceanogr 2:2–7
Stancheva M, Merdzhanova A, Dobreva DA (2017) Fat soluble vitamins, cholesterol, and fatty acid composition of wild and farmed black mussel (*Mytilus galloprovincialis*) consumed in Bulgaria. J Aquat Food Prod Technol 26(2):181–191
Stanley DW, Hultin HO (1982) Quality facters in cooked north Atlantic squid. Can Inst Food Sci Technol J 15:277–282
Stanley DW, Hultin HO (1984) Proteolytic activity in North American squid and its relation to quality. Can Inst Food Sci Technol J 17:163–716
Stratev D, Popova T, Zhelyazkov G, Vashin I, Dospatliev L, Valkov E (2017) Seasonal changes in quality and fatty acid composition of black mussel (*Mytilus galloprovincialis*). J Aquat Food Prod Technol 26(7):871–879
Stroud GD (1981) Handling and processing oysters. Torry Advisory Note No: 84. Ministry of Agriculture, Fisheries and Food. Torry Advisory Research Station
Su XQ, Dinh TT (2003) Omega-3 polyunsaturated fatty acid content in different edible portions of commerical scallop. Asia Pac J Clin Nutr 12(Suppl):S63

Su YC, Liu C (2014) Shellfish handling and primary processing. In: Boziaris IS (ed) Seafood processing technology, quality and safety. Wiley Blackwell, Chichester, pp 11–28

Su YC, Yang Q, Has C (2010) Refrigerated seawater depuration for reducing *Vibrio parahaemolyticus* contamination in pacific oyster (*Crassostrea gigas*). J Food Prot 73(6):1111–1115

Subasinghe MM, Jinadasa BKKK, Navarathne AN, Jayakodye S (2019) Seasonal variations in the total lipid content and fatty acid composition of cultured and wild *Crassostrea madrasensis* in Sri Lanka. Heliyon 5(2):e01238

Sukumar VK, Muruganantham D, Jeyasekaran G, Shanmıgam SA (2014) Studies on improvement in sensory quality and acceptability of frozen squid (*Sepıoteuthıs lessonıan*) upon treatment with imported chemicals. Int J Curr Res 6(12):10455–10459

Sukumaran S, Pittman K, Patterson WK, Dickson J, Yeend S, Townsend A, Broadbridge V, Price TJ (2010) A phase I study to determine the safety, tolerability and maximum tolerated dose of green-lipped mussel (*Perna canaliculus*) lipid extract, in patients with advanced prostate and breast cancer. Ann Oncol 21(5):1089–1093

Sungsri-in R (2010) Development of pink color of squid and the effect of chemical treatment on physico-chemical changes of squid during frozen storage. A thesis submitted to Prince of Songkla University, Food Science and Technology program, Hat-Yai, Thailand, 102 p

Sykes AV, Oliveira AR, Domingues PM, Cardoso CM, Andrade JP, Nunes ML (2009) Assessment of European cuttlefish (*Sepia officinalis* L.) nutritional value and freshness under ice storage using a developed Quality Index Method (QIM) and biochemical methods. LWT-Food Sci Technol 42(1):424–432

Tabakaeva OV, Tabakaev AV, Piekoszewski W (2018) Nutritional composition and total collagen content of two commercially important edible bivalve molluscs from the Sea of Japan coast. J Food Sci Technol 55(12):4877–4886

Takashi T (1965) Squid meat and it processing. In: Borgstrom G (ed) Fish as food, vol 4. Academic Press, New York, pp 339–354

Teerawut S, Arjpong R, Kwan-On P (2017) Microbiological and physical assessment of green mussel (*Perna viridis*) coated with green tea and ascorbic acid stored under modified atmosphere packaging. Thammasat Int J Sci Technol 22(2):1–9

Telahiguea K, Chetouia I, Rabeha I, Romdhaneb MS, El Cafsi M (2010) Comparative fatty acid profiles in edible parts of wild scallops fromthe Tunisian coast. Food Chem 122:744–746

Terzi G, Gücüklüoglu A, Cadırcı Ö, Kevenk TO, Alisarlı M (2013) Effects of Chitosan and lactic acid ımmersion on the mussels'quality changes during the refrigerated storage. Kafkas Univ J Vet Fac 19(2):311–317

Thanonkaew A, Benjakul S, Visessanguan W (2006) Chemical composition and thermal property of cuttlefish (*Sepia pharaonis*) muscle. J Food Compos Anal 19(2):127–133

Thomas MA (2016) Rapid cooling of eastern oysters (*Crassostrea virginica*) using an onboard icing unit to slow the growth of *Vibrio vulnificus* and *Vibrio parahaemolyticus*. A master thesis Submitted to the Graduate Faculty of the Louisiana State University and Agricultural and Mechanical College

Thomas MA, Hall SG, Supan J (2016) Effects of rapid cooling of eastern oysters *Crassostrea virginica* using an onboard chilling unit. Poster presentation at the World Aquaculture Society meeting, Las Vegas, NV, 22–26 Feb 2016

Tir M, Rebeh I, Telahigue K, Hajji T, Mejri H, El Cafsi M (2015) Seasonal variation of the chemical content and fatty acid composition of mantle and tentacle of male and female *Sepia officinalis*. J Am Oil Chem Soc 92:1643–1650

Torrinha A, Gomesa F, Oliveira M, Mendes CE, Delerue-Matosa C, Casal S, Morais S (2014) Commercial squids: characterization, assessment of potential health benefits/risks and discrimination based on mineral, lipid and vitamin E concentrations. Food Chem Toxicol 67:44–56

Tosun ŞY, Ucok-Alakavuk DU, Ulusoy Ş (2018) Quality changes of thermal pasteurized mussels (*Mytilus galloprovincialis*) during refrigerated storage at 4±1°C. Aquat Sci Eng 33(4):117–123

Tribuzi G, Laurindo JB (2015) Evaluation of different dehydration methods of cooked mussels. Blucher Chem Eng Proc 1(2):4169–4176

Tribuzi G, Aragao GMF, Laurindo JB (2015) Processing of chopped mussel meat in retort pouch. Food Sci Technol Campinas 35(4):612–619
Tsai CH, Sun Pan B, Kong MS (1991) Browning behavior of taurine and proline in model and dried squid systems. J Food Biochem 15:67–77
Tsangridis A, Sanchez P, Ioannidou D (2002) Exploitation patterns of *Octopus vulgaris* in two Mediterranean areas. Sci Mar 66(1):59–68
Turan H, Sönmez G, Çelik Y, Yalçin M, Kaya Y (2008) The effects of hot smoking on the chemical composition and shelf life of mediterranean mussel (*Mytılus galloprovıncıalıs* l. 1819) under chilled storage. J Food Process Preserv 32:912–922
Ueng Y, Chow C (1998) Textural and histological changes of different squid mantle muscle during frozen storage. J Agric Food Chem 46:4728–4733
Unal-Sengor GF, Ceylan Z, Dogruyol H (2018) Determination of the quality changes of whole cuttlefish (*Sepia officinalis*, L., 1758) stored at 2°C. Ege J Fish Aquat Sci 35(2):115–119
Unnikrishnan-Nair TS, Joseph KG, Muraleedharan V, Kalaimani N (1989) Preparation of mussel marinade. Fish Technol 26(2):119–121
Vairamani S (2010) Studies on biochemical composition, polysaccharides and collagen from *Sepiella inermis* Orbginy 1848. Ph.D Thesis, Khadhir Mohideen College, Tamil Nadu, India, p 223
Valencia-Perez AZ, Garcıa-Morales MH, Cardenas-Lopez JL, Herrera-Urbina JR, Rouzaud-Sandez O, Ezquerra-Brauer JM (2008) Effect of thermal process on connective tissue from jumbo squid (*Dosidicus gigas*) mantle. Food Chem 107:1371–1378
Vasakou A, Vareltzis K, Bloukas JG (2003) Effect of sodium lactate and potassium sorbate on quality characteristics and shelf life of Mediterranean mussel (*Mytilus galloprovincialis*) meat during chilled storage in pouches with water. Ital J Food Sci 15:359–370
Vasconcelos GJ, Lee JS (1972) Microbial flora of Pacific oysters (*Crassostrea gigas*) subjected to ultraviolet-irradiated seawater. Appl Microbiol 23:11–16
Vaz-Pires P, Barbosa A (2004) Sensory, microbiological, physical and nutritional properties of iced whole common octopus (*Octopus vulgaris*). LWT-Food Sci Technol 37:105–114
Vaz-Pires P, Seixas P (2006) Development of new quality index (QIM) schemes for cuttlefish (*Sepia officinalis*) and broadtail shortfin squid (*Index coidetii*). Food Control 17(12):941–949
Vaz-Pires P, Seixas P, Mota M, Lapa-Guimaraes J, Pickova J, Lindo A, Silva T (2008) Sensory, microbiological, physical and chemical properties of cuttlefish (Sepia officinalis) and broadtail shortfin squid (*Illex coindetii*) stored in ice. LWT-Food Sci Technol 41:1655–1664
Vecchione M (2002) Cephalopods. In: Carpenter KE (ed) The living marine resources of the Western Central Atlantic.Vol. 1. Introduction, molluscs, crustaceans, hagfishes, sharks, batoid fishes and chimaeras. FAO species identification guide for fishery purposes and American Society of Ichthyuologist and Herpetologists Special Publication No. 5. FAO, Rome, pp 149–244
Vega-Gálvez A, Miranda M, Clavería R, Quispe I, Vergara J, Uribe E, Paez H, Scala KD (2011) Effect of air temperature on drying kinetics and quality characteristics of osmo-treated jumbo squid (*Dosidicus gigas*). LWT-Food Sci Technol 44:16e23
Venkatesan V, Mohamed KS (2015) Gastropod classification and taxonomy. Summer school on recent advances in marine biodiversity conservation and management, Central Marine Fisheries Research Institute, Kochi, pp 38–41
Ventrella V, Pirini M, Pagliarani A, Trombetti F, Manuzzi M, Borgatti A (2008) Effect of temporal and geographical factors on fatty acid composition of *M. galloprovincialis* from the Adriatic sea. Comp Biochem Physiol B 149:241–250
Venugopal V (2006) Applications of enzymes in fish processing and quality control. In: Seafood processing. CRC Press, Boca Raton, pp 401–425
Vongsawasdi P, Nopharatana M, Khueankhancharoen J, Changyoug C (2011) Effect of modified atmosphere packaging on qualities and shelf life of precooked baby clam (*Paphia undulata*). Kasetsart J Nat Sci 45:530–538
Wakimoto T, Kondo H, Nii H, Kimura K, Egami Y, Oka Y, Yoshida M, Kida E, Ye Y, Akahoshi S, Asakawa T, Matsumura K, Ishida H, Nukaya H, Tsuji K, Kan T, Abe I (2011) Furan fatty acid

as an anti-inflammatory component from the green-lipped mussel *Perna canaliculus* format. Proc Natl Acad Sci U S A 108(42):17533–17537
Wang Y, Zhang M, Mujumdar AS (2011) Trends in Processing technologies for dried aquatic products. Dry Technol 29:382–394
Wang Q, Xue C, Li Z, Fu X, Xu J, Xu Y (2007) Changes in the Contents of ATP and Its Related Breakdown Compounds in Various Tissues of Oyster During Frozen Storage. Journal of Ocean University of China (Oceanic and Coastal Sea Research), 6(4):407–412
Waterman JJ (1963) Processing mussels, cockles and whelks. Ministry of Agriculture, Fisheries and Food Torry Research Station. Torry Advisory Note. No. 13
Webb NB, Thomas FB, Busta FF, Monroe RJ (1969) Variations in proximate composition of North Carolina scallop meats. J Food Sci 34:471–474
Wenstrom RT, Gorton TS (1985) Apparatus for Opening and Eviscerating Scallops. United States Patent No. 4,532,677
Wijsman JWM, Troost K, Fang J, Roncarati A (2019) Global production of marine bivalves. Trends and challenges. In: Smaal AC, Ferreira JG, Grant J, Petersen JK, Strand Ø (eds) Goods and services of marine bivalves. Springer, Cham, pp 7–27
Wikipedia (2019a) Mussel. https://en.wikipedia.org/wiki/Mussel#cite_note-:0-1. Accessed 4 Oct 2019
Wikipedia (2019b) Mediterranean mussel. https://en.wikipedia.org/wiki/Mediterranean_mussel. Accessed 4 Oct 2019
Wikipedia (2019c) Perna viridis. https://en.wikipedia.org/wiki/Perna_viridis. Accessed 4 Oct 2019
Wikipedia (2019d) Perna perna. https://en.wikipedia.org/wiki/Perna_perna. Accessed 4 Oct 2019
Wikipedia (2019e) Clam. https://en.wikipedia.org/wiki/Clam. Accessed 1 Nov 2019
Wikipedia (2019f) Veneidae. https://en.wikipedia.org/wiki/Veneridae. Accessed 1 Nov 2019
Wikipedia (2019g) Oyster. https://en.wikipedia.org/wiki/Oyster. Accessed 11 Nov 2019
Wikipedia (2019h) Scallop. https://en.wikipedia.org/wiki/Scallop. Accessed 25 Nov 2019
Wikipedia (2019i) Argopecten irridans. https://en.wikipedia.org/wiki/Argopecten_irradians. Accessed 26 Nov 2019
Wikipedia (2019j) Cephalopod. https://en.wikipedia.org/wiki/Cephalopod. Accesed 25 Dec 2019
Wikipedia (2019k) Squid. https://en.wikipedia.org/wiki/Squid. Accessed 25 Dec 2019
Wongso S, Ushio H, Ohshima T, Yamanak H (1998) Changes in content of octoplne, acidic opines, related amino acids and phosphoarginine in the adductor muscle of three species of scallop during storage. J Food Biochem 22:65–81
Xiong YL (1997) Protien denaturation and functionality losses. In: Erickson MC, Hung YC (eds) Quality in frozen food. NY Chapman and Hall, New York, pp 111–134
Yamanaka H (1989) Changes in polyamines and amino acids in scallop adductor muscle during storage. J Food Sci 54(5):1133–1135
Yamanaka H, Shiomi K, Kikuchi T (1987) Agmatine as a potential index for freshness of common squid (*Todarodes pacificus*). J Food Sci 52:936–938
Yap WG, Orano CF (1980) Preliminary studies on the holding of live mussels after harvest. SEAFDEC Aquaculture Department Quarterly Research Report 4(3):22–24
Yean YS, Pruthiarenun R, Doe P, Motohiro T, Gopakumar K (1998) Dried and smoked fish products. In: Doe PE (ed) Fish drying and smoking: production and quality. Technomic Publishing Co., Lancester, pp 47–88
Yi J, Qian X, Hu X, Dong P, Liao X, Zhang Y (2013) Shucking of bay scallop (*Argopecten irradians*) using high hydrostatic pressure and its effect on microbiological and physical quality of adductor muscle. Innov Food Sci Emerg Technol 18:57–64
Young JW, Olson RJ, Rodhouse PGK (2013) The role of squids in pelagic ecosystems: an overview. Deep-Sea Res II 95:3–6
Yuan P, Deng S, Hatab S, Yuan N, Huo J (2017) Comparative evaluation of the quality changes in squid (*Ommastrephes bartrami*) during flake and slurry ice storage. Emirates J Food Agricult 29(5):339–345

Zamir R, Qasim R, Khan MA (1999) Physiochemical changes in tissue of edible mussels *Perna viridis* at refrigerator temperature (7±2°C). Pak J Biol Sci 2(1):151–156

Zandee DI, Kluytmans JH, Zurburg W, Pieters H (1980) Seasonal variations in biochemical composition of *Mytilus edulis* (L.) with reference to energy metabolism and gametogenesis. Neth J Sea Res 14:1–29

Zhang GC, Mao Z (2004) Research advances of aquatic product drying technologies. Trans CSAE 20:297–300

Zhang X, Zhao C, Huang C, Duan H, Huan P, Liu C, Zhang X, Zhang Y, Li F, Zhang HB, Xiang J (2011) A BAC-based physical map of Zhikong scallop (*Chlamys farreri* Jones et Preston). PLoS One 6(11):e27612

Zhang H, Xia WS, Xu YS, Jiang OX, Wang CX, Wang WJ (2013) Effects of spray-drying operational parameters on the quality of freshwater mussel powder. Food Bioprod Process 91:242–248

Zhang J, Walton WC, Wang Y (2017) Quantitative quality evaluation of Eastern oyster (*Crassostrea virginica*) cultured by two different methods. Aquac Res 48:2934–2944

Zhao CC, Jiang GH, Eun JB (2017) Optimization of drying process for squid-laver snack by a combined method of fuzzy synthetic and response surface methodology. J Food Qual Article ID 9761356, 10 p

Zhu J, Wu S, Wang Y, Li J (2016) Quality changes and browning developments during storage of dried-seasoned squid (*Dosidicus gigas* and *Ommastrephes bartrami*). J Aquat Food Product Technol 25(7):1107–1119

Zlatanos S (2008) Seasonal variation in the fatty acid composition of the Mediterranean mussel (*Mytilus galloprovinciallis*) cultured in Greece. Electron J Environ Agric Food Chem 7(6):3015–3019

Zlatanos S, Laskaridis K, Feist C, Sagredos A (2006) Proximate composition, fatty acid analysis and protein digestibility-corrected amino acid score ofthree Mediterranean cephalopods. Mol Nutr Food Res 50:967–970

Chapter 4
Echinoderms

4.1 Introduction

Echinoderms are defined as literally spiny skinned animals. These are exclusively marine animals. They are the best-known group of marine invertebrates due to their frequent occurrence in coastal regions, their interesting shapes and volumes, and their macroscopic size. This group includes feather stars, sea stars, brittle stars, sea urchins, and sea cucumbers. The term echinoderm which means spiny skin refers to an endoskeleton composed of hard calcium-rich plates just beneath the delicate skin.

The name Echinoderm is a Greek word and means spiny skin. The spiny skins are benthic organisms with a few exceptions and can be seen at all depths from the shore. These marine invertebrates have two different sexes, and their reproduction is outside the egg and sperm released into the sea water. It is realized by fertilization. During the life cycle of these creatures, they transform from embryos with bilateral symmetry to planktonic larvae (Zito et al. 2005).

Except for a few species which inhabit brackish waters, all echinoderms are benthic organisms found in marine environments. Echinoderms inhabit depths ranging from shallow waters at tide lines to the deep sea (Mulcrone 2005).

Echinoderm species are abundant both in coral reef communities as in the intertidal and shallow subtidal areas around the world. Echinodermata members usually prefer muddy and sandy regions, spaces of stone and algae in shallow waters, areas where *Posidonia* is dense (Belbachir and Mezali 2018), under stones, hard and rocky sheltered areas, and grounds with alga (Hermosillo-Nunez et al. 2015). Echinoderms are one of the macrobenthic animals associated with seagrass. Echinoderms play an important role in the ecology of the marine ecosystem, they are primary consumers in the food chain cycle in these ecosystems, and echinoderms can act as a beach cleaner for marine organisms that consume organic materials that enter the water (Uneputty et al. 2017).

In 2016, the total production of echinoderms in the world was reported as about 105 thousand tons by capture and 217 thousand tons by aquaculture (FAO 2016).

N. Gökoğlu, *Shellfish Processing and Preservation*,
https://doi.org/10.1007/978-3-030-60303-8_4

Increasing global pressures for the collection of echinoderms for various commercial enterprises threatened these invertebrates and contributed to the ongoing concern of depletion of marine resources worldwide. Increasing global trade with the collection of echinoderms for home aquariums, souvenirs, and biomedical products as well as overfishing threatened some echinoderm species (Micael et al. 2009).

4.2 Sea Urchin

4.2.1 General Information About Sea Urchins

4.2.1.1 Biology of Sea Urchins

Sea urchins are members of a large group of marine invertebrates in the phylum Echinodermata. All sea urchins have a hard-calcareous shell called a test, which is covered with a thin epithelium and is usually armed with spines. The spines are used for locomotion, for protection, and for trapping drifting algae for food. Between the spines are tube feet that are used in food capture, for locomotion, and for holding on to the substrate. Sea urchins also have small pinchers, called pedicellariae, that are used for defence and for clutching food. The mouth is located on the underside. It consists of a complex array of skeletal elements, plates, and teeth arranged in five-symmetry called "Aristotle's lantern." The mouth leads to the digestive tract, which empties through the anus located on the top of the test. The sexes are generally separate in sea urchins. Females release up to several million eggs into the sea, where fertilization takes place (Price 2000). The reproduction periods of sea urchins vary according to their habitat, and the most common months of ovulation are July and August. The release of eggs and sperm into the sea water takes place through pores located on ambulacral plaques (King et al. 1994). After a multistage planktonic existence, the larvae settle and metamorphose to the characteristic form. In sea urchins, the body is spherical or flattened. There are no arms and the most prominent feature is that the limestone plates interlock each other and form a shell. This shell completely covers the other sides of the body, except for only two poles (Price 2000).

Sea urchins, which have a wide variety and spread, are very important organisms because they are the species that provide information about water pollution in their regions. In addition to being of serious ecological importance, the consumption of gonads in the food industry enables sea urchins to form an important place worldwide (Guidetti et al. 2004; Gianguzza et al. 2006).

4.2.1.2 Distribution of Sea Urchins

Sea urchins are widely distributed in polar, temperate, and tropical oceans, where they are conspicuous members of most benthic marine communities. They play an important ecological role as herbivorous grazers, and their ability to alter algal community states has made them the subject of numerous ecological studies. There are about 850 living species of sea urchins, and at least 17 of these are commercially valued as food (Harris and Eddy 2015).

4.2.1.3 Production and Trade of Sea Urchins

At first glance, the hard shell and long spines of sea urchins give the impression that they are not a creature that people will be overly interested in eating. Whereas, sea urchin or urchin roe is an important culinary taste and flavour in many European countries, Chile, North America, Asia and especially Japan, which make up about 90% of the worldwide demand (Sun and Chiang 2015). The market for sea urchins is very traditional with Japan consuming about 80–90% of the total current global supply. There is a domestic market in many sea urchins harvesting countries, especially in Chile, New Zealand, and the Philippines. In Europe, the market is also traditional and is mainly in continental Europe and in the Mediterranean countries, Italy, France, and Spain (Stefansson et al. 2017). In a report on the global marketing of sea urchins prepared by Stefansson et al. (2017), it has been reported that the global annual market demand for sea urchins can be estimated at 60,000–70,000 tons and it was difficult to obtain reliable figures on the size of the global market based on harvest figures reported by FAO in 2014. They also reported that it is difficult to determine the world harvest volume of sea urchins, especially in countries producing before the 1980s, since they were transmitted to FAO below estimates. Other echinoderms are sometimes collected with sea urchin data and FAO cannot easily filter differences.

About half of the global harvest consists of Chilean sea urchins. Russia, Japan, the United States and Canada are other main harvesters. *Strongylocentrotus intermedius*, *Strongylocentrotus franciscanus*, and *Strongylocentrotus droebachiensis* are mainly harvested species (Sun and Chiang 2015).

In most aquatic species, the term roe applies only to eggs, but in the case of sea urchins, roe refers to both male and female gonads. Urchin roe, or uni, is said by many to be an acquired taste. The texture is slimy, and the sweet-salty combination is rather unusual for seafood. Over time, however, consumers have spread the word about their fondness for urchin. In Norway, a sea urchin entrepreneur attempted to increase local demand by describing the taste as similar to that of Russian caviar. The world's largest wholesale sea urchin market is in Japan. As Japanese urchin stocks were exploited and overfished, leading to the collapse of native stocks in the early 1970s, the continued growth of uni consumption in Japan was increasingly met by other countries. Generally, Japanese domestic uni auctioned is considered to be of higher quality than imports, and the freshest urchin with the best uni flavour and quality is captured from the inshore waters of the Japanese island of Hokkaido, where almost half of the Japanese sea urchin harvest is landed. Because of its traditional role in Japanese cuisine and the relative scarcity of urchins, uni is considered a delicacy (Sun and Chiang 2015).

It has been reported that sea urchin roes are one of the popular taste and flavours of Mediterranean cuisine and Italians mix the sea urchin eggs with various sauces and serve them with pasta (Grisolia et al. 2012).Sea urchin is consumed in Mediterranean countries in a variety of ways such as blended into sauces, with pasta, breads or used as an ingredient in various dishes. Sea urchins have also found popularity outside the traditional markets with increasing ethnic populations and the growing popularity of sushi restaurants (Stefansson et al. 2017).

Sea urchin roes are also a popular product in the United States. The state of California is an important fishery centre for the red sea urchin and offers fresh roe to regional markets. California urchins are also exported to Asian markets in substantial amounts (Grisolia et al. 2012). On the United States Pacific Coast, red, purple, and green sea urchins are commercially harvested. On the Atlantic Coast, only green sea urchins are commercially harvested (Price 2000).

Many areas of Southern and Western Australia are home to the purple sea urchin *Heliocidaris erythrogramma*, generally considered to be the highest quality local urchin. As with urchins caught in the United States, Australian urchins of particularly high quality are often exported to the lucrative Japanese market. Moreover, like the United States, the development of an export market to Japan was followed by the growth of a larger domestic market in Australia for urchin roe than seen in decades past.

In Europe, France, Italy, and Spain are the leaders in terms of sea urchin consumption. The purple sea urchin species favoured in France, *Paracentrotus lividus*, is harvested from the Atlantic and Mediterranean areas as well as in Ireland. *Sphaerechinus granularis* (purple sea urchin) is mainly harvested in the Netherlands and France. *Psammechinus miliaris* is harvested from the Atlantic coast of France and *Strongylocenrotus droebachiensis* from Iceland, Norway (Monfort 2002).

Prices of sea urchins vary with some factors such as appearance, colour, flavour, texture, species, region of harvest, demand and distribution and processing (Sun and Chiang 2015; Stefansson et al. 2017).

4.2.1.4 Fishery of Sea Urchins

Sea urchins are commonly found in the mid to low intertidal zone, at depths up to about 50 m. They prefer rocky substrate and are often found in and around kelp beds. Scuba diving is most common technique used for fishing sea urchins around the world. It requires basic SCUBA equipment to dive to depths between 2 and 20 m and although depths as far as 40 m can be achieved shallower depths are most common. This type of diving can be made using single or double tanks and using basic compressed air or mixed gases to extend dive times. There are many restrictions in fishing of sea urchins by scuba diving such as requiring expensive equipment such as full diving team and boats and diving safety issues (James et al. 2018).

In United States, West Coast sea urchins are commercially harvested by divers using "hooka" diving gear, consisting of a low-pressure air compressor that feeds air through a hose from the vessel to the divers. Sea urchins are harvested from the ocean bottom with a hand-held rake or hook and put into a hoop-net bag or wire basket. The basket is winched onto the boat and emptied into a larger net bag (Price 2000). Since urchins are selected manually with this method, the fact that individuals of the desired size can be selected by divers and the undersized sample can be sent back without significant mortality contributes to the sustainability of sea urchins (James et al. 2018). Urchins are harvested with a drag in areas where diving is not feasible.

The average harvest size has decreased gradually due to over exploitation in countries where sea urchins are fished. For this reason, regulations such as limiting the harvest season and creating protection areas have encouraged the sea urchin culture (Sætra 2019).

4.2.1.5 Some Important Species of Sea Urchins

Various types of sea urchins are harvested for their roe globally. Green sea urchin (*Strongylocentrotus pulcheriius*), red sea urchin (*Strongylocentrotus franciscanus*), and purple sea urchin (*Strongylocentrotus intermedius*) are the species of highest demand globally. *Pseudocentratus depressus*, *Heliocidaris crassispina*, and *Strongylocentrotus droebachiensis* are also in good demand (Bledsoe et al. 2003).

Tripneustes ventricosus **(Lamarck, 1816) (White Sea Urchin)**
The White Sea urchin is a member of the sea urchin group and is large. As with all urchins, it has a spherical exoskeleton (test) consisting of fused calcareous plates placed in two separate groups containing five ambulacral and five inter-ambulacral series. The plate series extend from the mouth in the centre of the animal's ventral side to the anus in the centre of the dorsal side. They use their spines for defence against predators. They can do this while on the move. The reproductive system of the white Sea urchin consists of five gonads, sometimes more or less fused and suspended by mesenterial strands to the "roof" of both male and female urchins. The gonads are not only the source of eggs or sperm, which are referred to as roe, but also serve as the main nutrient storage organ. Gonads tend to be bright orange in colour in females and light yellow in males. The white sea urchin is found along the west coast of Africa from the Gulf of Guinea to Walvis Bay and along the western central Atlantic Ocean from Bermuda, to the Carolina coast of the United States of America, and the Caribbean to Brazil. White sea urchins live in various shallow water habitats such as rock rubble, moss rock circles and sea grass beds. Sometimes many White Sea urchins form in tidal pools. However, it tends to high death events due to very high water-temperatures and periodic subaerial exposure of such habitats, especially in tides (Pena et al. 2010).

Paracentrotus lividus **(Lamarck, 1816)**
P. lividus is distributed from the Mediterranean along the Northeast Atlantic coasts and in shallow waters where it can generally reach high density. They have a wide range of water temperatures in the winter season, in the range of 10–15 °C, and in the summer, in the range of 18–28 °C. They prefer grounds with sandy, rocky and sea meadows in sub littoral areas between 10 and 20 m depth. *P. lividus* are very sensitive to low and high salinity values. Usually, deaths occur when they stay in the salinity values of 0.15% and 0.39% for a long time (Boudouresque and Verlaque 2007).

Strongylocentrotus droebachiensis **(Müller, 1776) (Green Sea Urchin)**
The green sea urchin lives at a depth of 0–300 m, but is usually 0–50 m. They are mainly found on hard surfaces like rocks and on cobbles. They are mainly found on hard substrates like boulders and on cobble. They also occur on soft substrate, but in exposed areas they are prone to be crushed by stones thrown around by turbulent water, especially the smaller sea urchins. Sea urchins are often associated with *Laminaria* kelp and can often destroy the kelp bed when the sea urchin biomass reaches a critical point, and thus create a barren. The green sea urchin is an omnivore and feeds on algae, invertebrates, mussels, and other food sources it can scavenge. Food preference is a complex response to abiotic and biotic factors, such as environmental causes and nutritional status (Sætra 2019).

Stomopneustes variolaris **(Lamarck, 1816)**
Growing up to a test diameter of about 18 cm, the red sea urchin is one of the largest types of sea urchin in the world. It has a dark purple colour test and spine. Whether it is rarely, the test or spine can be reddish or light purple. Since the length of the vertebrae may be related to the habitat, sea urchins tend to have larger spines in deeper, calmer waters (Jinadasa et al. 2016). Red sea urchins are found on the west coast of North America. They range northward to Sitka and Kodiak, Alaska, and along the Asiatic coast as far south as the southern tip of Hokkaido Island, Japan (Kato and Schroeter 1985).

Strongylocentrotus purpuratus **(Stimpson, 1857)**
The purple sea urchin, usually occupy shallow waters, from the mid to low intertidal zones to depths more than 50 m. They prefer rocky substrates, particularly ledges and crevices, and avoid sand and mud.

4.2.1.6 Nutritional Composition of Sea Urchins

Sea urchin roe contains an assortment of nutrients. The nutritional composition of sea urchins varies depending on their nutritional and reproductive status. Sea urchin roe also contains calcium, phosphorus, iron, Vitamins A, B_1, B_2, B_{12}, nicotinic acid, pantothenic acid, folic acid, and carotenes (Kato and Schroeter 1985).

The average moisture, crude protein, crude fat and crude ash contents of *Stomopneustes variolaris* samples obtained from three different stations were determined as 69.28%, 14.99%, 7.99% and 2.28%, respectively (Jinadasa et al. 2016). In another study, moisture, crude protein, crude oil, and crude ash contents of *P. lividus* samples were found as 79.87%, 12.03%, 3.05% and 2.25%, respectively (Mol et al. 2008).

The chemical composition of sea urchins varies according to the season, feeding, temperature and breeding period. It has been reported that season and gender affected the chemical composition of *Evechinus chloroticus* gonads (Verachia et al. 2012) and *Paracentrotus lividus* (Dincer and Cakli 2007; Rocha et al. 2019; Martinez-Pita et al. 2010) and *Arbacia lixula* (Martinez-Pita et al. 2010). Similar results were reported for *Strongylocentrotus franciscanus* (McBride et al. 2004), *Arbacia lixula*. Gonads of *Paracentrotus lividus* showed the highest glycogen concentration in autumn and winter months compared to gonads during the spring and summer months having the lowest glycogen concentration due to spawning. The seasonal variations observed the biochemical composition of *P. lividus* gonads was related to the energy requirements for protein synthesis in gamete production (Montero-Torreiro and Garcia-Martinez 2003). The seasonal variations in the gonads of *P. lividus* were observed mostly for the proteins and lipids (Fernandez 1997, 1998).

The main component of the nutritional composition of sea urchins is protein. In particular, the protein content of the gonads is related to the reproductive period. Before ovulation, gonads contain significant levels of protein. There is an inverse relationship between protein storage and gametogenesis. Carbohydrate is used as the main energy source for the growth of gonads of sea urchins. When the gonad mass reached the highest level, the carbohydrate content is at the lowest level. Carbohydrate content increased in spring when food was abundant (Arafa et al. 2012).

Fatty acid composition of sea urchin varies with diet composition and reproductive status. The polyunsaturated fatty acids of *P. lividis* gonads collected from the intertidal zone of the Tunis Gulf constituted the highest rate among the total fatty acids. The highest PUFA level was determined in winter and spring, and the lowest level in summer (Arafa et al. 2012). Similarly, in *P. lividus* PUFAs were more abundant than SFAs and MUFAs (De La Cruz-García et al. 2000; Rocha et al. 2019). Mol et al. (2008) found that MUFA and PUFA contents of *P lividus* were higher than SFA. EPA and DHA were the major polyunsaturated fatty acids of sea urchin gonads (Arafa et al. 2012; Rocha et al. 2019).

4.2.2 Postharvest Quality Changes in Sea Urchins

Japanese buyers take several factors into account when purchasing uni. Colour, texture, presentation, and size are all important considerations, making for a highly competitive market. The best quality urchin roe is usually golden orange to yellow and has a distinct, sweet ocean taste, while poorer quality urchin tends to be bitterer or have brownish coloration. In markets such as the United States and Canada, where urchin roe is still a relatively new product, quality is not as important as in Japan (Sun and Chiang 2015).

Uni may be sold fresh, but is also steamed, baked, sautéed, or frozen. Uni used for sushi is brined and treated with alum. It may also be coloured. Colour and condition are important for determining grades of uni, and a bright orange product is the most desirable. The salted gonadal tissue from the sea urchin may also be fermented, producing a paste (neriuni). Also, a more watery preparation, mizuuni, is prepared using a dry cure process. Doro uni is made by washing the gonadal tissue with dilute alcohol, then draining it, and mixing with salt. Sea urchins are cultured in China, and some wild harvested animals are held in raceways where they are fed a high-nutrition diet. Proper husbandry and nutrition can yield over a 40% increase in gonadal production (Batista 2007).

Sea urchin processors have little control over the quality of extracted roe, as this mainly depends on where and when the sea urchins were harvested. Buyers for processing companies try to assert some quality control dockside by sampling the catch before they agree to purchase, which allows them to either reject the harvest outright or pay the fishermen a lower or higher price depending upon the quality (Sun and Chiang 2015).

In a study examining the quality and shelf life during refrigerated storage of fresh roes of *Paracentrotus lividus* sold in Palermo, Italy, it was determined that roes were sensorially acceptable for up to 72 h, *Listeria* and *Salmonella* were not isolated, *Vibrio* was isolated in 71.43% of the samples (Panebianco et al. 2011).

4.2.3 Postharvest Handling of Sea Urchins

4.2.3.1 Live Transport of Sea Urchins

Sea urchins are relatively hardy animals, and if handled correctly, they can survive considerably out of the water without suffering high mortality and quality loss, even after immersion in sea water. In a previous trial, green sea urchins collected were transported to a live holding system. They were held in this system until the beginning of the experiment (approximately 6 h). Two temperature regimes were applied (ambient and 4 °C up to 28 h) and (5 and 3 °C up to 56 h). They were held in an insulated plastic container. Inside the container the urchins were covered with hessian sacks that had been soaked in seawater. The maximum transport period for an average transport temperature of 14 °C was found 14 h, at 5.0 °C was 36 h, and at 3.0 °C was 44 h. If the transport period is longer then it becomes necessary to transport sea urchins immersed in cool, seawater. There are systems designed to carry live seafood in chilled shipping containers and although they are not specifically designed for sea urchins, they may also be suitable for transporting sea urchins for long distances and over extended periods (James and Evensen 2018). Holding and shipping of sea urchins in recirculating sea water may open options on keeping them alive for long periods especially if ammonia stripping is carried out. Still, such transport is challenging, expensive and difficult, due to factors such as oxygen supply, increase of carbon dioxide in the water, accumulation of toxic ammonia and changes in pH. For these challenges, a RAS holding, and transport system was developed by Technion, Israel Institute of Technology and BioFishency which not only recirculates the water, but additionally controls the pH and removes toxic ammonia. The survival of sea urchins held in the RAS system at 4 °C was high during the first 5 days. Eight days from catch the survival was only 80%, after 12 days about 50% and after 15 days, 10% (Stefansson and Olafsdottir 2019).

James et al. (2017) conducted a simulated transport trial for live sea urchins to investigate possible transport times to deliver sea urchins to a processor, markets, or restaurants in a live, whole, and high quality. Sea urchins collected were hold in the sea in catch bags until they were packed into 'isopro' (polystyrene) boxes. The urchins were randomly allocated to one of six boxes that were packed with ice gel packs. The boxes were then exposed to various transport periods. During the transport periods, the isopro boxes were stored in a cool room (approximately 10–13 °C). After 24, 34 and 44 h respectively, they were opened, and the condition of the sea urchins and any mortalities were recorded. According to the results of the experiment, it was determined that there was almost no death in the transfer times used in the trial. It was found that only one chestnut died during the transplant. When the sea urchins were taken out of the transport boxes, all the urchins looked in good condition. Very limited mortality was reported in seawater holding systems after 7 days of storage. According to the results of the experiment, it was determined that there was almost no death in the transfer times used in the trial. It was found that only one chestnut died during the transplant. When the sea urchins were taken out of the transport boxes, all the urchins looked in good condition. Very limited mortality was reported in seawater holding systems after 7 days of storage.

4.2.4 *Processing of Sea Urchins*

Processing and packaging of the sea urchins should be done with great care, especially if the product is intended for the Japanese market. Sea urchins can be shipped live to markets or processed to extract the roe (Sun and Chiang 2015). Whole sea urchins are transported to the processing facility and stored in the cold until they are processed there. Before processing, the shell should be opened, and the gonads should be removed with a small spoon. A knife is used to strike along the central line of the shell, separating it into two parts, the upper and lower. If carried out carefully, shell and spines will not be mixed with the gonads. The shell is shaken gently up and down, with the mouth side down, to get rid of some of the viscera and ingested food, then the gonads are removed with a spoon, bamboo scoop or fingernails (Kramer and Nordin 1979). Roes are placed in baskets after excess membranes are removed. The gonads are placed in a thin layer in a shallow basket for draining. This basket is put in a cool, dark place as warmth causes deterioration and oozing. Once the gonads are removed, they should be protected from direct sunlight (Kramer and Nordin 1979). The baskets come to the packaging line after immersion in salted ice water and potassium solution to maintain the firm texture of the roe. The highest quality roe should be of uniform size, firm texture, attractive colour, and optimum maturity. Some high-end roe is packed into plastic trays or cups destined for the Japanese supermarket trade. The highest quality roe is purchased by buyers from sushi bars and restaurants, who prepare the fresh roe and repackage and combine it with other products in a meal. Some of the higher quality urchins are sold by supermarkets and fish retailers serving household consumption. Sea urchins are generally purchased from supermarkets to consume on special occasions. There is also a relatively important gift market, where uni is packaged and processed in various ways and sold with catalogues for special occasions. These are given as gifts for special occasions such as the New Year, welcome parties, and visits to relatives or friends. It is sometimes sold mixed with other products. In such products, lower quality, less mature, smaller, less hard urchins are used. These low-quality roes are also used in dried, salted, baked and frozen products (Reynolds and Wilen 2000). Wholesalers play an important role in distributing sea urchins to various buyers, such as restaurants (mainly Japanese style restaurants), supermarkets, fish markets, and catering companies (Sun and Chiang 2015).

4.3 Sea Cucumbers

4.3.1 General Information About Sea Cucumbers

4.3.1.1 Biology of Sea Cucumbers

Sea cucumbers are echinoderms from the class Holothuroidea. They are marine animals with a leathery skin and an elongated body containing a single, branched gonad. Like all echinoderms, sea cucumbers have an endoskeleton just below the skin, calcified structures that are usually reduced to isolated microscopic ossicles joined by connective tissue. In some species these can sometimes be enlarged to flattened plates, forming armour. In pelagic species such as *Pelagothuria natatrix*, the skeleton is absent and there is no calcareous ring. Most sea cucumbers, as their name suggests, have a soft and cylindrical body, lengthened, rounded off and occasionally fat in the extremities, and generally without solid appendages. Their shape ranges from almost spherical for “sea apples” to serpent-like for Apodida or the classic sausage-shape, while others resemble caterpillars. The mouth is surrounded by tentacles, which can be pulled back inside the animal. The body of a holothurian is roughly cylindrical. It is radially symmetrical along its longitudinal axis and has weak bilateral symmetry transversely with a dorsal and a ventral surface. There are five ambulacra separated by five ambulacral grooves, the interambulacral. The ambulacral grooves bear four rows of tube feet but these are diminished in size or absent in some holothurians, especially on the dorsal surface. The two dorsal ambulacra make up the bivium while the three ventral ones are known as the trivium (Wikipedia 2020a).

Sea cucumbers are fed with detritus in the bottom structure and planktons in the water column. It is known that these creatures play a major role in maintaining the ecological balance due to their nutritive properties with the bottom structure of detritus. Species live in a variety of bottom types, from surface waters to 100 m deep, and often form dense populations in shallow sub-littoral sea grass meadows (Kazanidis et al. 2010).

Sea cucumbers are responsible for mixing and changing the substrate. Holothurians are sediment strainers that pass large amounts of sediment through their gut, assimilating bacteria, detritus and low-content organic matter, especially living diatoms. Sea cucumber improves environmental conditions by preventing the formation of anaerobic conditions with both nutrition and movement activity and causes an increase in species diversity. In some regions, the absence of sea cucumber can cause the sea floor to solidify, which can cause the destruction of the necessary habitat for other benthic and buried living creatures (Bruckner et al. 2003).

4.3.1.2 Distribution of Sea Cucumbers

Although sea cucumbers generally live close to the beach, they are creatures that can be found at any depth. In general, they prefer sandy, sandy-muddy grounds with algae and coral reefs. Sea cucumbers are an economically valuable aquatic creature found throughout the world, especially in the tropical West Pacific and Indian coasts, in the South Pacific and Asian countries, in the seas of Japan.

Sea cucumber production through cultivation worldwide is 205 thousand tons and it is stated that 99% of it is made in Asian countries, mostly in China and Indonesia. Sea cucumber is produced intensively in Asian countries, especially in Sri Lanka, Indonesia, and Korea, and recently in American and African countries. According to FAO data, 221 thousand tons of 238 thousand tons of production made by hunting and aquaculture are carried out in Asian countries (FAO 2016).

4.3.1.3 Production and Trade of Sea Cucumbers

The international trade structure for sea cucumbers is different from typical aquatic trades. The demand is limited to Chinese and other Asian consumers (often of Chinese origin), and the trade is also controlled by these groups. About 90% of the trade volume goes through Hong Kong and Singapore, while the Chinese mainland is the main consumer market. Sea cucumber is chiefly exported in dried form (beche-de-mer), whereas fresh and frozen products only take up a small proportion of the international trade (Xu et al. 2015).

The highest Holothuroidea capture landings in the 2000s were yielded by Indonesia, followed by the Philippines. On average basis, almost 47% of the world's Holothuroidea landings per annum, producing an average of 2572 t (wet weight) catches per year, were contributed together by the Philippines and Indonesia during 2000 and 2005. Japan as the largest capture fishery producer of the temperate species (*A. japonicus*), produced an average of 8101 t per year between the period 2000 and 2005 (Bordbar et al. 2011).

Sea cucumber is a premium product with large popularity in China, the largest producer of this seafood and its related products. The Chinese sea cucumber farming industry showed an overall stable trend in 2017. Farm-gate prices have continued an upward trend in the past 2 years. Production volumes and stocked amounts have also increased significantly. Total production of sea cucumber in China amounted to 200,000 tonnes in 2017. The provinces of Shandong, Liaoning, and Fujian led the national supply, contributing with 100,000 tonnes (50%), 70,000 tonnes (35%), and 20,000 tonnes (10%), respectively. The Hebei province produced 10,000 tonnes, while other provinces accounted for the remainder 5%. In 2017, the total output value of sea cucumber products reached nearly CNY 30 billion1 (USD 4.44 billion) in China (FAO-Globefish 2018).

Countries trading sea cucumbers export to one of the three main centers, where they are re-exported, mostly to Chinese consumers around the world (Bruckner et al. 2003).

4.3.1.4 Fishery of Sea Cucumbers

Harvesting sea cucumbers is done by hand from intertidal waters or diving from deeper waters. Fishing gears often damage the skin of sea cucumbers in deep water, which leads to poor quality of the final product (Mathews et al. 1990). The divers usually store the harvested sea cucumbers in sacks, while the reef collectors store them in containers or buckets with clean sea water.

The traditional dive-fishing method for sea cucumber is still being used in northern China, the only difference being that the diver's equipment and the containers for storing the sea cucumbers have been improved. In general, divers leave the boat, seek sea cucumber visually between 2 and 20 m depth, and then collect them by hand and store them in mesh bags tied to their waist. The bags are made of nylon or polypropylene rope, with a mesh size of 0.5–1.0 cm, a length of 60–80 cm, an opening diameter of 15–20 cm, and a bottom diameter of 40–50 cm. The divers work underwater for about 1–2 h, less in winter when lower water temperatures prevail (Liu et al. 2015).

Only larger individuals should be collected during harvest, while smaller ones should be left to mature and grow to the length required for harvest (Purcell et al. 2012).

Sea cucumber should be kept away from sunlight to prevent the skin from drying out, causing the end product to spoil. High-value species are usually kept on a flat surface in a single layer, away from sunlight, until treated. If stacked on top of each other, the body wall may break, and this will decrease its value (Ram 2017).

It has been reported that sea cucumber stocks are under the risk of excessive exploitation due to its limited and irregular distribution, easy collection, and slow recycling of overfishing (Bumrasarinpai 2006).

4.3.1.5 Some Important Species of Sea Cucumbers

Among the approximately 1200 known species of sea cucumber, some 70 are harvested worldwide. Sea cucumbers belong to the class Holothuroidea and so are also referred to as holothurians. Most species harvested commercially belong to the order Aspidochirotida, specifically to the families Holothuriidae and Stichopodidae, and are mostly tropical. A few species belonging to the order Dendrochirotida, family Cucumariidae, are also fished commercially. Species in the orders Apodida, Dactylochirotida, Elasipodida and Molpadida are mostly not fished (Purcell et al. 2012).

The main sea cucumber species grown in aquaculture are *A. japonicus* (temperate species), *Isostichopus fuscus* (tropical species), and *Holothuria scabra* (tropical species) (Xu et al. 2015).

Apostichopus japonicus **(Selenka, 1867)**

Apostichopus japonicus is a species of sea cucumber in the family Stichopodidae. It is found in shallow temperate waters along the coasts of south east Asia. The Japanese sea cucumber is found along the coast of Russia, China, Japan and Korea. The range extends from Alaska and Sakhalin Island to the Amami Islands, Japan. The Japanese sea cucumber sifts through the sediment on the seabed with its tentacles and feeds on detritus and other organic matter including plant and animal remains, bacteria, protozoa, diatoms, and faeces. It lives in temperate seas. In locations where the water heats up excessively in summer it undergoes aestivation, going into a state of dormancy. The threshold temperature is about 25 °C. The Japanese sea cucumber is used for food. The largest fishery is in Japan. Fishing methods include diving and hand collection at depths of up to 20 m and the use of trawls at greater depths (Wikipedia 2020b).

Halothurian edulis **(Lesson, 1830)**

Holothuria edulis, commonly known as the edible sea cucumber or the pink and black sea cucumber, is a species of echinoderm in the family Holothuriidae. *Holothuria edulis* is a medium-sized sea cucumber reaching a length of about 30 cm, is a common and widespread species in the Indo-Pacific Ocean. It lives on the seabed at depths down to 20 m. Its range extends from the Red Sea and East African coast to Sri Lanka, Japan, China, Indonesia, the Philippines, northern Australia, and various Pacific islands. It is found in several different habitats including on sandy or muddy substrates, on coral rubble and in seagrass meadows. It can be found on inner and outer reef flats, on back reef slopes and in lagoons (Wikipedia 2020c).

Actinopyga agassizii **(Selenka, 1867)**
Actinopyga agassizii can be found from 0 to 54 m deep. It forages on fine detrital sediments in algal turfs, seagrass beds and in rubble or sand-covered areas. Distribution of this species is in Caribbean coast of Florida (USA), Cuba, Mexico, Puerto Rico, Dominican Republic, Haiti, Jamaica, Belize, Guatemala, Honduras, Nicaragua, Costa Rica, Panama, Colombia (Atlantic), Venezuela (Bolivarian Republic of), the Bahamas, Barbados and the United States of America. The Japanese sea cucumber sifts through the sediment on the seabed with its tentacles and feeds on detritus and other organic matter including plant and animal remains, bacteria, protozoa, diatoms, and faeces (Purcell et al. 2012).

Actinopyga echinites **(Jaeger, 1833)**
Despite being named "deep-water" redfish, the species lives in shallow waters, mostly on flats (reefs and seagrass beds) down to 10 m depth with relatively high densities. This species has separate sexes and is reported to live more than 12 years. Spawning occurs in the dry season, size of maturity is reported at about 12 cm, or a weight between 45 and 90 g. It is found throughout the western central Pacific, Asia, Africa, and Indian Ocean region. Common in the Indo-Pacific, islands of western Indian Ocean, Mascarene Islands, East Africa, Madagascar, southeast Arabia, Sri Lanka, Bay of Bengal, East Indies, north Australia, the Philippines, China and southern Japan, South Sea Islands (Purcell et al. 2012).

Holothuria tubulosa **(Gmelin, 1791)**
Holothuria tubulosa, the cotton-spinner or tubular sea cucumber, is a species of sea cucumber in the family Holothuriidae. It is the type species of the genus Holothuria and is placed in the subgenus Holothuria, making its full name *Holothuria tubulosa*. *Holothuria tubulosa* is found in temperate regions of the eastern Atlantic Ocean as far north as the Bay of Biscay and in the Mediterranean Sea, where it is abundant. It is found on sandy seabeds, among seagrass (*Posidonia* spp.) and on muddy rocks to a depth of about 100 m (Wikipedia 2020d).

4.3.1.6 Nutritional Composition of Sea Cucumber

The body wall is the sea cucumber's major edible (and medicinal) organ. It consists mainly of epithelial and dermal connective tissues. Among the general biochemical components (related to nutrition), moisture content stands the highest, followed by crude protein, ash content and total carbohydrate, and finally crude fat content. The body wall of sea cucumber also contains polysaccharides, lipids, and collagen (Xia and Wang 2015).

Proteins are the most abundant chemical components and can account 40–60 wt% of sea cucumber dry matter (Wen et al. 2010). Most sea cucumber proteins are in collagen form. Collagens, which constitute the body tissues of sea cucumbers, contribute to the flavours of sea cucumbers as the main structural proteins (Liu et al. 2017).

Protein and lipid contents of sea cucumber vary depending on season. Seasonal variations in the proximate composition in body wall of *A. japonicus* have been reported (Gao et al. 2011). In that study, it was determined that the moisture content changed irregularly in all seasons, and the moisture content reached the lowest value in September with a slow decrease from June to September, and the highest in November. It was reported that the highest protein values were reached in September. Minimum carbohydrate values were found in September and November. The highest lipid contents were found in March. On the other hand, ash content did not change throughout the year (Gao et al. 2011). Ash content in sea cucumbers is higher than in other sea foods due to the many calcareous ossicles found in body wall tissues (Liu et al. 2015). The highest lipid content for *Isostichopus* sp. was found in May and the highest percentage of protein was determined in July (Vergara and Rodríguez 2016).

During the aestivation period, *A. japonicus* ceases to feed. Interestingly, crude protein, carbohydrate, and lipid contents of fresh body wall tissues were comparatively high when the moisture content was at its minimum. As *A. japonicus* loses 30–50% of its body weight during aestivation, it can be suggested that the main nutritional constituents of the body wall, including protein, carbohydrate, and lipid, may all participate in the energy supply. The results may also show that there are no special needs for lipid as a source of energy during the aestivation period (Liu et al. 2015).

Sea cucumbers contain an interesting combination of valuable amino acids; glycine being the major component in almost all species identified. Glutamic acid, aspartic acid, alanine, and arginine are prominent, among others. Another important feature of sea cucumber's amino acid composition is its low lysine/arginine ratio together with high score of essential amino acids (EAAs) due to presence of considerable amount of threonine, tyrosine, and phenylalanine (Wen et al. 2010; Bordbar et al. 2011). A total of 18 amino acids were identified in the dry body wall tissues, digestive tract, and respiratory tree of *A. japonicus*. They were alanine, valine, leucine, isoleucine, proline, phenylalanine, tryptophan, methionine, glycine, serine, threonine, cysteine, tyrosine, aspartic acid, glutamic acid, lysine, arginine, and histidine (Liu et al. 2015).

Moreover, sea cucumbers have Vitamin A, Vitamin B_1 (thiamine), Vitamin B_2 (riboflavin), Vitamin B_3 (niacin), and minerals, especially calcium, magnesium, iron, and zinc (Chen 2003; Ridzwan 2007).

There are very few studies on sea cucumber mineral composition. Eleven sea cucumber species were reported as excellent sources of macro-elements (Na, Cl, Ca, and Mg). Calcium is present in sea cucumbers in large quantities (Chang-Lee et al. 1989). It has been reported that Ca concentrations of sea cucumbers are significantly higher than that of other seafood products such as fish, shrimps, crabs. The fact that sea cucumbers are embedded in a bendable body wall, isolated, have the skeleton of microscopic ossicles consisting of $CaCO_3$, is shown as the reason for the high content of Ca (Wen and Hu 2010).

Some studies have shown that the body wall of the sea cucumber contains many types of active polypeptides. The pentapeptides that consist of leucine, proline, serine, and arginine are extracted from the epithelial tissue of sea cucumber transplantation tumor through immune mediation. The lipid composition of sea cucumber mainly consists of phospholipid, which makes up 90% of the total fat content. The cholesterol content of sea cucumber as a low cholesterol animal source is 1% (Yang and Bai 2015).

It has been reported that the composition of fatty acids varies by species and region and the feed has an important effect on the fatty acid profile since sea cucumbers are fed at the bottom. It is reported that sea cucumbers (*Holothuria tubulosa*, *Holothuria polii* and *Holothuria mammata*) (Aydin et al. 2011) and *Holothuria arenicola* (Haider et al. 2015) have higher PUFA content compared to MUFA and SFA, and especially deep sea species need higher PUFA content to maintain membrane fluidity at high pressure and low temperatures (Aydin et al. 2011). However, Gonzalez-Wanguemert et al. (2018) reported SFA as the dominant for *H. mammata*, while they reported SFA as the least concentrated fatty acid for *H. polii* and *H. tubulosa*. On the contrary, Fawzya et al. (2015) reported that SFA is dominant for *T. ananas* and *H. fuscogilva*. Wen et al. (2010) also reported dominance in SFA for seven different sea cucumber species (*S. herrmanni*, *T. ananas*, *T. anax*, *H. fus*cogilva, *H. fuscopunctata*, *A. caerulea*, *B. argus*) and Ridzwan et al. (2014) reported for *Holothuria scabra*, *Holothuria leucospilota*, *Holothuria atra*. Al Azad et al. (2017) reported for *Holothuria edulis* and *Holothuria scabra*. In studies with various sea cucumber species, the higher EPA content has been reported compared to the DHA content (Xiang et al. 2006; Zhong et al. 2007; Li et al. 2009). DHA was not detected in eight different sea cucumber species (*S. herrmanni*, *T. ananas*, *T. anax*, *H. fusco*gilva, *H. fuscopunctata*, *A. caerulea*, *B. argus*) by (Wen et al. 2010) and in three species (*Holothuria scabra*, *Holothuria leucospilota*, *Holothuria atra*) by Ridzwan et al. (2014).

On the other hand, sea cucumber is reported to contain bioactive substances that have not yet been identified as helping wound healing. It is thought that this effect is caused by fatty acids with branched chain structure (Fredalina et al. 1999; San Miguel-Ruiz and Garcia-Arraras 2007; Aydin et al. 2011; Ridhowati et al. 2018).

Traditionally, sea cucumbers are eaten by Chinese people more for their tonic value than for their seafood taste. Hence, the popular Chinese name for sea cucumber is haishen, which means, roughly, ginseng of the sea. Chinese commonly consume certain types of food as medicines for prevention and treatment of illness. From the health viewpoint, sea cucumber is an ideal tonic food (Chen 2004). The active ingredients in sea cucumber are believed to allay fatigue, boost the immune system, strengthen resistance to disease, treat injuries, prevent inflammation, and provide liver and blood vessel protection, among many other properties (Xia and Wang 2015). Sea cucumbers are known in Malaysia and Chinese culture as a tonic and traditional medicine to treat diseases such as asthma, rheumatism and cuts and burns. In addition, the anti-angiogenic, anticancer, anti-coagulant, anti-hypertension, anti-inflammatory, antimicrobial, antioxidant, anti-thrombotic, antitumor effects of the compounds extracted from sea cucumber species are also reported. These medicinal benefits and health functions of sea cucumbers are derived from bioactive compounds such as triterpene glycosides, chondroitin sulphates, glycosaminoglycan, sulphated polysaccharides, sterols, phenolics, peptides, cerebrosides and lectins (Bordbar et al. 2011). These bioactive compounds are naturally produced by sea cucumber as a defence against a competitive environment (Bhakuni and Rawat 2005). This type compounds are of interest in industrial applications such as nutraceuticals, pharmaceuticals, agrochemicals, and functional foods and in the development of new drugs (Venugopal 2009). In a study, a purified extract obtained from *Athyonidium chilensis* (Holothuria) was characterized by chromatographic and mass spectrometric techniques, and its bioactive potential was evaluated and two saponins were detected (Sottorff et al. 2013). In a study in which the antimicrobial effect of sea cucumber (*Holothuria atra*) was investigated, it was determined that the sea cucumber extract contained bioactive compounds such as phenols, terpenoids and saponins, and the *n*-hexane fraction and methanol extract showed high antibacterial activity against *Pseudomonas aeruginosa* (Sukmiwati et al. 2020), methanol extract from *Stichopus variegates* also inhibited *Escherichia coli* and *Aspergillus niger* (Shakouri et al. 2017). In another study, it was determined that saponins, terpenoids, and steroid compounds is found in *Stichopus vastus* extract and these have antimicrobial activity (Sukmiwati et al. 2018). Protein hydrolysates (Zou et al. 2016), Polysaccharides (Liu et al. 2012), polyphenols (Wu et al. 2014) and *Phospholipids* (Husni et al. 2009) from sea cucumbers showed antioxidant activities. Moreover, anti-coagulant and anti-thrombotic (Mou et al. 2017), antidiabetic (Hu et al. 2014), antiobesity (Xu et al. 2014) activities of sea cucumbers are also reported (Tables 4.1, 4.2 and 4.3).

4.3.2 Live Transport and Storage of Sea Cucumbers

Compared to other shellfish, sea cucumbers are difficult to keep and transport live. Because they do not have a protective exoskeleton and can autolyze when they are stressed or removed from sea water (Gianasi et al. 2016). Sea cucumbers do not have scales and a hard skeleton to protect them from various effects during their storage.

In a previous study, five different media (seawater ice, freshwater ice, freshwater ice + fish salt, and bagged freshwater ice) have been tried for live transport of *Cucumaria frondosa*. Iced seawater was found the most effective for live storage of *C. frondosa*, yielding 100% survival and largely intact body wall and muscles (Gianasi et al. 2016). In that study, it was stated that the use of ice sea water minimizes direct contact between ice and body wall, preventing burns, and also that ice sea water completely buffered the sea cucumbers against changes in environmental conditions and prevented exposure to the air causing stress.

Table 4.1 Proximate composition of different sea cucumber species (g/100 g)

Species	Moisture	Protein	Fat	Ash	Reference
Holothuria scabra	84.54–87.21	4.78–9.53	0.17–0.37	3.59–11.06	Ozer et al. (2004).
Bohadschia argus	85.82	9.74	0.19	2.87	Fawzya et al. (2015)
Holothuria fuscogilva	84.34	9.96	0.17	4.76	Fawzya et al. (2015)
Actinophyga lecanora	90.81	4.44	0.21	3.43	Fawzya et al. (2015)
Apostichopus japonicus	91.74	3.35	0.29	2.97	Li et al. (2018)
Stichopus variegatus	93.36	2.64	0.23	0.09	Ridhowati et al. (2018)
Acaudina molpadioides	77	12.94	0.03	1.03	Chen (2003)
Thelenota ananas	76.97	16.64	0.27	1.60	Chen (2003)
	87.78	8.0	0.21	2.81	Fawzya et al. (2015)
Parastichopus spp.	76.94	4.7	0.03	0.9	Chang-Lee et al. (1989)
Holothuria tubulosa	81.37–86.51	7.75–10.22	1.61–1.90	4.14–5.10	Kunili and Arik-Colakoglu (2019)
	86.76	12.30	0.22	0.72	Bilgin and Tanrikulu (2018)
Holothuria polii	46.3	7.37	0.32	13.4	Gonzalez-Wanguemert et al. (2018)
	81.24	8.66	0.15	7.85	Aydin et al. (2011)
Holothuria tubulosa	84.30	8.82	0.18	5.13	Aydin et al. (2011)
	81.2	3.01	0.21	14.7	Gonzalez-Wanguemert et al. (2018)
Holothuria mammata	85.24	7.88	0.09	5.13	Aydin et al. (2011)
	73.6	11.1	11.1	9.61	Gonzalez-Wanguemert et al. (2018)
Holothuria edulis	85.56	10.15	0.19	0.18	Al Azad et al. (2017)
Holothuria scabra	84.49	7.94	0.042	0.68	Al Azad et al. (2017)
Isostichopus sp.	83.96–86.92	2.74–6.63	0.07–0.35	3.16–3.81	Vergara and Rodríguez (2016)

4.3.3 *Postharvest Quality Changes of Sea Cucumbers*

Endogenous proteases cause autolysis and results in sensory changes in seafood meat. New compounds formed because of lipid oxidation and protein elution cause changes in colour, flavour, and taste. Exposure of sea cucumbers to various environmental conditions during transportation, storage and processing may cause softening of the body wall, as well as strong odour formation and even animal death. In addition, the digestive system releases enzymes that cause the hydrolysis of collagen, the main component of the body wall. Dead or unhealthy sea cucumbers exhibit deteriorated body walls that decrease the final products' quality and result in economic losses (Gianasi et al. 2016).

Table 4.2 Fatty acid compositions of different sea cucumber species (% of total fatty acids) (wet weight)

Species	SFA	MUFA	PUFA	n-3	EPA+DHA	References
Holothuria tubulosa	35.57–42.85	26.22–28.33	28.80–37.85	10.45–15.42	7.97–12.52	Kunili and Arik-Colakoglu (2019)
	15.48	13.92	57.76	28.8	18.55	Aydin et al. 2011
Holothuria polii	18.28	15.90	56.47	24.71	15.22	
Holothuria mammata	19.21	15.01	53.38	25.58	15.29	
Stichopus herrmanni	47.20	37.70	15.10	3.44	1.31	Wen et al. (2010)
Thelenota ananas	54.70	29.80	15.40	5.69	3.92	
Thelenota anax	61.60	27.0	11.40	3.81	3.10	
Holothuria fuscogilva	59.50	32.19	8.32	2.32	1.24	
Holothuria fuscopunctata	56.55	38.35	5.10	1.25	0.32	
Actinopyga mauritiana	31.23	45.64	23.33	6.11	3.84	
Actinopyga caerulea	49.54	36.32	14.14	4.82	3.53	
Bohadschia argus	47.40	36.60	16.0	3.25	1.96	
Apostichopus japonicus	17.11–30.89	29.94–35.45	25.87–42.93	16.09–28.74	9.65–18.9	Gao et al. (2011)
Holothuria tubulosa	19.73	21.40	26.02	14.20	13.19	Bilgin and Tanrikulu (2018)

Body wall thickness of BDM greatly influences commercial value of sea cucumber. The thicker body wall flesh generally resulting in better texture and improved eating quality. The body wall of sea cucumbers contains a high proportion of collagen that has a major influence on dried sea cucumber (bêche-de-mer (BDM)) firmness and texture quality, and therefore, product value (Ram 2017).

4.3.4 *Processing of Sea Cucumbers*

If harvested sea cucumbers are not to be processed directly, they should be kept in a cool and shady place on the boat. It can also be cooked after being removed from the sea to preserve the freshness of sea cucumbers. As soon as the sea cucumber is hauled up on board, a 2–3 cm slit is made near the cloaca. The body wall near the oral region is pressed to induce the animal to eviscerate, water is also squeezed out from the body by pressing. Immediately the sea cucumber are to be transferred to plastic boxes, with smooth inner surface and without drain holes. To prolong holding

Table 4.3 Mineral contents of different sea cucumber species (mg/100 g dry weight)

	Macro elements					References
	Na	K	Ca	Mg	P	
Holothuria mammata	6650	3860	4110	1270		Gonzalez-Wanguemert et al. (2018)
Holothuria polii	5690	3270	14,500	2140		Gonzalez-Wanguemert et al. (2018)
Holothuria tubulosa	7520	5580	9580	1660		Gonzalez-Wanguemert et al. (2018)
Holothuria arenicola	4750	520	5700	4750		Haider et al. (2015)
Actinopyga mauritiana	6220	620	2610	1870		Haider et al. (2015)
Parastichopus californicus	8800	400	2500	1400	500	Bechtel et al. (2012)
	Micro elements					**References**
	Mn	Cu	Zn	Fe	Cr	
Holothuria mammata	0.45	–	1.05	3.37	0.087	Gonzalez-Wanguemert et al. (2018)
Holothuria polii	0.46	–	0.89	4.06	0.077	Gonzalez-Wanguemert et al. (2018)
Holothuria tubulosa	8.66	–	22.7	440	1.52	Gonzalez-Wanguemert et al. (2018)
Holothuria arenicola	5.23	0.95	4.28			Haider et al. (2015)
Actinopyga mauritiana	5.85	5.11	5.23			Haider et al. (2015)
Parastichopus californicus	4.36	0.35	4.04	18.42	0.67	Bechtel et al. (2012)

time in fish boxes sea water is added to the boxes and the water is to be changed every 12 h (Gurmani and Krishnamurthy 1994).

The processing methods of different sea cucumber types are different due to differences in the thickness and softness of the body walls and the degree of water retention after cooking. There may be differences in the processing of commercial processors according to their own experience (Purcell et al. 2016). The cut made on sea cucumbers (to remove guts and aid drying) should be in different locations on the body for different species. Both black teatfish (*Holothuria whitmaei*) and white teatfish (*H. fuscogilva*) should be cut on the dorsal side, while other species that attain large sizes (i.e. *H. fuscopunctata*, *Thelenota ananas* and *T. anax*), should be cut on the ventral side. In both instances, the cut should stop short of the anus and mouth. For most other species, the general practice among processors is to make a small longitudinal cut (i.e. 2–3 cm) on the ventral surface at the anus (Purcell et al. 2016).

4.3.4.1 Salting of Sea Cucumber

Salting is a method used to protect sea cucumbers. It is one of the oldest methods used in the preservation of foods. In this method, salt or salt water is used. It is reported that the estimated ratio of salt to sea cucumber is usually 1–3 (Purcell 2014). Salting is carried out for dehydration in sea cucumbers prior to drying process. Because in salting, water is removed from sea cucumbers by osmotic dehydration (Khan 2012). Salt removes water from the body of sea cucumbers, making it heavier after cooking and drying, thus making more money on sale. It is reported that the salting process provides a good appearance to dried sea cucumbers (Purcell 2014). Coarse salt should be used in salting sea cucumbers, because it has a porous structure and slowly penetrates the sea cucumber tissue. Fine particles can damage the skin of sea cucumber (Chong et al. 2015).

There are two general methods for salting of sea cucumbers, wet salting, and dry salting.

4.3.4.2 Boiling of Sea Cucumber

One of the pre-treatments applied to sea cucumbers after harvesting is cooking. This process is carried out with water and salt. Cooking process is applied to sea cucumbers before drying. If cooking is done under unfavourable conditions, sea cucumber skin may be damaged, and the appearance quality may decrease. Cooking time affects the softening and breakdown of muscle fibres (Duan et al. 2010).

Eviscerated sea cucumbers are transferred to the boiling sea water one by one and heated for 45 min with stirring until each piece has attained elasticity like a rubber ball. During the process, scum, mud, and slime are removed and the pan is refilled with clean sea water and reheated (Gurmani and Krishnamurthy 1994).

4.3.4.3 Drying of Sea Cucumber

Sea cucumbers and are either consumed raw, dried, or boiled in many tropical and subtropical countries. Dried sea cucumber is known as bêche-de-mer (BDM) and it is consumed as a delicacy and for its perceived medicinal value. There is high demand for BDM and sea cucumber products in south-east Asian countries and at least 15,000 tonnes of BDM are traded annually in South East Asia.

Bêche-de-mer processing is done by boiling, cleaning, smoke drying and sun drying before storage. The cooking time is 15–25 min at 50 °C. If cooked at 100 °C, the skin will be damaged. This time is 10–15 min for even thin species. Sea cucumbers are cooled depending on the species after the first cooking, and then their viscera are removed. Valuable species are cut from the dorsal side while low or medium-value species are cut from the anal side and sliced (Ram 2017).

Drying Methods

While drying of seafood, various dryers such as vacuum-assisted solar dryers, cabin or tunnel dryers are used; sun drying, solar cabinet and tunnel drying methods are used for sea cucumber (Ram 2017). New drying technology is needed for sea cucumber to obtain better quality dried sea cucumber.

Sun Drying

Traditionally, sea cucumbers are dried using the sun drying technique. However, it is not possible to use this method continuously. Because it is weather dependent method and improper weather conditions can cause the product to deteriorate (Ram 2017). In this method, which is based on drying sea cucumber under the sun, capital and operating costs are low. There are no energy, labour or machine costs. Free solar energy is used and does not require experienced labour. Water is usually removed by evaporation (Chong et al. 2015). In the traditionally used sun drying method, sea cucumber is boiled and then salted, and dried for 3–5 days with solar radiation. However, this process has negative aspects such as being time consuming, contaminated, and causing loss of many active ingredients (Duan et al. 2007). Moreover, there is a risk of being invaded by various animals (Chong et al. 2015).

Hot Air Drying

Due to the above-mentioned disadvantages of the sun drying method, the hot air-drying method is a simple and popular method used as an alternative to the sun drying method. However, it has been reported that hot air-drying causes large product deformations and thermally induced deterioration (Duan et al. 2010; Ratti 2001). Also in hot air drying, hot air accelerates the drying of the sea cucumber surface and causes a surface hardening and so this hardening prevents the transfer of moisture from the body wall and can cause serious cracking of the surface and poor sea cucumber quality (Moon et al. 2014).

Freeze Drying

The greatest advantage of freeze-dried sea cucumbers is that they are ready to eat after simple rehydration. Moreover, they can be kept fresh without any additive or preservative. The whole processing procedure is carried out below the freezing point under vacuum (no oxygen), with minimal loss of bioactive components, and lower final moisture content. With their colour, shape, and nutrients preserved, these sea cucumbers will have the same taste as fresh sea cucumbers after rehydration (Mao et al. 2015).

In a study, it has been shown that freeze drying method can be used for seafood (Yun et al. 2006). Although this method is an excellent method, it has been reported that takes a long processing time and energy consumption and investment expenses are high (Ratti 2001). Duan et al. (2010) developed a microwave freeze drying technique for sea cucumber (*Stichopus japonicus*). They reported that the microwave freeze dryer reduces the drying time by half the conventional freeze drying process and offers a quality product similar to freeze-dried and for this reason can replace the microwave freeze drying method with the conventional freeze drying method. Although microwave freeze drying can significantly increase the drying speed, in this technology, it is reported that the naturally in homogeneous distribution of the microwave area causes an uneven temperature distribution in the drying material (Duan et al. 2010; Dolan and Scott 1994).

Cabinet Drying

Cabinet dryer is used with other drying technologies. This type of drying is more flexible as it provides drying without focusing on the problem of air, season or animals. It is a typical cabin filled with a shallow tray. To ensure optimum drying, hot air flow occurs throughout the area inside the cabinet (Chong et al. 2015).

Combined Methods

A combination of drying methods can be used to avoid the disadvantages of a single drying method and combine the advantages of various drying methods. Duan et al. (2007) examined the effects of the combined use of different drying methods on the rehydration and textural properties of the product. For this purpose, drying of sea cucumber (*Stichopus japonicus*) was tried with the combination of freeze drying and microwave frying, and the combination of freeze drying and microwave vacuum drying. With the combination of freeze drying and microwave vacuum drying, more acceptable quality and better drying efficiency were achieved.

Combination of infrared radiation and freeze-drying pre-treatment was found to be effective for drying the fresh sea cucumbers. The drying time of sea cucumbers with infrared radiation pre-treatment is shorter than that of raw materials without pre-treatment (Mamatov et al. 2019).

Far Infrared Radiation Drying

In far-infrared radiation drying, which electromagnetic wave energy is absorbed directly by dried food, the temperature distribution in the product (FIRD) is relatively uniform. This kind of uniform temperature distribution accelerates the drying rate and reduces the internal resistance in the surface hardening seen in the air-drying method. Due to the high drying rate, a higher quality product is obtained and consumes less energy. It is an energy-saving method (Moon et al. 2014). Far-infrared radiation drying method has been used to dry other foods such as mushroom (Yeon et al. 2004), blueberry (Shi et al. 2008), banana (Nimmol et al. 2007), onion (Marinos-Kouris and Maroulis 1995) and barley (Afzal and Abe 2000). A limited number of studies have been carried out on drying sea cucumber with this method. Moon et al. (2014) examined the effects of far-infrared radiation drying temperature on sea cucumber (*Stichopus japonicus*) drying rate and compared them with the hot air-drying method. At the end of this study, it has been found that the far infrared radiation drying reduces the drying time and a higher quality product is obtained compared to the air-drying method. Zhao et al. (2018) tried to dry sea cucumber using far infrared drying method. They found that drying temperatures had a significant effect on relative moisture content, drying rate, shrinkage rate and rehydration rate.

Quality of Dried Sea Cucumber

The quality of Bêche-de-mer is classified according to the species, size, colour, odour, moisture content and foreign matter content of the sea cucumber. The larger dried cucumber is considered higher quality and higher price. The quality dried product should have a smooth surface, a homogeneous shape, a clean body wall and a pleasant odour (McElroy 1990). Improper processing methods can cause nutrient losses in the product. Ozer et al. (2004) reported that the nutritional composition of sea cucumber (*Holothuria scabra*) processed by two different methods was affected by the processing method. In this study, sea cucumbers harvested in different months were processed in two ways. The first method was gutting by cutting the anus, then squeezing the body, removing the viscera. The second method was to remove the viscera by cutting along the length of the body. Samples prepared in both ways were boiled in clean sea water for 45 min and then immersed in fresh water. All samples were boiled the same way a second time. The collected samples were dried at 37 °C for 1 day and then at 70 °C for 2 days. Different boiling temperatures have been reported to affect the collagen of sea cucumber. Collagen is the main structural component of the major edible part of sea cucumber, thus changes of collagen during cooking will affect the physicochemical properties of the whole product (Dong et al. 2011).

Since BDM is subjected to multi-step processing involving multiple cooking, salting, and drying, the texture and nutritional composition of the product may vary during processing (Ram 2017).

4.3.4.4 Smoking of Sea Cucumber

Smoking sea cucumber and other food materials that is initially intended to prolong the product durability has developed to obtain certain appearance and taste of the food material. Several studies have indicated that smoking is a preserving method that does not only increase the durability but also can give desired taste and colour of the smoked product due to the presence of phenolic and carbonyl compounds (Salindeho 2017).

Smoking is not a common method for sea cucumbers. If salting or drying cannot be done, smoke curing is traditionally used. Smoking process of sea cucumbers are carried out after evisceration and cutting. Eviscerated and cut products are salted before boiling and smoking (Ram et al. 2014).

References

Afzal TM, Abe T (2000) Simulation of moisture changes in barley during far infrared radiation drying. Comput Electron Agric 26:137–145

Al Azad S, Shaleh SRM, Siddiquee S (2017) Comparison of fatty acid and proximate composition between *Holothuria edulis* and *Holothuria scabra* collected from coastal water of Sabah, Malaysia. Adv Biosci Biotechnol 8:91–103

Arafa S, Chouaibi M, Sadok S, El Abed A (2012) The Influence of season on the gonad index and biochemical composition of the sea urchin *Paracentrotus lividus* from the Golf of Tunis. Sci World J 2012:815935, 8 p

Aydin M, Sevgili H, Tufan B, Emre Y, Kose S (2011) Proximate composition and fatty acid profile of three different fresh and dried commercial sea cucumbers from Turkey. Int J Food Sci Technol 46:500–508

Batista I (2007) By-catch underutilized species and underutilized fish parts as food ingredients. In: Shahidi F (ed) Maximising the value of marine by products. CRC Press, Boca Raton, pp 171–191

Bechtel PJ, Oliveira ACM, Demir N, Smiley S (2012) Chemical composition of the giant red sea cucumber, *Parastichopus californicus*, commercially harvested in Alaska. Food Sci Nutr 1(1):63–73

Belbachir NE, Mezali K (2018) Food preferences of four aspidochirotid holothurians species (Holothuroidea: Echinodermata) inhabiting the *Posidonia oceanica* meadow of Mostaganem area (Algeria). SPC Beche-de-mer Inf Bull 38:55–59

Bhakuni DS, Rawat DS (2005) Bioactive marine toxins. In: Bhakuni DS, Rawat DS (eds) Bioactive marine natural products. Springer, New Delhi, pp 151–207

Bilgin S, Tanrikulu HO (2018) The changes in chemical composition of *Holothuria tubulosa* (Gmelin, 1788) with ambient-drying and oven-drying methods. Food Sci Nutr 6:1456–1461

Bledsoe GE, Bledsoe CD, Rasco B (2003) Caviars and fish roe products. Crit Rev Food Sci Nutr 43(3):317–356

Bordbar S, Anwar F, Saari N (2011) High-value components and bioactives from sea cucumbers for functional foods—a review. Mar Drugs 9:1761–1805

Boudouresque C, Verlaque M (2007) Ecology of *Paracentrotus lividus*. In: Lawrence JM (ed) Edible sea urchins: biology and ecology. Elsevier Science, pp 243–285

Bruckner AW, Johnson KA, Field JD (2003) Conservation strategies for sea cucumbers: can a CITES Appendix II listing promote sustainable international trade? SPC Beche-de-mer Inf Bull 18:24–33

Bumrasarinpai R (2006) Overview of issues on concerned aquatic species under the convention on international trade in endangered species of wild fauna and flora (CITES). Regional technical consultation on international fisheries related issues WP03

Chang-Lee MV, Price RJ, Lampila LE (1989) Effect of processing on proximate composition and mineral content of sea cucumbers (*Parastichopus* spp.). J Food Sci 54:567–568

Chen J (2003) Overview of sea cucumber farming and ranching practices in China. Beche-de-mer Inf Bull 18:18–23

Chen J (2004) Present status and prospects pf sea cucumber industry in China. In: Conand C, Purcell S, Uthicke S, Hamel JF, Mercier A (eds) Advances in sea cucumber aquaculture and management. Food and Agriculture Organization of the United Nations, Rome, pp 25–39

Chong NVW, Pindi W, Chye FY, Shaarani SM, Lee JS (2015) Effects of drying methods on the quality of dried sea cucumbers from Sabah—a review. Int J Nov Res Life Sci 2(4):49–64

De La Cruz-García C, Lopez-Hernandez J, González-Castro MJ, Rorrihez-Bernaldo De Quiros ARB, Simal-Lozano J (2000) Protein, amino acid and fatty acid contents in raw and canned sea urchin (Paracentrotus lividus) harvested in Galicia (NW Spain). J Sci Food Agric 80(8):1189–1192

Dincer T, Cakli S (2007) Chemical composition and biometrical measurements of the Turkish Sea Urchin (*Paracentrotus Lividus*, Lamarck, 1816). Crit Rev Food Sci Nutr 27:21–26

Dolan JP, Scott EP (1994) Microwave freeze-drying of aqueous solutions. Advances in heat and mass transfer in biological systems. ASME–Heat Transfer Division, HTD 288:91–98

Dong X, Zhu B, Sun L, Zheng J, Jiang D, Zhou D, Wu H, Murata Y (2011) Changes of collagen in sea cucumber (*Stichopus japonicus*) during cooking. Food Sci Biotechnol 20(4):1137–1141

Duan X, Zhang M, Mujumdar AS (2007) Study on a combination drying technique of sea cucumber. Dry Technol 25(12):2011–2019

Duan X, Zhang M, Mujumdar AS, Wang S (2010) Microwave freeze drying of sea cucumber (*Stichopus japonicus*). J Food Eng 96:491–497

FAO (2016) Yearbook of fishery statistics summary tables. Fisheries and Aquaculture Department, Rome, 200 p. http://www.fao.org/3/i9942t/I9942T.pdf

FAO-Globefish (2018) Globefish highlights. A quarterly update on world seafood markets. FAO, Rome, p 31

Fawzya YN, Januar HI, Susilowati R, Chasanah E (2015) Chemical composition and fatty acid profile. Squalen Bull Mar Fish Postharvest Biotechnol 10(1):27–34

Fernandez C (1997) Effect of diet on the biochemical composition of Paracentrotus livid (Echinodermata: Echinoidea) under natural and rearing conditions (Effect of diet on biochemical composition of urchins). Comp Biochem Physiol A 118(4):1377–1384

Fernandez C (1998) Seasonal changes in the biochemical composition of the edible sea urchin *Paracentrotus livid*us (Echinodermata: Echinoidea) in a lagoonal environment. Mar Ecol 19(I):1–1

Fredalina BD, Ridzwan BH, Zainal Abidin AA, Kaswandi MA, Zaiton H, Zali I, Kittakoop P, Jais AM (1999) Fatty acid compositions in local sea cucumber, *Stichopus chloronotus* for wound healing. Gen Pharmacol 33:337–340

Gao F, Xu Q, Yang H (2011) Seasonal biochemical changes in composition of body wall tissues of sea cucumber *Apostichopus japonicus*. Chin J Oceanol Limnol 29(2):252–260

Gianasi BL, Hamel JF, Merciera A (2016) Experimental test of optimal holding conditions for live transport of temperate sea cucumbers. Fish Res 174:298–308

Gianguzza P, Chiantore M, Bonaviri C, Cattaneo-Vietti R, Vielmini I, Riggio S (2006) The Effects of recreational *Paracentrotus lividu*s fishing on distribution patterns of sea urchins at Ustica Island MPA (Western Mediterranean, Italy). Fish Res 81:37–44

Gonzalez-Wanguemert M, Roggatz CC, Rodrigues MJ, Barreira L, Moreira da Silva M, Custodi L (2018) A new insight into the influence of habitat on the biochemical properties of three commercial sea cucumber species. Int Aquat Res 10:361–373

Grisolia JM, Lopez F, de Dios Ortuzar J (2012) Sea urchin: from plague to market opportunity. Food Qual Prefer 25:46–56

Guidetti P, Terlizzi A, Boero F (2004) Effects of the edible sea urchin, *Paracentrotus lividus*, fishery along the Apulian rocky coast (SE Italy, Mediterranean Sea). Fish Res 66:287–297

Gurmani ON, Krishnamurthy S (1994) Some aspects of processing and quality control of Beche-De-Mer for export. Bull Cent Mar Fish Res Inst 46:81–84

Haider MS, Sultana R, Jamil K, Lakht-e Z, Tarar OM, Shirin K, Afzal W (2015) A study on proximate composition, amino acid profile, fatty acid profile and some mineral contents in two species of sea cucumber. J Anim Plant Sci 25(1):168–175

Harris LG, Eddy SD (2015) Sea urchin ecology and biology. In: Brown NP, Eddy SD (eds) Echinoders aquaculture. Wiley, New Jersey, pp 1–24

Hermosillo-Nunez B, Rodríguez-Zaragoza F, Ortiz M, Galvan-Villa C, Cupul-Magana A, Rios-Jara E (2015) Effect of habitat structure on the most frequent echinoderm species in habiting coral reef communities at Isla Isabel National Park (Mexico). Community Ecol 16(1):125–134

Hu S, Xia G, Wang J, Wang Y, Li Z, Xue C (2014) Fucoidan from sea cucumber protects against high-fat high-sucrose diet-induced hyperglycaemia and insulin resistance in mice. J Funct Foods 10:128–138

Husni A, Shin I-S, You S, Chung D (2009) Antioxidant properties of water and aqueous ethanol extracts and their crude saponin fractions from a far-eastern sea cucumber, *Stichopus japonicus*. Food Sci Biotechnol 18(2):419–424

James P, Evensen T (2018) Sea Urchin transportation systems for land, sea and air. NOFIMA Report 12, Tromsö, 32 p

James P, Evensen T, Samuelsen A (2017) Commercial scale sea urchin roe enhancement in Norway: enhancement, transport and market assessment. Nofima Technical Report No:7. Tromsö, 22 p

James P, Noble C, Siikavuopio S, Sloan R, Hannon C, Thorarinsdöttir G, Ziemer N, Lochead J (2018) Sea Urchin Fishing Techniques Report (Activity A3.1.1 of the NPA URCHIN project), Nofima, Trömsö, 21 p

Jinadasa BKKK, De Zoysa HKS, Jayasinghe GDTM, Edirisinghe EMRKB (2016) Determination of the biometrical parameters, biochemical composition and essential trace metals of edible sea urchin (*Stomopneustes variolaris*) in Sri Lanka. Cogent Food Agric 2:1143343

Kato S, Schroeter SC (1985) Biology of the red sea urchin, *Strongylocentrotus franciscanus*, and its fishery in California. Mar Fish Rev 47(3):1–20

Kazanidis G, Antoniadou C, Lolas AP, Neofitou N, Vafidis D, Chintiroglou C, Neofitou C (2010) Population Dynamics and reproduction of *Holothuria tubulosa* (Holothuroidea: Echinodermata) in the Aegean Sea. J Mar Biol Assoc UK 90:895–901

Khan M (2012) Osmotic dehydration technique for fruit preservation—a review. Pak J Food Sci 22(2):71–85

King CK, Hoegh-Guldberg O, Byrne M (1994) Reproductive cycle of *Centrostephanus rodgersii* (Echinoidea), with recommendations for the establishment of a sea urchin fishery in New South Wales. Mar Biol 120:95–106

Kramer DE, Nordin DMA (1979) Studies on the handling and processing of sea urchin roe. I. Fresh product. Fisheries and Marine Service Technical Report No. 870, Vancouver, 36 p

Kunili IE, Arik-Colakoglu F (2019) Chemical and nutritional characteristics of Holothuria tubulosa (Gmelin, 1788); A seasonally comparative study. J Aquat Food Prod Technol 28(7):716–728

Li D-T, Chang Y-Q, Wu Z-H, Chen W, Wang J-Y, Chen G-D (2009) Analysis of nutritive composition of body wall in wild sea cucumber *Apostichopus japonicus* Selenka at Zhangzi Island in Spring and Autumn. Fish Sci 28:365–369

Li M, Qi Y, Mu L, Li Z, Zhao Q, Sun J, Jiang Q (2018) Effects of processing method on chemical compositions and nutritional quality of ready-to-eat sea cucumber (*Apostichopus japonicus*). Food Sci Nutr 7:755–763

Liu X, Sun Z, Zhang M, Meng X, Xia X, Yuan W, Liu C (2012) Antioxidant and antihyperlipidemic activities of polysaccharides from sea cucumber *Apostichopus japonicus*. Carbohydr Polym 90:1664–1670

Liu G, Sun J, Liu S (2015) From fisheries towards aquaculture. In: Yang H, Hamel JH, Mercier A (eds) The sea cucumber *Apostichopus japonicus*. History, biology and aquaculture. Elsevier, London, pp 1–23

Liu F, Zamora L, Jeffs A, Quek SY (2017) Biochemical composition of the Australasian sea cucumber, *Australostichopus mollis*, from a nutritional point of view. Forum Nutr 42(1):12

Mamatov S, Zhang M, Jia A, Liu X, Liu C (2019) Effect of infrared radiation (IR) pre-treatment on the freeze-drying of sea cucumber. Int J Innov Technol Explor Eng 8(12):475–478

Mao Y, Huang Y, Liu Q (2015) Processing and cuisine. In: Yang H, Hamel JH, Mercier A (eds) The sea cucumber *Apostichopus japonicus* history, biology and aquaculture. Elsevier, London, pp 367–382

Marinos-Kouris D, Maroulis ZB (1995) Transfer properties in the drying of solids. In: Mujumdar AS (ed) Handbook of İndustrial drying. Marcel Dekker, New York, pp 133–155

Martinez-Pita I, García FJ, Pita ML (2010) Males and females gonad fatty acids of the sea urchins *Paracentrotus lividus* and *Arbacia lixula* (Echinodermata). Helgol Mar Res 64:135–142

Mathews V, Kookesh M, Bosworth R (1990) Subsistence harvest and use of sea cucumbers in southeast Alaska, Division of Subsistence, Alaska Department of Fish and Game. Technical paper no. 190

McBride SC, Price RJ, Tom PD, Lawrence JM, Lawrence AL (2004) Comparison of gonad quality factors: colour, hardness and resilience, of *Strongylocentrotus franciscanus* between sea urchins fed prepared feed or algal diets and sea urchins harvested from the Northern California fishery. Aquaculture 233:405422

McElroy S (1990) Beche-de-mer species of commercial species an update. SPC Bêche-demer Inf Bull 2:2–7

Micael J, Alves MJ, Costa VAC, Jone MB (2009) Exploitation and conservation of echinoderm. Oceanogr Mar Biol Annu Rev 47:191–208

Mol S, Baygar T, Varlik C, Tosun SY (2008) Seasonal variations in yield, fatty acids, amino acids and proximate compositions of sea urchin (*Paracentrotus lividus*) Roe. J Food Drug Anal 16:68–74

Monfort MC (2002) Fish Roe in Europe: supply and demand conditions. Food and Agriculture Organization of the United Nations, GLOBEFISH, Fishery Industries Division, Italy

Montero-Torreiro MF, Garcia-Martinez P (2003) Seasonal changes in the biochemical composition of body components of the sea urchin, *Paracentrotus lividus*, in Lorbé (Galicia, northwestern Spain). J Mar Biol Assoc U K 83:575581

Moon JH, Kim MJ, Chung DH, Pan CH, Yoon WB (2014) Drying characteristics of sea cucumber (*Stichopus japonicas* Selenka) using far infrared radiation drying and hot air drying. J Food Process Preserv 38:1534–1546

Mou J, Wang C, Li W, Yang J (2017) Purification, structural characterization and anticoagulant properties of fucosylated chondroitin sulfate isolated from Holothuria mexicana. Int J Biol Macromol 98:208–215

Mulcrone R (2005) "Echinodermata" (On-line). Animal Diversity Web. https://animaldiversity.org/accounts/Echinodermata/. Accessed 12 Mar 2020

Nimmol C, Devahastin S, Swasdisevi T, Soponronnarit S (2007) Drying and heat transfer behaviour of banana undergoing combined low-pressure superheated steam and far-infrared radiation drying. Appl Therm Eng 27:2483–2494

Ozer NP, Mol S, Varlık C (2004) Effect of the handling procedures on the chemical composition of sea cucumber. Turk J Fish Aquat Sci 4:71–74

Panebianco A, Giuffrida A, Tasca D, Atanossova V, Ziino G (2011) Shelf-life and microbiological properties of refrigerated sea urchin (*Paracentrotus lividus*) roe. Arch Leb 62(1):11–15

Pena MH, Oxenford HA, Parker C, Johnson A (2010) Biology and fishery management of the white sea urchin, *Tripneustes ventricosus*, in the eastern Caribbean. FAO Fisheries and Aquaculture Circular. No. 1056. Rome, 43 p

Price RJ (2000) Other aquatic life of economic significance: sea urchins. In: Martin RE, Carter EP, Flick GJ Jr, Davis LM (eds) Marine and freshwater products handbook. Technomic Publishing Co. Inc., Lancherster-Basel, pp 264–273

Purcell SW (2014) Processing sea cucumbers into bêche-de-mer: a manual for Pacific Island fishers. Southern Cross University, Lismore, and the Secretariat of the Pacific Community, Noumea. 44 p

Purcell SW, Samyn Y, Conand Y (2012) Commercially important se cucumbers of the world. FAO Species Catalogue for Fishery Purposes No. 6, Rome

Purcell SW, Ngaluafe P, Aram KT, Lalavanua W (2016) Variation in postharvest processing of sea cucumbers by fishers and commercial processors among three Pacific Island countries. SPC Beche-de-mer Inf Bull 36:58–66

Ram R (2017) Influence of processing techniques on quality and nutritional composition of the tropical sea cucumber *Holothuria scabra*. PhD thesis, James Cook University

Ram R, Chand RV, Southgate PC (2014) Effect of harvest and post-harvest processing methods on quality of Bêche-De-Mer in Fiji Islands. J Mar Sci Res Dev 4:153

Ratti C (2001) Hot air and freeze-drying of high-value foods: a review. J Food Eng 49(4):311–319

Reynolds JA, Wilen JE (2000) The sea urchin fishery: harvesting, processing and the market. Mar Resour Econ 15:115–126

Ridhowati S, Chasanah E, Syah D, Zakaria F (2018) A study on the nutrient substances of sea cucumber *Stichopus variegatus* flour using vacuum oven. Int Food Res J 25(4):1419–1426

Ridzwan BH (2007) Sea cucumbers, a Malaysian heritage, 1st edn. Research Centre of International Islamic University Malaysia (IIUM), Kuala Lumpur Wilayah Persekutuan, pp 1–15, 89–128

Ridzwan BH, Hanita MH, Nurzafirah M, Norshuhadaa SMP, Hanis ZF (2014) Free fatty acids composition in lipid extracts of several sea cucumbers species from Malaysia. Int J Biosci Biochem Bioinform 4(3):201–207

Rocha F, Baião LF, Moutinho S, Reis B, Oliveira A, Arenas F, Maia MRG, Fonseca AJM, Pintado M, Valente LMP (2019) The effect of sex, season and gametogenic cycle on gonad yield, biochemical composition and quality traits of *Paracentrotus lividus* along the North Atlantic coast of Portugal. Sci Rep 9:2994

Sætra IM (2019) Quality aspects of wild caught and enhanced sea urchins (*Strongylocentrotus droebachiensis*). Master thesis in Fisheries and Aquaculture Science

Salindeho N (2017) Application of nutmeg shell pyrolysis-based liquid smoke for sea cucumber (*Holothuria scabra*) processing. Int J ChemTech Res 10(12):278–283

San Miguel-Ruiz JE, Garcia-Arraras JE (2007) Common cellular events occur during wound healing and organ regeneration in the sea cucumber *Holothuria glaberrima*. BMC Dev Biol 7:1–19

Shakouri A, Shoushizadeh MR, Nematpour F (2017) Antimicrobial activity of sea cucumber (*Stichopus variegatus*) body wall extract in Chabahar Bay, Oman Sea. Jundishapur J Nat Pharm Prod 12:e32422

Shi J, Pan Z, Mchugh TH, Wood D, Hirschberg E, Olson D (2008) Drying and quality characteristics of fresh and sugar-infused blueberries dried with infrared radiation heating. LWT–Food Sci Technol 41:1962–1972

Sottorff I, Aballay A, Hernández V, Roa L, Munoz LX, Silva M, Becerra J, Astuya A (2013) Characterization of bioactive molecules isolated from sea cucumber *Athyonidium chilensis*. Rev Biol Mar Oceanogr 48(1):23–35

Stefansson G, Olafsdottir A (2019) Holding of sea urchins and scallops in a RAS transport system. Icelandic Food and Biotech R & D Report No: 25-19

Stefansson G, Kristinsson H, Ziemer N, Hannon C, James P (2017) Markets for sea urchins: a review of global supply and markets. Internal Matís Report: Skýrsla Matís, 10–17 Oct 2017, 45 p

Sukmiwati M, Diharmi A, Mora E, Susanti E (2018) Antimicrobial activity of sea cucumber (*Stichopus vastus*) from Natuna Coastal Sea of Riau Island. JPHPI-Indonesian Fish Process J 21(2):328–335

SukmiwatiM,IlzaM,PutriAE,SidaurukSW(2020)Antibacterialactivityofseacucumber(*Holothuria atra*) against Pseudomonas aeruginosa, vol 404. IOP Conf. Series: Earth and Environmental Science, IOP Publishing, p 012047. https://doi.org/10.1088/1755-1315/404/1/012047

Sun J, Chiang FS (2015) Use and exploitation of sea urchins. In: Brown NP, Eddy SD (eds) Echinoderm aquaculture. Wiley-Blackwell, New Jersey, pp 25–47

Uneputty PA, Tuapattinaja MA, Pattikawa JA (2017) Density and diversity of echinoderms in seagrass bed, Baguala Bay, Maluku, Eastern Indonesia. Int J Fish Aquat Stud 5(2):311–315

Venugopal V (2009) Marine habitat and resources. In: Venugopal V (ed) Marine products for healthcare: functional and bioactive nutraceutical compounds from the ocean. CRC Press Taylor & Francis Group, Boca Raton, pp 23–50

Verachia W, Sewell MA, Niven B, Leus M, Barker MF, Bremer PJ (2012) Seasonal changes in the biochemical composition of *Evechinus chloroticus* gonads (Echinodermata: Echinoidea). N Z J Mar Freshw Res 46(3):39941

Vergara W, Rodríguez A (2016) Nutritional composition of sea cucumber Isostichopus sp. Nat Resour 7:130–137

Wen J, Hu C (2010) Elemental composition of commercial sea cucumbers (holothurians). Food Addit Contam Part B 3(4):246–252

Wen J, Hua C, Fana S (2010) Chemical composition and nutritional quality of sea cucumbers. J Sci Food Agric 90:2469–24745

Wikipedia (2020a). https://en.wikipedia.org/wiki/Sea_cucumber. Accesed 12 April 2020

Wikipedia (2020b) *Apostichopus japonicus*. https://en.wikipedia.org/wiki/Apostichopus_japonicus. Accessed 18 Apr 2020

Wikipedia (2020c) *Holothuria edulis*. https://en.wikipedia.org/wiki/Holothuria_edulis. Accessed 18 Apr 2020

Wikipedia (2020d) *Holothuria tubulosa*. https://en.wikipedia.org/wiki/Holothuria_tubulosa Accessed in 18 April 2020

Wu F-J, Xue Y, Liu X-F, Xue C-H, Wang J-F, Du L et al (2014) The protective effect of eicosapentaenoic acid-enriched phospholipids from sea cucumber *Cucumaria frondosa* on oxidative stress in PC12 cells and SAMP8 mice. Neurochem Int 64:9–17

Xia S, Wang X (2015) Nutritional and medicinal value. In: Yang H, Hamel JH, Mercier A (eds) The sea cucumber *Apostichopus japonicus* history, biology and aquaculture. Elsevier, London, pp 353–365

Xiang Y-H, Su X-R, Dong M-M (2006) The composition of amino acids and fatty acids in body walls and alimentary canals of sea cucumber Apostichopus japonicus. Fish Sci 25:280–282

Xu H, Wang J, Chang Y, Xu J, Wang Y, Long T, Xue C (2014) Fucoidan from the sea cucumber *Acaudina molpadioides* exhibits anti-adipogenic activity by modulating the Wnt/β-catenin pathway and down-regulating the SREBP-1c expression. Food Funct 5:1547–1555

Xu D, Su L, Zhao P (2015) *Apostichopus japonicus* in the worldwide production and trade of sea cucumbers. In: Yang H, Hamel J-F, Mercier A (eds) The sea cucumber *Apostichopus japonicus*. History, biology and aquaculture. Academic, pp 383–398

Yang H, Bai Y (2015) *Apostichopus japonicus* in the life Chinese people. In: Yang H, Hamel JH, Mercier A (eds) The sea cucumber *Apostichopus japonicus*. History, biology and aquaculture. Elsevier, London, pp 1–23

Yeon GS, Kim MH, Han CS, Cho SC, Kang TH, Lee HC, Kim CB, Kim JG (2004) Drying characteristics of oak mushroom using conveyer far infrared dryer-down draft air flow type. Korean J Biosyst Eng 29:37–44

Yun X, Han XH, Nong SZ, Yang H (2006) Technological condition of vacuum freeze-drying in Stichopus japonicus. J Fish Sci China 13(4):662–665

Zhao H-B, Dai J-A, Qiao L-M, Zhang L-L, Wu K (2018) Far infrared drying of sea cucumber. Mod Food Sci Technol 34(3):101–107

Zhong Y, Khan MA, Shahidi F (2007) Compositional characteristics and antioxidant properties of fresh and processed sea cucumber (*Cucumaria frondosa*). J Agric Food Chem 55:1188–1192

Zito F, Costa C, Sciarrino S, Cavalcante C, Poma V, Matranga V (2005) Cell adhesion and communication: a lesson from echinoderm embryos for the exploitation of new therapeutic tools. In: Matranga V (ed) Echinodermata. Springer, Berlin, pp 7–36

Zou S, Pan R, Dong X, He M, Wang C (2016) Physicochemical properties and antioxidant activities of two fucosylated chondroitin sulfate from sea cucumber *Acaudina molpadioidea* and *Holothuria nobilis*. Process Biochem 51:650–658

Chapter 5
Shellfish Safety

5.1 Introduction

Safe food is a food substance that is physically, chemically, and microbiologically suitable for consumption and has not lost its nutritional value. Safe food is food that does not pose a health hazard to the consumer.

A biological, chemical, or physical agent or condition that has the potential to have a negative health effect for the consumer is defined as a hazard. The estimated size of the hazard that is likely to occur is also defined as the risk.

Bacteria, viruses, parasites and naturally occurring toxins can cause foodborne diseases in fresh and processed shellfish. Shellfish also can be contaminated by substances or organisms introduced into the environment through animal and human pollution or agricultural runoff. Other factors that can increase the risk of illness include environmental conditions in the growing waters, harvesting methods, processing operations, and handling during marketing. Many molluscan shellfish (primarily oysters and mussels) are eaten raw or with minimal heating. This is a major concern to national and international health agencies because bacteria in these products have caused food-borne illnesses and deaths (Flick Jr and Granata 2010). Shellfish can concentrate contaminants in their tissues and become unsafe to eat. Shellfish contaminants fall into three categories: bacteria and viruses, marine biotoxins, and chemicals. Shellfish-associated hazards can be broadly grouped as environmental, intrinsic, and process-related, categorizes hazards associated with various shellfish products, in the order of decreasing risk. Shellfish items, which are consumed raw without any cooking, are the most hazardous, while products consumed soon after thorough heat processing pose minimum microbial hazards (Venugopal and Gopakumar 2017).

Bivalve molluscs are fed microscopic size particles from the water by filtering. This filter feeding mechanism can cause public health concern in relation to the pollutants in the waters pumped by bivalves with a relatively large amount. Particulate ingestion is the manner which shellfish concentrate viruses and bacteria from the

N. Gökoğlu, *Shellfish Processing and Preservation*,
https://doi.org/10.1007/978-3-030-60303-8_5

water. Viruses and bacteria are too small for the shellfish's gills to detect or reject, but due to their small size in relation to such food items as algae and small particles suspended in the water, they tend to attach to these surfaces. Other pollutants of public health concern that shellfish concentrate basically via this mechanism are many types of pesticides, toxic algae, and ions of heavy metals attached to particulate matter (Croonenberghs 2000). Bivalves are minimally processed, and traditionally consumed raw or lightly cooked as a whole. The hazards of contamination of bivalve by harmful microorganisms arise from the lack of traditional cooking procedures sufficient to ensure consumer safety. These conditions make them an important carrier of foodborne diseases (Oliveira et al. 2011).

5.2 Microbial Risks

Microbial hazards take the first place among food-borne risks. Microbial risks are the most difficult to control hazards. The most common of food-borne diseases are of bacterial origin, progressing faster and occur relatively quickly. Microorganisms contaminated with food at any stage of production can pose a risk when it comes to process errors, such as inadequate temperature applications. After bacterial contamination occurring in food for any reason reaches a certain level, it may cause infection or intoxication when food is consumed (Tayar 2010).

Microbial contaminants are pathogenic microorganisms that cause disease in shellfish-consuming humans. Included are faecal-borne viral, bacterial, and protozoan pathogens, and naturally occurring bacterial pathogens. The indicator bacteria used within the shellfish industry are generally faecal (and total) coliforms and *Escherichia coli*. Enterococci, *Clostridium perfringens* and other indicators are also useful in helping to elucidate the sources of faecal-borne contamination in shellfish waters. Faecal-borne bacteria are present in nearly all coastal waters where human activities and animals contribute faecal contamination. Faecal-borne bacterial species have been the cause of significant outbreaks of shellfish-borne diseases (Jones 2009). Bivalves living in polluted waters are often subject to contamination from domestic sewage and land run-off, which typically contain pathogenic bacteria such as Salmonella spp. and Vibrio spp. (Gosling 2015). Faecal-borne bacterial pathogens account for a significant fraction of shellfish-borne disease in some areas. The major microbial contaminants include viruses, vibrios, and to a lesser extent faecal borne bacterium.

5.2.1 Bacterial Pathogens

Bacteria are everywhere in the marine environment. Shellfish can be a source of bacteria for human pathogens. Raw and partially cooked mollusc shellfish have long been vectors of infectious agents (Hariharan and Amadi 2016).

It is reported that 20% of all outbreaks of shellfish include *Vibrio* species, while *Salnonella*, *Shigella* and *Listeria* species are other species (Potasman et al. 2002).

5.2.1.1 *Vibrio* Species

Various species of *Vibrio*, specific to both marine and river mouth environments have been identified as factors of human diseases associated with shellfish. These include *Vibrio parahaemolyticus*, *Vibrio vulnificus*, *Vibrio cholerae*, *Vibrio fluvialis*, *Vibrio hollisae*, *Vibrio mimicus*, *Vibrio alginolyticus* and unidentified or various *Vibrio* species (Hariharan and Amadi 2016). Vibrios are among the fastest growing bacteria in nature. If they are not cooled immediately after harvest, they easily grow in molluscan shellfish. *Vibrio cholerae*, *Vibrio vulnificus* and *Vibrio parahaemolyticus* are the most important pathogens. The pathogens, *V. vulnificus* and *V. parahaemolyticus*, can cause waterborne diseases, but are particularly dangerous when combined with a filter-feeding vector, such as molluscan shellfish. Filter-feeding molluscs pump the surrounding water over their gills, simultaneously obtaining oxygen and food. *Vibrio* spp. are often found attached to particles and as these particulates are passed over the sieve-like gills of filter-feeding molluscs, they are strained out of the water and retained (Raszl et al. 2016).

It has been reported that in naturally contaminated bivalve molluscs, vibrios cannot be easily removed by depuration (Vasconcelos and Lee 1972; Rodrick and Schneider 1991; Wright et al. 2009) compared to faecal bacterial indicators such as *E. Coli* (Lee and Rangdale 2008). It has been reported that artificial depuration is not reliable to eliminate *Vibrio parahaemolyticus* in oysters (Croci et al. 2005), and even depuration may cause cross contamination of *V. parahaemolyticus* in other oysters (Ramos et al. 2012).Therefore, such processing methods may not be sufficient to maintain the necessary public health in the harvested product contaminated with vibrios (Lee and Rangdale 2008). Eyles and Davey (1984) also reported that depuration did not reduce *V. parahaemolyticus* levels in shellfish. It is reported that another method of purification involving the transport of shellfish from a limited harvesting area to an open area where natural cleaning can occur is available, but it will not be relied upon to completely eliminate *V. vulnificus* in shellfish (Drake et al. 2007).

All vibrios are heat sensitive. It is reported that in the heat treatment applied to shellfish, a heating that will increase the internal temperature to 60 °C will be sufficient to eliminate pathogenic vibrios, and chilling and refrigeration are critical control measures to prevent the growth of these microorganisms (Dalsgaard et al. 2001). It is reported that *V. vulnificus* is known to grow between 8–43 °C and *V. parahaemolyticus* grows within the 5–45.3 °C range (Baker 2016).

The heat shock process applied for shucking the oysters was determined to reduce the level of *V. vulnificus* in oyster meats. Additional reductions were found immediately after the heat shock during the washing and cooling phases (Hesselman et al. 1999). Heating oysters for 10 min at 50 °C in water was determined to be sufficient to reduce *V. vulnificus* to an undetectable level (Cook and Ruple 1992). It has been shown that combination of hot-water/cold-shock processes reduced *V. parahaemolyticus* in oysters to nondetectable levels within 22 min at 50–52 °C (Andrews et al. 2000).

Ionising radiation is another way to eliminate vibrios in shellfish. In another study, live oysters (*Crassostrea virginica*) with naturally incurred and artificially inoculated pathogenic vibrios, were exposed to 0–3 kGy dose Cobalt-60 gamma radiation. *Vibrio vulnificus* was reduced from 10^6 cfu/g oyster meat to non-detectable levels with 0.75–1.0 kGy irradiation exposure. *Vibrio parahaemolyticus* required 1.0–1.5 kGy for reduction to non-detectable levels (Andrews et al. 2003). In a study aiming to determine the radiation decimal reduction dose (D10) of toxigenic *Vibrio cholerae* in pure culture, it is found that a dose of 1.0 kGy was optimal for choro mussels and abanico clams, whereas 2.0 kGy produced the best results when treating common clams (Torres et al. 2001). Lopez (2001) stated that 1.0 kGy would be enough to render Uruguayan mussels Vibrio-safe (Lopez 2001). Rashid et al. (1992) found that the gamma-radiation dose needed to reduce by 10^4 the number of *Vibrio* isolates is about 3 kGy in frozen shrimps. It has been reported that combination of sodium hypochlorite (NaClO) and gamma irradiation reduce *V. parahaemolyticus* in shucked oysters and clams (Park and Ha 2018).

Chilling can be effective in reducing the number of vibrios in shellfish products. *Vibrio vulnificus* in cold stored shellstock oysters and shucked oyster meats reached undetectable levels (MPN <3/g) within 14 days and 21 days respectively (Cook and Ruple 1992). Bradshaw et al. (1974) reported that *V. parahaemolyticus* inoculated onto the surface of cooked shrimp or crab gradually decline in numbers at incubation temperatures of 10 °C and below and multiply if held at 18.3 °C. In shellstock oysters stored at 0, 2 and 4 °C, *V. vulnificus* levels have been reported 1-log unit decreased after 3 days of storage, while remain unchanged for 14 days at 2 and 4 °C. However, after 10 days at 0 °C, a 2.5-log unit decrease was detected (Kasper and Tamplin 1993). However, others stated quite the reverse reports. Thomas (2016) found that the onboard icing using ice-slurry immediately after harvest did not change the levels of *V. parahaemolyticus* and *V. vulnificus* in oysters (*Crassostrea virginica*). Similarly, Melody et al. (2008) reported that on-board and dockside icing did not reduce the levels of *V. vulnificus* or *V. parahaemolyticus* in oysters (*Crassostrea virginica*).

Liu et al. (2009) reported that freezing and frozen storage is a widely used method to protect the product quality by preventing the growth of bacteria, and that *V. parahaemolyticus* can provide a certain degree of reduction in oyster meat. They determined a reduction of a 0.22 log MPN/g in population of *V. parahaemolyticus* in Pacific oysters cryogenically frozen. Parker et al. (1994) reported that freezing oysters under vacuum-packaged conditions at −20 °C is significantly effective in reducing loads of *V. Vulnificus*. A reduction of 3–4-logs in *Vibrio vulnificus* population in oysters inoculated with *V. vulnificus* and frozen at −20 °C was found in that study. The frozen storage of oysters at −18 and −24 °C has been reported to be effective in the inactivation of *V. parahaemolyticus* in oyster homogenates (Muntada-Garriga et al. 1995). Freezing of oysters at −40 °C and then storage for 8–10 weeks achieved a 4- to 5-log reduction in the *V. vulnificus* population (Cook and Ruple

1992). The greater log reduction in population of V. *vulnificus* frozen by cryogenic freezing compared to air blast freezing was determined by Espinoza Rodezno et al. (2013). However, it was determined that both freezing methods is not effective on reduction of *V. parahaemolitycus* population. This is thought to be since the fact that *V. parahaemolyticus* tolerates low temperatures more than *V. vulnificus* because of rapid and severe morphological changes during cold stress.

The application of pressure is reported to inactivate vibrios in shellfish. The application of pressure is reported to inactivate vibrios in shellfish. Vibrios are more sensitive to pressure than most other bacteria. Because they have a cell membrane that can be inactivated at pressure levels ranging from 200 to 350 MPa (Espinoza 2013). It was reported that *V. vulnificus*, *V. parahaemolyticus* and *V. cholerae* are sensitive to high hydrostatic pressure treatment at pressure levels between 200 and 300 MPa (Berlin et al. 1999). The initial *V. parahaemolyticus* population (8.4 × 105–3.4 × 107 cfu/g) reduced to non-detectable levels in Pacific oysters (*Crassostrea gigas*) after 345 MPa for 30 and 90 s (Calik et al. 2002). A reduction of 5-log of *V. parahaemolyticus* and *V. vulnificus*, achieved with high pressure processing of 300 MPa for 180 s and 250 MPa for 120 s respectively has been reported (Cook 2003).

5.2.1.2 *Salmonella*

Another bacterium related to human health is *Salmonella*. *Salmonella* spp. are Gram-negative, rod-shaped enteric bacteria, human and animal wastes are sources of *Salmonella* and are therefore ubiquitous in polluted water environments. Since bacteria are resistant to physicochemical factors, they are present in surface and groundwater (Soniat 2009). Microorganisms belonging to the genus *Salmonella* are not natural inhabitants of the aquatic environment, but several *Salmonella* serovars are widely distributed in the sea, river mouth and river waters, and in various seafood such as molluscs, shrimp, oysters, and various fish species. The presence of these microorganisms in aquatic products and environments where they grow is mainly due to hygiene errors during production (Li et al. 2009; Novoslavskij et al. 2016). *Salmonella*, introducing into the marine environment, can contaminate the fauna, especially molluscan shellfish products. Although it has been reported to be isolated from fish, shrimp, clams, mussels, oysters, crabs, lobsters, squid, cuttlefish, and octopus (Kumar et al. 2009), there are few studies on *Salmonella* in various shellfish there are few studies on *Salmonella* in various shellfish (Martinez-Urtaza et al. 2004). It has been reported that salmonellosis is one of the most important bacteria in food poisoning epidemics in the world, in most cases poisoning is associated with seafood, especially shellfish consumption, and the incidence of salmonellosis caused by the consumption of contaminated raw shellfish is the main concern of the public health institutions and the seafood industry (Correa et al. 2006). *Salmonella* is among the most important causes of gastrointestinal disease worldwide, and many countries that import seafood will restrict the import of raw food products containing these pathogens. *Salmonella* occurs both in water and in contaminated coastal areas or ponds and fresh fish from these areas (Feldhusen 2000). *Salmonella enteritidis*, *Salmonella paraty*phi, and *Salmonella typhimurium* are important species and are responsible for high mortality rates (Amagliani et al. 2012). Historically, sewage contamination of shellfish harvest beds led to large shellfish-associated outbreaks of *Salmonella s*erotype *typhi* infections (Iwamoto et al. 2010). *Salmonella enterica* serovar *Enteritidis* and serovar *Typhimurium* are the most common *Salmonella* that cause infection and death (Butt et al. 2004).

Salmonella isolation was reported in 133 bivalve samples harvested from a polluted shellfish growing area (Monfort et al. 1994). In a study that examined 100 fish and shellfish samples, *Salmonella* was detected in 59% of shrimp samples and 30% of oysters (Shabarinath et al. 2007). Martinez-Urtaza et al. (2003) reported the isolation of *Salmonella serovars* in live molluscs from the Galician region of Spain. In Shrimp, oyster and mussel samples collected from different markets in Alexandria, Egypt, *Salmonella* was determined as 14.0%, 8% and 4 8%, respectively (Bakr et al. 2011). In a study conducted on the presence of *Salmonella enterica* subsp. *enterica* in different species of bivalve molluscs and seawater, it was reported that the presence of *Salmonella enterica* subsp. *enterica* in seawaters and in bivalve molluscs varied by bivalve species and classification areas (Rubini et al. 2018). A study studied on the presence of *Salmonella* spp. in cockles and carpet shells collected from a class B growing natural bed in Sardinia (Italy), the fact that no *Salmonella* has been detected confirms this (Marceddu et al. 2017).

The most common factors contributing to salmonellosis outbreaks are improper cooking, inadequate storage, cross-contamination and use of raw ingredients in the preparation of seafood. Main post-harvest critical control points for *Salmonella* control in seafood, irrespective of whether the primary source is a marine or an aquaculture product, include: primary chilling immediately in an ice-water slurry on vessels and at harvest site; in cooked products, applying time–temperature regimes to give log reductions of contamination levels at sites of microbiological concern; rapid chilling after cooking; plate freezing, followed by frozen storage (Amagliani et al. 2012).

5.2.1.3 *Escherichia coli*

Escherichia coli has been reported to serve as a useful indicator of faecal pollution due to its relationship with the digestive system of warm-blooded animals, and this Gram-negative, rod-shaped, constitute to coliform population. Although *E. coli* is not generally considered pathogenic, a few of pathogenic strains have been identified. Pathogenic types include enterotoxigenic, enteropathogenic, enteroinvasive, and hemorrhagic strains. The first three are usually associated with human faecal contamination, whereas the latter is most often associated with farm animals (Soniat 2009).

The *E. coli* outbreak due to consumption of butterfly shrimp in sushi restaurants in Nevada (USA), which is thought to be caused by poor food processing practices and infected food processors, has been reported (Jain et al. 2008).

It has been reported that the use of the depuration system as an effective sanitation strategy towards control of *E. coli* contamination in mussels. In a research that study the performance of a shellfish depuration system for elimination of *E. coli*, freshly harvested mussels were artificially contaminated with *E. coli* and were kept immersed in an isothermal tank containing recirculated clean seawater. The results showed an at least 2 log reduction of *E. coli* in mussels after 24 h of treatment (Andritsos et al. 2016).

Shellfish are purified to reduce the risk of microbiological contamination. Microbiological purification consists of immersing live shellfish in tanks continuously fed clean seawater for a period that is sufficient to eliminate microbiological contaminants and render the shellfish suitable for human consumption (Gueguen et al. 2011). In a study, a 95% reduction in the number of *E. coli* in depuration process of 48 h was determined (Schwab et al. 1998).

Microbial contamination significantly affects the shellfish industry. Therefore, to protect public health in the world, shellfish harvest areas are limited. Harvest is limited in areas where crustaceans are subjected to microbial contamination. It is considered as the best protection strategy to harvest from regions with good water quality where there is no microbial contamination or at minimum levels (Jones 2009). Shellfish safety issues continue to revolve around these two categories: the quality of the waters in which shellfish are grown, and the conditions under which shellfish are harvested, processed, and distributed (Wittman and Flick 1995).

5.2.2 Viruses

Viruses, unlike bacteria, have been reported to be the main cause of shellfish outbreaks as a result of pollution because sewage is now treated more thoroughly than in the past and viruses survive better than bacteria in sanitation procedures (Croonenberghs 2000). Human pathogenic viruses enter the marine environment by direct discharge of treated or untreated sewage wastewater. Failure of current water treatment applications to provide virus free wastewater wastes causes human pathogen to enter the sea and river mouth waters continuously. Bivalve molluscs fed by filtering the water accumulate contaminants in polluted waters in their edible tissues and cause public health problems, especially due to virus contamination of shellfish products such as raw consumed oysters and clams and other molluscs that have undergone little treatment (Bosch and Guyader 2010). Viruses are species-specific intracellular parasites and are the leading cause of shellfish-borne disease in humans. Although human pathogenic viruses are readily taken up and accumulated, they do not infect or grow in molluscan shellfish. They can persist for extensive periods in the marine environment and in shellfish (Jones 2009). It has been reported that most of the shellfish-borne viral outbreaks are limited to norovirus and hepatitis A virus, and norovirus outbreaks associated with raw oyster consumption are common in the world, and infection of the hepatitis A virus is the most serious viral infection due to shellfish consumption (Baker et al. 2010; Bosch and Guyader 2010). Bellou et al.

(2013) conducted a review to investigate viral outbreaks from shellfish and their distribution in different countries and to determine whether different shellfish and virus species are involved. In this study, they examined various databases from 1980 to 2012, and reported that 359 shellfish-origin viral outbreaks were detected and most of the reported outbreaks were in East Asia, followed by Europe, America, Oceania, Australia, and Africa.

5.2.2.1 Norovirus (NoV)

Noroviruses (NoVs) belong to the family Caliciviridae and are the leading cause of human gastroenteritis worldwide. Human NoV infection is transmitted mainly through the faecal-oral route and has often been associated with the consumption of contaminated foods (Li et al. 2011). Seafood plays an important role in the emergence of Norovirus infections. Virus contamination of seafood takes place in two main ways: contamination of the waters where the seafood is found with human and animal waste and contamination that occurs during the preparation and serving of food by infected people (Cheng et al. 2005). It is realized by the consumption of shellfish that are grown in waters contaminated by sewage. During the processing, packaging, and serving of shellfish that are grown in clean waters for consumption, the products can be contaminated with norovirus by sick people and outbreaks can occur (Hassard et al. 2017).

5.2.2.2 Hepatitis A Virus (HAV)

Hepatitis is the most serious viral disease associated with the consumption of raw shellfish, and the hepatitis A virus (HAV) is the most common agent (Croonenberghs 2000). HAV is responsible for human acute viral hepatitis. Only one serotype of HAV has been identified worldwide (La Bella et al. 2017). Hepatitis A infection is a mild illness characterized by sudden onset of fever, malaise, nausea, anorexia, and abdominal discomfort, in humans (Di Pinto et al. 2003).Shellfish are not only considered high risk food if eaten raw or cooked lightly but also recognized as a potential vehicle of food-borne diseases of HAV (Potasman et al. 2002). It has been reported that mussels and cockle are an important risk factor for HAV because they are consumed raw or lightly cooked, and HAV may remain infectious in the mussel tissues unless thoroughly cooked (Croci et al. 2005; Pinto et al. 2009). In a study that a total of 213 shellfish (52 oysters, 69 cockles and 92 mussels) from a culture farm and two retailed markets were investigated for HAV contamination, it was found that 3.8% of the shellfish and 2.9 and 6.5% of the cockle and mussel, respectively, showed positive for HAV (Namsai et al. 2011).

It is reported that viruses can remain in shellfish for more than a few days and there is not enough intervention beyond cooking for bivalve shellfish contaminated with enteric viruses (Kingsley et al. 2007). Depuration alone has a limited effect on reducing the level of viruses in shellfish and is not suitable for shellfish harvested from more heavily contaminated areas (Singh and Nagalakshimi 2013). Other studies have shown that it is more difficult to remove enteric viruses such as hepatitis A virus (HAV) and norovirus (NoV) by purification from the infected oysters (Hernroth and Allard 2007; Ueki et al. 2007; Mcleod et al. 2009; Correa et al. 2012; Neish 2013).

High pressure treatment has been reported to be promising for hepatitis A virus (HAV) and norovirus inactivation, and treatment of 460 MPa at room temperature has been found to be sufficient to inactivate the 7 log10 PFU virus stock of HAV (Kingsley et al. 2002). Inactivation effects of HHP on human viruses found in the tissues of oysters, mussels, and clams have been shown (Prapaiwong et al. 2009; Kingsley 2013, 2014). Under 350–400 MPa high pressure processing (HPP) at 8.7–10.3 °C for 1 min resulted in a 6-log reduction of hepatitis A (Calci et al. 2005).

Enteric viruses are more resistant to inactivation in water sources and are removed slowly, or not at all, from bivalves by depuration process (La Bella et al. 2017). Methods such as depuration to remove bacterial pathogens in bivalve have been used successfully, whereas had little effect on viral contamination. In a study, after a 48-h depuration, the bacterial levels of the oyster decreased by 95%, while the Norwalk virus levels only decreased by 7% (Schwab et al. 1998).

Many enteric human viruses, such as Hepatitis A and Norwalk, have been reported to be heat resistant and that viruses can survive when mild cooked (Millard et al. 1987; Peterson et al. 1978). However, there are studies reported that an effective cooking process inactivates viruses. Flannery et al. (2014) reported that domestic cooking practices based on shell opening alone do not inactivate infectious virus in mussels, however, cooking mussels at high temperatures is effective to reduce infectious virus concentrations and the risk of illness in consumers. In another study, cooking for 3 min at >85 °C was found to be sufficient to inactivate poliovirus and hepatitis A in clams (Millard et al. 1987). Lees (2000) reported that cooking shellfish is an effective method for norovirus decontamination and at least 90 s of heat treatment at 90 °C will be sufficient. However, a balanced heat treatment should be applied to protect the sensory quality of shellfish (Richards et al. 2010).

5.3 Biotoxins

Some species of Cyanobacteria (blue-green algae) and Pyrrophyta (dinoflagellates) produce toxic compounds which are not poisonous for fish but for human. Filter-feeding shellfish such as clams and mussels accumulate the poisons from the dinoflagellates when feeding on them. They are normally eaten whole, including the intestinal tract, and raw or following a very mild heat treatment (Lees 2000). Marine biotoxins are a naturally occurring phenomenon that is not associated with sewage contamination of coastal waters but is produced by Dinoflagellates and diatoms (Gosling 2015). The marine biotoxins comprise many distinct compounds, all produced by species of naturally occurring marine algae. The algae are at the bottom of the marine food chain. Consequently, the biotoxins produced by some algae are collected and concentrated through levels of the food chain (e.g., molluscs, crustaceans, and finfish) and ultimately are consumed by humans (Ward and Hart 1996).

Algae, one of the natural parts of the aquatic food chain, are photosynthetic organisms ranging from single cell to complex multicellular structures. Algae reproduce enough to accumulate on the water surface in the presence of favourable ambient conditions for their reproduction and this disrupts the aquatic balance. This rapid increase in algae population is called algae bloom and causes the water to turn into red, green, or brown. As a result of excessive reproduction and toxin production of some algae species, a condition called harmful algae bloom arises that threaten human health. In some of the harmful algae reproduction, various poisons are synthesized depending on the type of microalgae that show excessive reproduction. Although these poisons exhibit different properties in fresh-brackish water and marine species, they are all temperature resistant compounds and do not degrade at cooking temperature (He 2015) This means that even well-cooked bivalves might still present a risk to consumer's safety. Accumulation of toxic marine algae in raw or light cooked shellfish has been associated to Paralytic Shellfish Poisoning (PSP), Diarrhetic Shellfish Poisoning (DSP), Neurotoxic Shellfish Poisoning (NSP), Amnesic Shellfish Poisoning (ASP) and Azaspiracid Poisoning (AZP) occurrences (Oliveira et al. 2011).

Bivalve molluscs, especially oysters, are consumed raw in most countries, and many are marketed as live bivalve molluscs. Because it is difficult to preserve live shellfish outside their natural environment, this also results in the need for rapid assessment of their sanitary safety. Factors affecting the occurrence and accumulation of toxic algae cannot be controlled, and the prediction of toxic algae has severe limitations. Therefore, and because there is a very strong need to market shellfish as a live product, management practices for safe bivalve mollusc production are very specific. Official control of shellfish safety is mostly conducted on live shellfish, i.e. through regular, continuous surveillance of shellfish growing areas (Lawrence et al. 2011).

Digestive gland, mantle, gonad, and gill tissues all retain the toxins although the levels vary between tissues and between species (FAO 2004). The ability of shellfish to retain and accumulate toxins varies greatly according to the shellfish species (Hegaret and Shumway 2009). For example, mussels accumulate more toxin, and more rapidly, than most of the other bivalve species. They actively feed on toxic cells and accumulate high levels of toxins because they have nerves insensitive to PSP toxins. Oysters accumulate low levels of toxins because they are sensitive to PSP toxins. The accumulation of PSP toxins in bivalves varies according to external factors such as temperature, microbiota, presence, and concentration of toxic dinoflagellates (Bricelj and Shumway 1998).

5.3.1 *Paralytic Shellfish Poisoning (PSP)*

PSP is caused by eating bivalves that contain saxitoxin (STX) and its derivatives. STX and its analogues are synthesized by *Alexandrium tamarense* and *Alexandrium cantenella* dinoflagellates in both marine and freshwater environments (Faber 2012). The toxins, a group of about 20, are bacterial catabolites in dinoflagellate cells of the genera, *Alexandrium*, *Gymnodinium* and *Pyrodinium*. The toxins produced by these organisms are the most common and widespread of the shellfish toxins and are among the most potent neurotoxins known. They act by blocking the passage of sodium ions through cell membranes, thus inhibiting nerve impulse transmission (Gosling 2015). The basic effective toxin is saxitoxin (Pomati et al. 2006). Saxitoxin is an inhibitory neurotoxin that induces flaccid paralysis by actively suppressing excitation of neuronal impulses. In humans, PSP may develop after consuming contaminated raw or cooked shellfish containing high concentrations of STX. Since Saxitoxin is a heat resistant and very strong neurotoxin, it can cause adverse effects in humans even at low doses. There is no antidote and detoxification pathway either (Faber 2012).

Bivalve shellfish such as mussels, clams, oysters, and scallops pose a particularly high risk as they feed by filtration providing toxin accumulation (Turnbull et al. 2013). During the process of filtration, the dinoflagellate cells and cysts are transported to the oesophagus and the stomach of the bivalve molluscs. The digestion takes place in the stomach and the diverticula whereby the PSP toxins are released and enter the digestive organs. The particular toxin mixture retained in soft tissues of the shellfish varies in concentration and over time and is determined by the species and strains of the dinoflagellates and shellfish as well as by other factors like environmental conditions (FAO 2004). While the toxin is found quite intensely in the digestive glands in clams, its concentration in the gills, ovaries and adductor muscles is less than 80 μg/100 g. Since toxin is not easily found in the adductor muscles in oysters, even if they are exposed to toxic algae, these parts are not dangerous for public health (Cambella et al. 1993). In a study with mussels, Kwong et al. (2006) listed the accumulation of toxins in organs as hepatopancreas, internal organs, gills, legs, and adductor muscles. It has been reported that different types of shellfish accumulate toxins at different rate, even in individuals of the same species and even under the same environmental conditions, it is thought that these differences may result from the difference in shell size (Setala et al. 2014; Morono et al. 2001).

Symptoms of PSP can include tingling, numbness, or burning in the extremities or mouth; a lack of coordination; drowsiness; fever; and rash. Individuals may also experience nausea, vomiting and diarrhoea. In severe cases, PSP can result in respiratory arrest and death.

Saxitoxin cannot be removed from shellfish by cooking, freezing, or other post-harvesting processes. Toxins do not alter the taste, smell, or appearance of shellfish. The most obvious detoxifying method for shellfish contaminated with PSP is to transfer the shellfish to waters free from toxic organisms. Although this is a suitable method for some shellfish products, some species have been reported to remain toxic for a long time (Shumway et al. 1995).

Although toxins cannot be destroyed by normal cooking procedures, it is reported that coagulation of proteins in shellfish tissues may cause redistribution in the organs of shellfish and some toxins can be released into cooking liquids (FAO 2016). It is reported that even if cooking does not eliminate the danger of poisoning, in case of low toxicity can reduce toxicity to safe levels. Some researchers have determined that heat treatment and cooking are effective in reducing PSP toxicity. These effects were probably caused by the release of toxin into cooking water. MacDonald (1970) reported the effectiveness of pan-frying cooking in shellfish products in reducing PSP toxicity. Noguchi et al. (1980a, b) identified a sharp decrease in toxin levels after sterilization of infected scallops. Prakash et al. (1971) found that cooking caused a 70% reduction in the toxicity of soft-shell clam. Berenguer et al. (1993) also concluded that the industrial canning process caused a decrease in the PSP levels of the Mediterranean cockles. The standard canning process provided a reduction (over 50% of toxicity level in raw material) in PSP toxicity of mussel meat (Vieites et al. 1999). A significant reduction (approximately 65%) in toxicity of lobster hepatopancreas was observed after boiling or steaming (Lawrence et al. 1994).

Reboreda et al. (2010) determined that thermal treatment on PSP contaminated mussels, oysters and oysters reduces PSP levels below the detection limit.

5.3.2 *Diarrhetic Shellfish Poisoning (DSP)*

Diarrhetic Shellfish Poisoning is caused by toxins (ocadaic acid, dinofisis toxin, yessotoxin pectotoxin) produced by dinoflagellates (Aune and Yndstad 1993). Diarrhetic shellfish toxins (DST) are a group of phycotoxins that include Okadaic acid (OA) and structurally related toxins. DSP is caused primarily by okadaic acid (OA) and its derivatives, the dinophysistoxins (DTXs), produced by dinoflagellate species of *Dinophysis* and *Prorocentrum* (Gosling 2015). The first group, acidic toxins, includes okadaic acid (OA) and its derivatives named dynophysistoxins (DTXs). The second group, neutral toxins, consists of polyether-lactones of the pectenotoxin group (PTXs). The third group includes a sulphated polyether and its derivatives the yessotoxins (YTXs) (FAO 2004).

It has been reported that the shellfish causing DSP in Japan are mussels (*Mytilus edulis* and *M. coruscum*) scallops (*Patinopecten yessoensis* and *Chlamys nipponesis akazara*) and short-necked clams (*Tapes japonica* and *Gomphina melaegis*) and in the European Atlantic coast are *Mytilus edulis* and *Ostrea* sp. (FAO 2004). Although DSP is predominantly seen in Europe and Japan, it is reported to be seen in North and South America, Australia, Indonesia, and New Zealand due to the spread of toxic dinoflagellates globally (Scoging and Bahl 1998).

The main symptoms of DSP are cramps, severe diarrhoea and vomiting and patients usually recover within 3–4 days (Gosling 2015). Consumption of shellfish contaminated with high levels of OA-type toxins will result in adverse effects such as gastrointestinal disorder, diarrhea, abdominal cramps, nausea, and vomiting. The onset of symptoms, which are never lethal, ranged from 30 min to a few hours after ingestion of the toxic shellfish, with complete recovery within 3 days. The intensity of the symptoms in humans depends upon the amount of toxin ingested. Diarrhoeic Shellfish Poisoning (DSP) in humans is caused by the ingestion of contaminated bivalves such as mussels, scallops, oysters, or clams. The fat-soluble DSP toxins accumulate in the fatty tissue of the bivalves (FAO 2004).

In a study, ozone treatment was applied to homogenized mussels and whole unshelled mussels at 4 °C and it was found that ozone treatment was treatment effective in reducing the amount of OA and its derivatives. The authors also found more OA reduction in homogenized mussel samples than in whole mussel samples (Louppis et al. 2011). Reboreda et al. (2010) showed that thermal processing of mussels does not change the DSP toxin levels compared to the controls in the whole body of the mussel.

5.3.3 Amnesic Shellfish Poisoning (ASP)

ASP is caused when bivalves are eaten that contain domoic acid (DA) and/or its analogues. The toxins are secreted by the diatom *Pseudo-nitzschia* multiseries, and the red alga, *Chondria armata*. DA is an analogue of glutamic acid, a neurotransmitter in the brain (Gosling 2015). Currently, 45 species *Pseudo-nitzschia* genus, 19 of which produce domoic acid have been discovered. These toxigenic species are found worldwide. *Nitzschia navis-varingica*, another diatom, also produces domoic acid (Schroeder et al. 2015). After consuming contaminated bivalves, there is nausea, vomiting and diarrhoea, confusion, memory loss, disorientation and even seizures, coma and death (Gosling 2015). While irreversible conditions such as loss of memory may be encountered in the treatment of delayed cases, success can be observed in the early treatment of domoic acid-related toxicities (Schroeder et al. 2015).

It has been reported that domoic acid is accumulates in a wide variety of shellfish species such as cockles, crabs, mussels, razor clams and scallops, they do this by filtering the plankton directly or feeding it directly with contaminated organisms and thus, the highest concentrations are found in the digestive glands compared to other tissues (Jeffery et al. 2004). However, accumulation rate varies between species of shellfish. In Canada in late 1987 there was an outbreak of an acute illness characterized by gastrointestinal symptoms and unusual neurologic abnormalities among persons who had eaten cultivated mussels. Perl et al. (1990) investigated this outbreak and found that the poisoning is caused by the consumption of mussels contaminated by domoic acid. In a study investigating the domoic acid concentrations in the different samples of molluscs, mussels, oysters, and scallops, it was found that the level of toxin was the highest in scallops while the lowest in mussels (Andjelkovic et al. 2012). In a similar study, it was found that domoic acid was the most prevalent in blue mussel (*Mytilus galloprovincialis*), followed by the European flat oyster (*Ostrea edulis*), the Mediterranean scallop (*Pecten jacobaeus*) and proteus scallop (*Flexopecten proteus*) (Ujevic et al. 2010). The presence of domoic acid in tissues of octopus (*Eledone cirrhosa*, *E. moschata* species, *Octopus vulgaris*) (Costa et al. 2004, 2005a), cuttlefish (*Sepia officinalis*) (Costa et al. 2005b) and swimming crab (*Polybius henslowii*) (Costa et al. 2003) collected from the Portuguese coast showed that toxin was transported through the sea food chain.

Novaczek et al. (1992) reported that industrial depuration may provide a means of removing domoic acid toxin from blue mussels (*Mytilus edulis*). These researchers transported mussels containing up to 50 µg/g domoic acid from a Prince Edward Island estuary into controlled laboratory conditions. They found that 50% of toxin is eliminated within 24 h.

Although cooking due to the heat stability of domoic acid does not destroy the toxin, it has been reported that normal home cooking processes such as boiling and steaming can reduce the amount of DA in shellfish due to the partial leakage of the toxin into the cooking liquids (Andjelkovic et al. 2012). The effect of cooking on the concentration of domoic acid in the Manila clam (*Ruditapes philippinarum*) and cockle (*Cerastoderma edule*) was investigated. In that study, these molluscs were cooked by steaming and boiling. They found that cooking provided different effect for these two species, a significant part of its domoic acid content was lost by cooking in the cockle, while the clam did not. The authors stated that cooking affects the toxin concentration in bivalves in a way that is species specific (Vidal et al. 2009). It has been determined that the cooking process provides more than 50% reduction in the concentration of DA in swimming crab (*Polybius henslowii*) (Costa et al. 2003). Reboreda et al. (2010) studied on reduction of ASP in scallops. They tried different elimination methods including evisceration, thermal processing, freezing and their combination. They found that the most effective method to reduce toxin level to the legal limit is ablation of the hepatopancreas. It was found insignificant reduction with the other methods.

5.3.4 *Neurotoxic Shellfish Poisoning (NSP)*

Neurotoxic Shellfish Poisoning is a little common intoxication, which has not been documented as a fatal intoxication in humans, and results from consumption of molluscan shellfish contaminated with brevetoxins (Zaccaroni and Scaravelli 2008). NSP is caused by a group of toxins, the brevetoxins (BTXs), secreted by marine dinoflagellates of the genus Karenia. The main symptoms of NSP are headaches, dizziness, muscle and joint pain and difficulty in breathing, and full recovery generally occurs within 48 h (Gosling 2015). In the first one symptom are both neurological and gastro-intestinal and appear within 1–3 h after mussels' ingestion. Neurological syndrome includes paraesthesia of area around the mouth, the face and throat, muscular ache, ataxia, inversion of thermal perception, bradycardia, mydriasis. Gastro-intestinal syndrome includes abdominal pain, nausea, diarrhoea.

5.3.5 *Azaspiracid Shellfish Poisoning (AZP)*

These toxins accumulate in bivalve molluscs that feed on toxic microalgae of the genus *Protoperidinium*. Azaspiracids cause azaspiracid poisoning, an illness like diarrhetic shellfish poisoning. The toxin may accumulate in the hepatopancreas (digestive glands), gonads and the adductor muscle. Such contamination can happen during the sample pre-treatment stage, freezing or cooking may damage the walls of internal organs, thus facilitating the toxin to leach into different tissues (Hess et al. 2005). An unusual biotoxin was first observed in mussels (*Mytilus edulis*) from Ireland in 1995 after an illness similar to diarrhetic shellfish poisoning (DSP) was observed. With the identification of AZA as the causative agent, the illness was named azaspiracid poisoning (AZP). Following the discovery in Risk Assessment of Azaspiracids (AZAs) in Shellfish: A report of the Scientific Committee of the FSAI August 2006 mussels, several other bivalve molluscs have been identified as containing AZAs including oysters (*Crassostrea gigas*, *Ostrea edulis*), scallops (*Pecten maximus*), clams (*Tapes phillipinarium*), cockles (*Cardium edule*) and razor fish (*Ensis siliqua*) (Anonymous 2006).

5.4 Chemical Contaminants

5.4.1 *Toxic Metals*

As a potential source of chemical contaminants in the aquatic environment, industrial wastes and mining activities have accumulation properties and, due to these properties, can damage the ecosystem and therefore the species diversity in the seas. The main sources of contaminants are of human origin. Metals (trace elements) are naturally present in many rocks and minerals (Gueguen et al. 2011). Several sources of metals can contribute to metal accumulation in marine shellfish, including metal in the water (or the dissolved phase), metal from prey (such as phytoplankton, small protozoans, bacteria), and metal in sediment (Wang and Hong Kong University of Science and Technology 2009). Due to natural weathering of the earth's crust, they are found in all environmental compartments, including seawater. Some trace elements that are absorbed by living organisms accumulate in the food chain, and therefore present a risk to humans, who are the final consumers at the top of the food chain. Metals, such as mercury, cadmium, lead, copper, and zinc can easily concentrated in shellfish meat (Gueguen et al. 2011).

Marine bivalves are known to pump significant amounts of seawater and thereby to process particles because of their extremely high filtration activity. Thus, there is a very close interaction between bivalves and the chemicals in the water. These contaminants are accumulated through both dissolved phase uptake and particulate ingestions by shellfish. Bivalve molluscs are exposed to large volumes of water during respiration and feeding. Metals and organic contaminants will thus inevitably enter the bivalves through the gills and other exchange surfaces (Wang and Hong Kong University of Science and Technology 2009). The ability of mollusks to accumulate metals from the water they live in enables them to be used as a useful tool in marine environmental research.

Lead, cadmium, and mercury are more toxic among heavy metals, even in traces. Pollution from lead, cadmium and mercury poses a serious problem due to their toxicity and ability to accumulate in biota (Nguyen et al. 2014).

5.4.1.1 Cadmium (Cd)

Cadmium present in atmosphere, water, or food when exposed to human in low concentration cause serious health problems and probably the death. Sources of cadmium human exposures are fossil fuels, iron and steel production, cement non-ferrous metals production, waste incineration, smoking, fertilizers, etc. Activities like volcanic eruption, mining and use of phosphate fertilizers provides cadmium exposures indirectly as toxin from earth crust (Sharma et al. 2016). It is used industrially in batteries, accumulators, coatings for protection of various metals against corrosion, alloys, plastic as an endurance enhancer and industry (Landrigan 1982). Cadmium, which is the element with the highest water solubility feature compared to other heavy metals, is transmitted to plants and animals through air, soil, and water, and to people with nutrients. For this reason, it has a high propagation rate in nature; it is taken into biological systems by plant and sea creatures and has the property of accumulation (Dalcorso et al. 2008). In marine environments, Cd normally occurs in lower concentrations than in open oceans water and in higher concentrations in coastal and estuarine environments due to the intensive industrial discharge, port activity and mining activity in rivers distributed in the sediment, particulate matter, water and marine organisms (Tamele and Loureiro 2020).

Differences in metal concentrations have been reported in bivalve mollusc species. In a study examining the metal concentrations of black mussels and oysters, it was reported that the concentration of Class B type or borderline metals in oysters was higher than mussels (Wang and Wong 2006). In a study examining cadmium (Cd) concentration in various bivalve species, the highest concentration of Cd was determined in scallops (*Chlamys varia*), while the lowest concentration was in clams (*Macoma balthica*) (Pigeot et al. 2006). In a study which investigated the behaviors of three bivalve species (*Mytilus galloprovincialis*, *Callista chione* and *Venus verrucosa*) taken from their natural environment and relatively polluted marine environments, when they were transferred to clean sea water in the laboratory environment and then exposed to Cd concentrations, the highest Cd concentration was found in *M. galloprovincialis* (2.60 mg), followed by *V. verrucosa* (0.55 mg/kg) and *C. chione* (0.31 mg/kg). Cd bioaccumulation behaviors of even different tissues of *M. galloprovincialis* and *C. chione* have changed, the highest concentration was determined in the gills of these two species, followed by the body and the mantle, respectively (Chalkiadaki et al. 2014). In another study investigating the heavy metal concentrations of various bivalve species (*Anadara granosa*, *Perna viridis*, *Crassostrea gigas*, *Meretrix lyrata* and *Amusium pleuronectes*) captured from the northern shores of Central Java (Indonesia), while the highest Cd concentration was found in *Amusium pleuronectes* the lowest was in *Perna viridis* (Yulianto et al. 2005). Cadmium levels of Indian white prawn (*Fenneropenaeus indicus*) and blue crab (*Portunus pelagicus*) obtained from Tok Bali Port in Kelantan, Malaysia were determined as 0.04 µg/g and 0.09 µg/g respectively (Salam et al. 2019).

5.4.1.2 Mercury (Hg)

Mercury is a toxic heavy metal released into the environment from both natural and anthropogenic sources. It is of great interest to consumers as to whether it can cause neurological effects at low dose levels. Mercury is distributed throughout the world, and cycles naturally through the earth's crust, atmosphere, oceans, and life forms, with trace amounts in fish, plants, and animals. Human activity increases the amount of mercury emitted into water, air, and soil, through coal-combustion (used to power utility plants); mining; waste disposal; preparation of fungicide seed dressings in agriculture; and chlor/alkali plants. Mercury is used in products such as batteries, vapor discharge lamps, fluorescent bulbs, switches, and thermometers. Many of these products end up in landfills or incinerators. Mercury enters the food chain with the help of aquatic microorganisms. Released into the environment in inorganic form, mercury is methylated by bacteria in water and converted to an organic form, usually methylmercury (Hellberg et al. 2005).

Mercury levels were measured in commercially crustacea caught from the Central Adriatic and Tyrrhenian coasts of Italy and differences in mercury levels were determined between species. The lowest levels of mercury were detected in shallow aquatic decapods, karamote shrimp (*Penaeus kerathurus*), wart crab (*Eriphia verrucosa*) and European spider crab (*Maja squinado*). Deepwater rose shrimp (*Parapenaeus longirostris*), blue and red shrimp (*Aristeus antennatus*) and Norway lobster (*Nephrops norvegicus*), which are in close contact with the bottom sediments, have been reported to exceed the limit values set by the European commission (Di Lena et al. 2018). Mercury levels of eight bivalve species collected from a local market in Poland were examined. For this purpose Dog cockle (*Glycymeris glycymeris*), manila clam (*Ruditapes philippinarum*), Atlantic jackknife clam (*Ensis directus*), blue mussel (*Mytilus edulis*), Pacific oyster (*Crassostrea gigas*), great scallop (*Pecten maximus*), common cockle (*Cardium edule*), and hard clam (*Mercenaria mercenari*a) were collected. Mercury concentrations were determined between 0.033–0.577 (mg/kg dry matter) in these selected bivalve species and the average highest mercury level was found in Pacific oyster (Szkoda et al. 2015). In a study examining mercury concentrations in eight cephalopod species purchased from the major fish landing ports and wet markets on the Peninsular Malaysia, mercury concentrations were determined as 0.099–715 mg/kg DW in squids, 0.233–0.486 mg/kg DW in cuttlefishes and 0.208 mg/kg DW in octopus (Ahmad et al. 2015).

5.4.1.3 Lead (Pb)

Lead is a ubiquitous and typical toxic contaminant released into the environment from natural and anthropogenic sources. However, the environment has been enriched by industrial processes, and man-made sources such as mining, industries, motor vehicle exhaust and batteries (Yingliang et al. 2014). Mining and smelting, soldering, battery manufacturing, ammunition, metal water pipes, paint and petrol are reported as anthropogenic sources. Pb occurs in the environment both organic and inorganic forms, organic lead is the most toxic lead form (Tamele and Loureiro 2020). Lead levels of Indian white prawn (*Fenneropenaeus indicus*) and blue crab (*Portunus pelagicus*) obtained from Tok Bali Port in Kelantan, Malaysia were determined as 0.008 μg/g and 0.015 μg/g respectively (Salam et al. 2019). Mok et al. (2014) determined heavy metal concentrations of some molluscs, crustaceans and cephalopods collected from major fish markets on the coast of Korea. Lead concentration was found to be 0.117 μg/g in oyster, 0.154 μg/g in mussel, 0.018 μg/g in bay scallop, 0.007 μg/g in Japanese flying squid, 0.021 μg/g in small octopus, 0.012 μg/g in snow crab and 0.007 μg/g in blue crab. They reported that these values are lower than the regulatory limits set by various countries. In a study investigating seasonally heavy metal concentrations of cultured oysters (*Crassostrea gigas*) in coast of the Gulf of California, in the state of Sinaloa, Mexico, the highest Pb concentration was found to be 3.46 mg/g in April (Gongora-Gomez et al. 2017).

It is reported that oysters can accumulate very high levels in zinc (Zn) and copper (Cu) in their tissues (Rainbow and Phillips 1993).

5.4.1.4 Zinc (Zn)

Zinc, which is a trace element in sea water, plays an important role in biological events. High zinc concentration in marine environments is parallel to human activities and urbanization, and can be caused by paint industry, mining, electro-coating, and synthetic fibre production. Zinc is one of the most ubiquitous and mobile among the heavy metals and is transported in natural waters in both dissolved forms and is associated with suspended particles. In seawater, much of the zinc is found to be in the dissolved form as inorganic and organic complexes (Dhinamala et al. 2017). In various studies examining heavy metal concentrations in shellfish, the highest concentration among the studied metals was determined in zinc. Some of them are shrimp (*Penaeus monodon*) (Hashmi et al. 2002; Meshram et al. 2014; Dhinamala et al. 2017), miscellaneous shrimp and crab species (Krishnamurti and Nair 1999), oyster (*Crassostrea virginica*) (Vazquez et al. 1995) and *Pincdata radiata* (Gokoglu et al. 2006).

5.4.1.5 Copper (Cu)

Cu naturally occurs in the aquatic environment in low concentrations. Seawater Cu concentrations are generally less than 1 ppb. Elevated aquatic Cu concentrations primarily occur near copper mining and smelting facilities and in urbanized areas. Aquatic habitats are susceptible to Cu pollution because they are the ultimate receptor of industrial and urban wastewater, storm water runoff, and atmospheric deposition. Cu is toxic at higher concentrations (Woody and O'Neal 2012).

Copper concentrations of *Ruditapes decussutus* and *Ruditapes philippinarum* collected in different sites of Southern Spanish Atlantic coast were reported to be 1.6–3.7 mg/kg and 1.4–2.8 ng/kg, respectively (Usero et al. 1997).

Heavy metal concentrations in shellfish vary with season. In a study where seasonal concentrations of shellfish (bivalves, crustaceans, gastropods, cephalopods) that are popular in Nha Trang, Vietnam are examined, lead, cadmium and mercury concentrations in bivalves, crustacea and gastropods in rainy season (September, November, and January) were significantly higher than the dry season (May, July) (Nguyen et al. 2014). Gongora-Gomez et al. (2018) reported in pen shell (*Atrina maura*) higher concentrations of metal (chromium, cadmium, nickel, lead, arsenic, zinc, and mercury) in the dry season. Gokoglu et al. (2006) reported seasonal changes in Cd, Cu and Zn concentrations of pearl oyster (*Pinctada radiata*). They found the highest concentration among the metals was zinc. Seasonal variation in concentrations of some metals has been reported to be associated with gonad maturation (Latouche and Mix 1981; Paez-Osuna and Marmolejo-Rivas 1990; Paez-Osuna et al. 1995), food supply (Bryan 1973) and run-off of particulate metal to the sea due to high precipitation (Fowler and Oregioni 1976).

5.4.2 Agricultural Chemicals

Many chemicals produced to facilitate human life pose a global threat to human health and the environment, from the production stage to the consumption stage. Chemicals used to increase human life quality and especially pesticides used in agricultural production; As a result of uncontrolled, unconscious, and unnecessary use, it has become a serious threat to the life and environment of the individual. Every year, many people in the world get poisoned or die due to pesticides. Today, chemicals called pesticides that fight pests are used to increase the product. However, these chemicals stay in water and soil for a long time, causing pollution and reaching people through the food chain. These chemicals, which are suitable for their purpose and benefit when applied in required doses and durations, can affect non-targeted organisms as a result of their careless and intensive use, on the other hand, they can contaminate the soil and aquatic ecosystem. They are transported into aquatic systems through surface run-offs, spray drifts, agricultural returns, and groundwater intrusions. The residues of the pesticides used for intensive agriculture practices can contaminate the water (surface runoff and surface drainage) generally within a few weeks after the application (Mensah et al. 2014).

Bioaccumulation of hydrophobic organic pollutants is associated with the lipid content of biota. Lipids and fats change throughout the seasons in relation to an animal's physiological condition, such as spawning and gametogenesis. Thus, the accumulated concentrations in biota of lipophilic contaminants, including organochlorine contents, can be influenced by seasonal changes and age-grade (Choi et al. 2016). Organochlorine pesticides (OCP) have low solubility in sea water and are lipophilic. Thus, fish and shellfish are important sources of OCP intake. These chemicals usually accumulate in tissues with high fat content or in muscles or specific organs (Harvey et al. 2008). Polychlorinated biphenyls (PCBs) can come from sources such as the environment, atmosphere, water, and food to biota. PCB concentration in the tissues of organisms is associated with their lipophilicity (Salem et al. 2014).

5.5 Shellfish Safety Assurance

The principles for controlling hazards in shellfish will include the identification of hazard, control of the hazard and monitoring of the effectiveness of the controls. The responsibility should be shared between governments, industry, and consumers, each having an important role to play in the protection of human health (Erondu and Anyanwu 2005). The most important point in prevention of hazards related to shellfish is the place and time of capture. Capture point should be clear of microorganisms, parasites, biotoxins and not contaminated with heavy metals, antibiotics, and agrochemicals.

Sea shores, the places close to marine or industrial locations are all hazardous for aquaculture products. Heavy metals, drugs, antibiotics, agrochemicals are all located in these areas and accumulates. There are two ways of avoiding this man-made risk. The first is limiting or banning the usage of these chemicals and obeying the regulations. The second is periodically control of contaminants, not using as harvest area and taking care of withdrawal periods.

Continuous monitoring is necessary to ensure the safety of food and to protect the marine environment. The key elements of shellfish biosecurity include adequate diagnostic and detection methods to monitor pathogens, disinfection and pathogen eradication methods, applications of the Hazard Analysis Critical Control Point (HACCP) system, Good Manufacturing Practices (GMP), practical guidance, and appropriate legislative controls (Venugopal and Gopakumar 2017).

The HACCP system is a management system which food safety is conducted through the analysing and controlling of biological, chemical, and physical hazards from raw material production, supplying and handling to processing, distribution, and consumption of the final product. HACCP has emerged as a system that eliminates the need to test the finished product by identifying the safety risks existing in the nature of the product and taking preventive measures. HACCP is a protective system in food control. The system is used to control the areas and points that may cause any hazard in food production. This system basically has adopted the principle of determining the protective measures that will prevent the hazards that may arise because of control delays or irreversible negativities in the final product (Jo 2010).

HACCP system has seven basic principles. These principles include hazard analysis, critical control point (CCP) identification, establishing critical limits, monitoring procedures, corrective actions, verification procedures and record keeping and documentation. In this system, if a deviation occurs indicating that control has been lost, the deviation is detected, and appropriate steps are taken re-establish control in a timely manner to ensure that potentially hazardous products do not reach consumer (NACMCF 1998).

For proper use of shellfish according to HACCP and GMP, the following are recommended (Mortimore and Wallace 1994; Gould 1994: Tzouros and Arvanitoyannis 2000).

- Only live shellfish may be processed.
- Clams, mussels, and oysters that are used for canning shall not contain excessive green algae and shall be free from sand gravel, pearls, discoloration, and shell pieces.
- Fresh and frozen lobster meat shall be free of the stomach, the intestinal tract, the gills, cartilage, shell, particles, liver, roe, and any other part that is not flesh and that might be physically and/or microbiologically harmful to consumers.

Other recommendations are as follows:

- Keep shellfish cool after harvesting. If the temperature of shellfish can rise, bacteria will grow, and the shellfish will become unsafe to eat
- All fresh shellfish should be stored in an open container in the refrigerator. Never store shellfish in water. They will die and may spoil.
- To ensure proper food safety, shellfish must be cooked to an internal temperature of at least 145 °F (62.77 °C).

5.5.1 *Protection from Toxin Hazards*

Harmful algal bloom can be followed by visual inspections and harvest time can be arranged in case of biotoxin presence. Temperature, pH, salinity, up-welling, down-welling interactions can be observed to avoid biotoxin disorders such as paralytic, neurological, diarrhetic and amnesic. Periodic sampling should be organized to detect changes in the composition of the plankton containing toxins and the geographical distribution. The interaction between filter feeding shellfish and biotoxins is essential for an efficient risk management. European Union member states are required to monitor shellfish for the presence of these toxins under Directive 91/492/EEC (EC 1991). Monitoring of harvest areas with the periodical water and soil analyses should be taking into consideration and chosen non-contaminated areas.

The points that can be intervened to prevent biotoxin hazard are suggested as follows (Hegaret and Shumway 2009):

- The harvest should be stopped when potentially toxic phytoplankton species are present
- Shellfish populations should be sampled before harvest and analysed toxicity in the shellfish tissues
- Specific tissues containing phycotoxins should be removed after harvest
- Advisories to cook harvested shellfish should be issued thoroughly in cases wherein phycotoxins are heat labile.
- Moreover, monitoring programs and regulations need to be established to limit the risk of intoxication of humans by toxic shellfish. To be efficient, the monitoring programs must allow rapid detection of any toxic algae or presence of toxic shellfish, to be able to limit the harvest and avoid contamination, as well as to prevent unnecessary disposal of toxic shellfish already harvested. Thus, many countries have monitoring programs screening the phytoplankton present in the water, but also the flesh of the shellfish on a regular basis. Moreover, a good network needs to be developed to be able to transfer the information very rapidly and efficiently to seafood harvesters, distributors, and consumers, as well as public health and medical professionals.

5.5.2 Protection from Microbial Hazards

Conventional commercial processes employed to purge the microbial contamination of live bivalves are depuration, performed in tanks, and relaying, performed in the natural environment. Detailed information about the depuration is given in Chap. 3 (Molluscan shellfish). There are many factors affecting depuration process. Some these factors are water quality, oxygenation and flow rates of water, temperature, water to bivalve ratio, salinity, removal, and deposition, of faecal material. Some disinfection methods such as use of chlorine, UV light, ozone and iodophors also affect depuration. In this case the most effective depuration conditions should be constituted to provide consumer safety.

Apart from depuration and relaying, post-harvest use of various techniques such as irradiation, pasteurization, high hydrostatic pressure (HHP) and freezing may help reduce the risk of infection by pathogens (Soniat 2009).

One of the most widely used management approaches is the regular microbiological monitoring of shellfish harvesting areas/shellfish tissues and classifying the areas (Rees et al. 2010). For this purpose, areas approved by various countries and authorities have been determined. Shellfish from these areas can be used for human consumption without further processing. The classification of a production area determines the treatment required before living bivalve molluscs (shellfish) may be marketed for human consumption. Shellfish production and relay areas are classified according to the levels of *E. coli* detected in shellfish flesh. Based on the numbers of E. coli in samples of shellfish flesh, harvesting areas are classified as follows (SEAFISH 2017):

Class A—where bivalve molluscs can be harvested for direct human consumption.

Class B—where bivalve molluscs must be:

1. Purified (cleansed of bacteria through an approved depuration unit).
2. Relayed in an approved Class A relaying area.
3. Heat treated by an approved method before they can be sold for human consumption.

Class C—where bivalve molluscs must be relayed (for a minimum of 2 months) to meet Class A or B, or be heat treated.

Prohibited area—area from which bivalve molluscs cannot be harvested for human consumption.

The classifications are based on the following criteria:

Class A—an area in samples of live bivalve molluscs must not exceed, in 80% of samples collected during the review period, 230 E. coli per 100 g of flesh and intravalvular liquid. The remaining 20% of samples must not exceed 700 E. coli per 100 g of flesh and intervalvular liquid.

Class B—an area where 90% of samples have less than 4600 E. coli per 100 g of flesh; the remaining 10% of samples must not exceed 46,000 E. coli per 100 g of flesh.

Class C—an area where molluscs must contain less than 46,000 E. coli per 100 g of flesh.

Prohibited area—an area where levels are higher than 46,000 E. coli per 100 g of flesh.

Protection of water resources is another measure to ensure shellfish safety. Management practices that improve contaminated water source will protect shellfish from contamination. For this, sources of contamination should be placed away from water bodies. Waste should not be mixed with water. Sewage discharged into water bodies should be treated and sewage systems should be well designed. Natural buffers or filters (such as wetlands, mangroves, settling ponds, riparian or littoral filters) could be created/restored between contaminant sources and at-risk waters. Such a strategy should include controls over wastewater or bilge dumping from commercial and recreational vessels in the vicinity of shellfish waters. It is essential that management efforts between watersheds and/or river basins and coastal zones are co-ordinated. This will require communication between local, state, and federal agencies and defining regulatory standards that can achieve both water and shellfish fishing goals (Rees et al. 2010).

5.5.3 *Protection from Chemical Hazards*

Present quantitative risk assessment procedures used by government agencies should be improved. Each chemical user should not exceed the determined limits and should know that an overdose will threaten his health. Feed may include chemicals and drugs consciously or contaminated. Each supplier or fish farmer must check their materials. Sale and supply of organotin antifouling paints are also be checked if they are used in the limitations of regulations. Proper records of aquaculture related hazards should be updated to advance preventive or precautionary measures.

Soil and water analyses should be performed periodically for chemical; pesticide and herbicide residues and suspected fish should be analysed for contamination. All the levels are expected to be under the determined governmental limits. Feed that exceeds dosage limit or is unapproved should not be used and should not be sold until withdrawal period has been met. Despite their low solubility in water, the level of polychlorinated biphenyls can be decreased by boiling in water.

References

Ahmad NI, Noh MFM, Mahiyuddin WRW, Jaafar H, Ishak I, Azmi WNFW, Veloo Y, Mokhtar FA (2015) The mercury levels in crustaceans and cephalopods from Peninsular Malaysia. Environ Sci Pollut Res 22(17):12960–12974

Amagliani G, Brandt G, Schiavano GF (2012) Incidence and role of Salmonella in seafood safety. Food Res Int 45:780–788

Andjelkovic M, Vandevijvere S, Van Klaveren J, Van Oyen H, Van Loco J (2012) Exposure to domoic acid through shellfish consumption in Belgium. Environ Int 49:115–119

Andrews LS, Park DL, Chen YP (2000) Low-temperature pasteurization to reduce the risk of Vibrio infections from raw shell-stock oysters. Food Addit Contam 19:787–791

Andrews LS, Jahncke M, Millikarjunan K (2003) Low-dose gamma irradiation to reduce pathogenic Vibrios in live oysters (*Crassostrea virginica*). J Aquat Food Prod Technol 12:71–82

Andritsos ND, Moschonas G, Roukas D (2016) Elimination of *Escherichia coli* from mussels during treatment in a shellfish depuration system. In: 2nd international congress on applied ichthyology and aquatic environment, Messolonghi, Greece, 10–12 Nov 2016

Anonymous (2006) Risk assessment of Azaspiracids (AZAs) in shellfish: a report of the scientific committee of the Food Safety Authority of Ireland (FSAI), August 2006, 39 p.

Aune T, Yndsted M (1993) Diarrhetic shellfish poisoning. In: I.R., Falconer (Ed), Algal Toxins in Seafood and Drinking Water. Academic Press, London, 87–104

Baker GL (2016) Food safety impacts from post-harvest processing procedures of molluscan shellfish. Foods 5:29

Baker K, Morris J, McCarthy N, Saldana L, Lowther J, Collinson A et al (2010) An outbreak of norovirus infection linked to oyster consumption at a UK restaurant. J Public Health 3(2):205–211

Bakr WMK, Hazzah WA, Abaza AF (2011) Detection of Salmonella and Vibrio species in some seafood in Alexandria. J Am Sci 7(9):663–668

Bellou M, Kokkinos P, Vantarakis A (2013) Shellfish-borne viral outbreaks: a systematic review. Food Environ Virol 5:13–23

Berenguer JA, Gonzalez L, Jimenez TM, Legarda JB, Olmedo JB, Burdaspal PA (1993) The effect of commercial processing on the paralytic shellfish poison (PSP) content of naturally-contaminated *Acanthocardia tuberculatum* L. Food Addit Contam 10:217–230

Berlin DL, Herson DS, Hicks DT, Hoover DG (1999) Response of pathogenic Vibrio species to high hydrostatic pressure. Appl Environ Microbiol 65:2776–2780

Bosch A, Guyader SFL (2010) Introduction: viruses in shellfish. Food Environ Virol 2(3):115–116

Bradshaw JG, Francis DW, Twedt RM (1974) Survival of *Vibrio parahaemolyticus* in cooked seafood at refrigeration temperatures. Appl Microbiol 27(4):657–661

Bricelj VM, Shumway SE (1998) A review of paralytic shellfish poisoning toxins in bivalve molluscs: occurrence and transfer kinetics. Rev Fish Sci 6:315–383

Bryan GW (1973) The occurrence and seasonal variation of trace metals in the scallops *Pecten maximus* (L) and *Chlamys opercularis* (L). J Mar Biol Assoc U K 53:145–166

Butt AA, Aldridge KE, Sanders CV (2004) Infections related to the ingestion of seafood part I: viral and bacterial infections. Lancet Infect Dis 4(4):201e212

Calci KR, Meade GK, Tezloff RC, Kingsley DH (2005) High-pressure inactivation of hepatitis A virus within oysters. Appl Environ Microbiol 71(1):339–343

Calik H, Morrissey MT, Reno PW, He A (2002) Effect of high-pressure processing on *Vibrio parahaemolyticus* strains in pure culture and Pacific oysters. J Food Sci 67:1506–1510

Cambella AD, Shumway SE, Lewis NI (1993) Anatomical distribution and spatio-temporal variation in paralytic shellfish toxin composition in two bivalve species from the Gulf of Maine. J Shellfish Res 12(2):389–403

Chalkiadaki O, Dassenakis M, Paraskevopoulou V, Lydakis-Simantiris N (2014) Experimental study of cadmium bioaccumulation in three Mediterranean marine bivalve species: correlation with selected biomarkers. Pure Appl Chem 86(7):1189–1204

Cheng PK, Wong DK, Chung TW, Lim WW (2005) Norovirus contamination found in oysters worldwide. J Med Virol 76(4):593–597

Choi JY, Yang DB, Hong GH, Kim K, Shin K-H (2016) Ecological and human health risk from polychlorinated biphenyls and organochlorine pesticides in bivalves of Cheonsu Bay, Korea. Environ Eng Res 21(4):373–383

Cook DW (2003) Sensitivity of Vibrio species in phosphate-buffered saline and in oysters to high-pressure processing. J Food Prot 66:2276–2282

Cook DW, Ruple AD (1992) Cold storage and mild heat treatment as processing aids to reduce the numbers of *Vibrio vulnificus* in raw oysters. J Food Prot 55:985–989

Correa AA, Toso J, Albarnaz JD, Simoes CMO, Barardi CRM (2006) Detection of salmonella typhimurium in oysters by PCR and molecular hybridization. J Food Qual 29:458–469

Correa AA, Rigotto C, Moresco V, Kleemann CR, Teixeira AL, Poli CR, Simoes CMO, Barardi CRM (2012) The depuration dynamics of oysters (*Crassostrea gigas*) artificially contaminated with hepatitis A virus and human adenovirus. Mem Inst Oswaldo Cruz 107(1):11–17

Costa PR, Rodrigues S, Botelho MJ, Sampayo MAM (2003) A potential vector of domic acid: the swimming crab *Polybius henslowii* Leach (Decapoda–Brachyura). Toxicon 42:135–141

Costa PR, Rosa R, Sampayo MAM (2004) Tissue distribution of the amnesic shellfish toxin, domoic acid, in *Octopus vulgaris* from the Portuguese Coast. Mar Biol 144(5):971–976

Costa PR, Rosa R, Pereira J, Sampaya MAM (2005a) Detection of domoic acid, the amnesic shellfish toxin, in the digestive gland of *Eledone cirrhosa* and *E. moschata* (Cephalopoda, Octopoda) from the Portuguese coast. Aquat Live Resour 18:395–400

Costa PR, Rosa R, Duarte-Silva A, Brotas V, Sampayo MA (2005b) Accumulation, transformation and tissue distribution of domoic acid, the amnesic shellfish poisoning toxin, in the common cuttlefish, *Sepia officinalis*. Aquat Toxicol 74(1):82–91

Croci L, De Medici D, Di Pasquale S, Toti L (2005) Resistance of hepatitis A virus in mussels subjected to different domestic cookings. Int J Food Microbiol 105 (2005):139–144

Croonenberghs RE (2000) Contamination in shellfish-growing areas. In: Martin RE, Carter EP Jr, Flick GJ, Davis LM (eds) Marine and freshwater products handbook. Technomic Publishing Co., Inc., Lancaster, Basel, pp 995–695

Dalcorso G, Farinati S, Maistri S, Furini A (2008) How plants cope with cadmium: staking all on metabolism and gene expression. J Integr Plant Biol 50:1268–1280

Dalsgaard A, Hoei L, Linkous D, Oliver JD (2001) Vibrio vulnificus. In: Hui YH, Pierson MD, Gorham JR (eds) Foodborne disease handbook, vol 1. Bacterial pathogens, 2nd edn. Marcel Dekker, Inc., New York, pp 439–470

Dhinamala K, Shalini R, Deepa P, Pushpalatha M, Arivoli S, Samuel T, Raveen R (2017) Zinc accumulation in gills and muscles of shellfish species from Pulicat lake, Tamil Nadu, India. Int J Zool Stud 2(5):1–7

Di Lena G, Casini I, Caproni R, Orban EO (2018) Total mercury levels in crustacean species from Italian fishery. Food Addit Contam Part B Surveill 11(3):175–182

Di Pinto A, Forte VT, Tantillo GM, Terio V, Buonavoglia C (2003) Detection of hepatitis A virus in shellfish (*Mytilus galloprovincialis*) with RT-PCR. J Food Prot 66(9):1681–1685

Drake SL, DePaola A, Jaykus L-A (2007) An overview of vibrio vulnificus and Vibrio parahaemolyticus. Compr Rev Food Sci Food Saf 6:120–144

EC (European Commission) (1991) Council Directive 91/492/EEC of 15 July 1991 laying down the health conditions for the production and the placing on the market of live bivalve molluscs. Official Journal of the European Communities L 268, 24/09/1991, pp 0001–0014

Erondu ES, Anyanwu PE (2005) Potential hazards and risks associated with the aquaculture industry. Afr J Biotechnol 4(13):1622–1627

Espinoza LA (2013) Development of safe and ready to eat frozen oyster products using microwave steam-venting technology. Louisiana State University Doctoral Dissertations, p 1339

Espinoza Rodezno LA, Sundararajan S, Solval KM, Chotika A, Li J, Zhang J, Alfaro L, Bangston JD, Sathivel S (2013) Cryogenic and air blast freezing techniques and their effect on the quality of catfish fillets. LWT-Food Sci Technol 54:377–382

Eyles MJ, Davey GR (1984) Microbiology of commercial depuration of the Sydney rock oyster, *Crassostrea commercialis*. J Food Prot 47:703–706

Faber S (2012) Saxitoxin and the induction of paralytic shellfish poisoning. J Young Investig 23(1):1–7

FAO (2004) Marine biotoxins. Food and Agriculture Organisation of the United Nations, Rome, 279 p

FAO (2016) Technical paper on toxicity equivalency factors for marine biotoxins associated with bivalve molluscs. Rome, 108 p

Feldhusen F (2000) The role of seafood in bacterial foodborne disease. Microbes Infect 2:1651–1660

Flannery J, Rajko-Nenow P, Winterbourn JB, Malham SK, Jones DL (2014) Effectiveness of cooking to reduce Norovirus and infectious F-specific RNA bacteriophage concentrations in Mytilus edulis. J Appl Microbiol 117:564–571

Flick GJ Jr, Granata LA (2010) Biological safety of fresh and processed shellfish. Southern Regional Aquaculture Center (SRAC) Publication No. 4901 April 2010

Fowler SW, Oregioni B (1976) Trace metals in mussels from the NW Mediterranean. Mar Pollut Bull 7:26–29

Gokoglu N, Gokoglu M, Yerlikaya P (2006) Seasonal variations in proximate and elemental composition of pearl oyster *Pinctada radiata* from the Gulf of Antalya, Turkey. J Sci Food Agric 86(13):2161–2165

Gongora-Gomez AM, García-Ulloa M, Muñoz-Sevilla NP, Domínguez-Orozco AL, Villanueva-Fonseca BP, Hernandez-Sepulveda JA, Ortega Izaguirre R (2017) Heavy-metal contents in oysters (*Crassostrea gigas*) cultivated on the south-eastern coast of the Gulf of California, Mexico. Hidrobiologica 27(2):219–227

Gongora-Gomez AM, Dominguez-Orozco AL, Villanueva-Fonseca BP, Munoz-Sevilla NP, Garcia-Ulloa M (2018) Seasonal levels of heavy metals in soft tissue and muscle of the pen shell *Atrina maura* (sowerby, 1835) (bivalvia: Pinnidae) from a farm in the south-eastern coast of the gulf of California, Mexico. Rev Int Contam Ambient 34(1):57–68

Gosling E (2015) Marine bivalve molluscs, 2nd edn. Wiley-Blackwell, Hoboken, pp 478–508

Gould WA (1994) Current good manufacturing practices food plant sanitation, 2nd edn. ITP, New York

Gueguen M, Amiard JC, Arnich N, Badot PM, Claisse D, Guerin T, Vernoux JP (2011) Shellfish and residual chemical contaminants: hazards, monitoring, and health risk assessment along french coasts. Rev Environ Contam Toxicol 213:55–111

Hariharan H, Amadi V (2016) Shellfish as reservoirs of bacterial pathogens. J Coast Life Med 4(4):253–258

Harvey J, Harwell L, Summers JK (2008) Contaminant concentrations in whole-body fish and shellfish US estuaries. Environ Monit Assess 137:403–412

Hashmi MI, Mustafa S, Tariq SA (2002) Heavy metal concentrations in water and tiger prawn (*Penaeus monodon*) from grow-out farms in Sabah, North Borneo. Food Chem 79:151–156

Hassard F, Sharp JH, Taft H, LeVay L, Harris JP, McDonald JE, Tuson K, Wilson J, Jones DL, Shelagh K, Malham SK (2017) Critical review on the public health impact of norovirus contamination in shellfish and the environment: a UK perspective. Food Environ Virol 9:123–141

He P (2015) Harmful algal blooms. In: Sahoo D, Seckbach J (eds) The alg world. Springer, London, pp 339–355

Hegaret H, Shumway SE (2009) Biotoxin contamination and shellfish safety. In: Shumway SE, Rodrick GE (eds) Shellfish safety and quality. Woodhead Publishing, Cambridge, pp 43–68

Hellberg RS, Morrissey MT, Nettleton JA (2005) Review of mercury in seafood. J Aquat Food Prod Technol 14(4):71–100

Hernroth B, Allard A (2007) The persistence of infectious adenovirus (type 35) in mussels (*Mytilus edulis*) and oysters (*Ostrea edulis*). Int J Food Microbiol 113:296–302

Hess P, Nguyen L, Aasen J, Keogh M, Kilcoyne J, McCarron P, Aune T (2005) Tissue distribution, effects of cooking and parameters affecting the extraction of azaspiracids from mussels, *Mytilus edulis*, prior to analysis by liquid chromatography coupled to mass spectrometry. Toxicon 46(1):62–71

Hesselman DM, Motes ML, Lewis JP (1999) Effects of a commercial heat-shock process on Vibrio vulnificus in the American oyster, *Croassostrea virginica*, harvested from the Gulf Coast. J Food Prot 62:1266–1269

Iwamoto M, Ayers T, Mahon BE, Swerdlow DL (2010) Epidemiology of seafood-associated infections in the United States. Clin Microbiol Rev 23(2):399–411

Jain S, Chen L, Dechet A, Hertz AT, Brus DL, Hanley K, Wilson B, Frank J, Greene KD, Parsons M, Bopp CA, Todd R, Hoekstra M, Mintz ED, Ram PK (2008) An outbreak of enterotoxigenic *Escherichia coli* associated with sushi restaurants in Nevada. Clin Infect Dis 47:1–7

Jeffery B, Barlow T, Moizer K, Paul S, Boyle C (2004) Amnesic shellfish poison. Food Chem Toxicol 42:545–557

Jo KB (2010) Review of food safety assurance and application of HACCP in production of sea frozen squid in DPR of Korea. United Nations University Fisheries Training Programme, Iceland [final project]. https://www.grocentre.is/static/gro/publication/206/document/jo09prf.pdf

Jones S (2009) Microbial contamination and shellfish safety. In: Shumway SE, Rodrick GE (eds) Shellfish safety and quality. Woodhead Publishing Limited, Cambridge, pp 3–28

Kasper CW, Tamplin ML (1993) Effects of temperature and salinity on the survival of Vibrio vulnificus in seawater and shellfish. Appl Environ Microbiol 59:2425–2429

Kingsley DH (2013) High pressure processing and its application to the challenge of virus-contaminated foods. Food Environ Virol 5:1–12

Kingsley DH (2014) High pressure processing of bivalve shellfish and HPP's use as a virus intervention. Foods 3:336–350

Kingsley DH, Meade GK, Richards GP (2002) Detection of both hepatitis A virus and Norwalk-like virus in imported clams associated with food-borne illness. Appl Environ Microbiol 68:3914–3918

Kingsley DH, Holliman DR, Calci KR, Chen H, Flick GJ (2007) Inactivation of a norovirus by high-pressure processing. Appl Environ Microbiol 73(2):581–585

Krishnamurti JA, Nair RV (1999) Concentration of metals in shrimps and crabs from Thane-Bassein creek system, Maharastra. Indian J Mar Sci 28:92–95

Kumar R, Surendran PK, Thampuran N (2009) Distribution and genotypic characterization of Salmonella serovars isolated from tropical seafood in cochin, India. J Appl Microbiol 106:515–524

Kwong R, Wang V, Lamb P, Yu P (2006) The uptake, distribution and elimination of paralytic shellfish toxins in mussels and fish exposed to toxic dinoflagellates. Aquat Toxicol 80(1):82–89

La Bella G, Martella V, Basanisi MG, Nobili G, Terio V, La Salandra G (2017) Food-borne viruses in shellfish: investigation on norovirus and HAV Presence in Apulia (SE Italy). Food Environ Virol 9:179–186

Landrigan PJ (1982) Occupational and community exposures to toxic metals: lead, cadmium, mercury and arsenic. West J Med 137:531–539

Latouche YD, Mix MC (1981) Seasonal variation in soft tissue weights and trace metal burdens in the Bay Mussel, Mytilus edulis. Bull Environ Contam Toxicol 27:821–828

Lawrence JF, Maher M, Watson-Wright W (1994) Effect of cooking on the concentration of toxins associated with paralytic shellfish poison in lobster hepatopancreas. Toxicon 32(1):57–64

Lawrence J, Loreal H, Toyofuku H, Hess P, Iddya K, Ababouch L (2011) FAO fisheries and aquaculture assessment and management of biotoxin risks in bivalve molluscs. FAO Fisheries and Aquaculture Technical Paper, No. 551, Rome

Lee RJ, Rangdale RE (2008) Bacterial pathogens in seafood. In: Borresen T (ed) Improving seafood products for the consumer. Woodhead Publishing, Cambridge, 612 p

Lees D (2000) Viruses and bivalve shellfish. Int J Food Microbiol 59:81–116

Li TH, Chiu CH, Chen WC, Chen CM, Hsu YM, Chiou SS, Chiou CS, Chang CC (2009) Consumption of groundwater as an independent risk factor of Salmonella Choleraesuis infection: a case control study in Taiwan. J Environ Health 72:28–31

Li D, Stals A, Tang Q-J, Uyttendaele M (2011) Detection of noroviruses in shellfish and semi processed fishery products from a Belgian seafood company. J Food Prot 77(8):1342–1347

Liu C, Lu J, Su Y-C (2009) Effects of flash freezing, followed by frozen storage, on reducing *Vibrio parahaemolyticus* in Pacific raw oysters (*Crassostrea gigas*). J Food Prot 72(1):174–177

Lopez C (2001) Evaluation of the natural prevalence of vibrio spp. in Uruguayan mussels (*Mytilus* sp.) and their control using irradiation. Irradiation to control vibrio infection from consumption of raw seafood and fresh produce. Results of a co-ordinated research project organized by the Pan American Health Organization and the Joint FAO/IAEA Division of Nuclear Techniques in Food and Agriculture, IAEA, Vienna, pp 71–77

Louppis AP, Katikou P, Georgantelis D, Badeka AV, Kontominas MG (2011) Effect of ozonation and γ-irradiation of post-harvest decontamination of mussels (*Mytilus galloprovincialis*) containing diarrheic shellfish toxins. Food Addit Contam 28:1735–1744

MacDonald EM (1970) The occurrence of paralytic shellfish poison in various species of shore animals along the Strait of Juan de Fuca in the state of Washington. M.S. thesis, University of Washington, Seattle, 65 p

Marceddu M, Lamon S, Consolati SG, Ciulli S, Mazza R, Mureddu A, Meloni D (2017) Determination of *Salmonella* spp., *E. coli* VTEC, *Vibrio* spp., and Norovirus GI-GII in bivalve molluscs collected from growing natural beds in Sardinia (Italy). Foods 6:88

Martinez-Urtaza J, Montserrat S, Gustavo HC, Antonio L, Oscar GM, Joaquin E (2003) Identification of Salmonella serovars isolated from live molluscan shellfish and their significance in the marine environment. J Food Prot 66(2):226–232

Martinez-Urtaza J, Saco M, De Novoa J, Perez-Pineiro P, Peıteado J, Lozano-Leon A, Garcıa-Martın O (2004) Influence of environmental factors and human activity on the presence of *Salmonella serovars* in a marine environment. Appl Environ Microbiol 70:2089–2097

McLeod C, Hay B, Grant C, Greening G, Day D (2009) Inactivation and elimination of human enteric viruses by Pacific oysters. J Appl Microbiol 107(6):1809–1818

Melody K, Senevirathne R, Janes M, Jaykus LA, Supan J (2008) Effectiveness of icing as a post-harvest treatment control of *Vibrio vulnificus* and *Vibrio parahaemolyticus* in the Eastern oyster (*Crassostrea virginica*). J Food Prot 71(7):1475–1480

Mensah P, Palmer C, Muller W (2014) Lethal and sublethal effects of pesticides on aquatic organisms: the case of a freshwater shrimp exposure to Roundup®. In: Larramendy ML, Soloneski S (eds) Pesticides—toxic aspects. InTech, pp 163–185

Meshram LN, Udawant SM, Pawar S, Mishra PS (2014) Bioaccumulation of heavy metals (Zn, Pb, Cd, AND Ni) in tissues of Penaeus monodon (Fabricius, 1798) from India. Int J Adv Res 2(3):548–555

Millard J, Appleton H, Parry JV (1987) Studies on heat inactivation of hepatitis A virus with special reference to shellfish. Part 1. Procedures for infection and recovery of virus from laboratory-maintained cockles. Epidemiol Infect 98:397–414

Mok JC, Kwon JY, Choi WS, Kang TS, Ha NY (2014) Contents and risk assessment of heavy metals in marine invertebrates from Korean coastal fish markets. J Food Prot 77(6):1022–1030

Monfort P, Le Gal D, Le Saux JC, Pielet G, Raguenes P, Boulben S, Plusquellec A (1994) Improved rapid method for isolation and enumeration of Salmonella from bivalves using Rambach agar. J Microbiol Methods 19:67–79

Morono A, Franco J, Miranda M, Reyero MI, Blanco J (2001) The effect of mussel size temperature, seston volume, food quality, and volume-specific toxin concentration on the uptake rate of PSP toxins by mussels (*Mytilus galloprovincialis* Lmk). J Exp Mar Biol Ecol 257:117–132

Mortimore S, Wallace C (1994) HACCP: a practical approach. Springer, Boston, xviii, 296 p

Muntada-Garriga JM, Rodriguez-Jerez JJ, Lopez-Sabater EI, Mora-Ventura MT (1995) Effect of chill and freezing temperatures on survival of *Vibrio parahaemolyticus* inoculated in homogenates of oyster meat. Lett Appl Microbiol 20:225–227

NACMCF (National Advisory Committee on Microbiological Criteria for Foods) (1998) Hazard analysis and critical control point principles and application guidelines. J Food Prot 61(9):1246–1259

Namsai A, Louisirirotchanakul S, Wongchinda N, Siripanyaphinyo U, Virulhakul P, Puthavathana P, Myint KS, Gannarong M, Ittapong R (2011) Surveillance of hepatitis A and E viruses' contamination in shellfish in Thailand. Lett Appl Microbiol 53:608–613

Neish A (2013) Investigative trials on the purification of oysters to identify ways of reducing norovirus. CEFAS (Centre for Environment, Fisheries & Aquaculture Science) contract report C5224, UK, 25 p

Nguyen TA, Roudot A-C, Massin DP (2014) Risk assessment of heavy metals in shellfish for the population in Nha Trang City, Vietnam. J Environ Health 76:56–64

Noguchi T, Ueda Y, Onoue Y, Kono M, Koyama K, Hashimoto K, Seno Y, Mishima S (1980a) Reduction in toxicity of PSP infested scallops during canning process. Bull Jpn Soc Sci Fish 46:1273

Noguchi T, Ueda Y, Onoue Y, Kono M, Koyama K, Hashimoto K, Takeuchi T, Seno Y, Mishima S (1980b) Reduction in toxicity of highly PSP-infested scallops during canning process and storage. Bull Jpn Soc Sci Fish 46:1339

Novaczek I, Madhyastha MS, Ablett RF, Donald A, Johnson G, Nijjar MS, Sims DE (1992) Depuration of domoic acid from live blue mussels (*Mytilus edulis*). Can J Fish Aquat Sci 49:312–318

Novoslavskij A, Terentjeva M, Eizenberga I, Valciņa O, Bartkevics V, Berzins AN (2016) Major food-borne pathogens in fish and fish products: a review. Ann Microbiol 66:1–15

Oliveira J, Cunha A, Castilho F, Romalde JL, Pereira MJ (2011) Microbial contamination and purification of bivalve shellfish: crucial aspects in monitoring and future perspectives—a mini review. Food Control 22:805–816

Paez-Osuna F, Marmolejo-Rivas C (1990) Trace metals in tropical coastal lagoon bivalves: Crassostrea corteziensis. Bull Environ Contam Toxicol 45:538–544

Paez-Osuna F, Frías-Espericueta MG, Osuna-López JL (1995) Trace metal concentration in relation to season and gonadal maturation in the oyster *Crassostrea iridescens*. Mar Environ Res 40:19–31

Park SY, Ha S-D (2018) Inactivation of Vibrio parahaemolyticus in shucked raw oyster (*Grassostrea gigas*) and clam (*Venerupis phillippinarum*) by using a combination of NaClO and gamma irradiation. Food Sci Technol Int 24(1):43–52

Parker RW, Maurer EM, Childers AB, Lewis DH (1994) Effect of frozen storage and vacuum packaging on survival of *Vibrio vulnificus* in Gulf Coast oysters (*Crassostrea virginica*). J Food Prot 57:604–606

Perl TM, Bedard L, Kosatsky T, Hockin JC, Todd ECD, Remic RS (1990) An outbreak of toxic encephalopathy caused by eating mussels contaminated with domoic acid. N Engl J Med 322:775–1780

Peterson DA, Wolfe G, Larkin EP, Deinhardt FW (1978) Thermal treatment and infectivity of hepatitis A virus in human feces. J Med Virol 2:201–206

Pigeot J, Miramand P, Guyot T, Sauriau P-G, Ficher D, Le Moine O, Huet V (2006) Cadmium pathways in an exploited intertidal ecosystem with chronic cadmium inputs (Marennes-Oleron, Atlantic coast, France). Mar Ecol Prog Ser 307:101–114

Pinto RM, Costafreda MI, Bosch A (2009) Risk assessment in shellfish-borne outbreaks of hepatitis A. Appl Environ Microbiol 75:7350–7355

Pomati F, Kellmann R, Cavalieri R, Burns BP, Neilan BA (2006) Comparative gene expression of PSP-toxin producing and non-toxic *Anabaena circinalis* strains. Environ Int 32(6):743–748

Potasman I, Paz A, Odeh M (2002) Infectious outbreaks associated with bivalve shellfish consumption: a worldwide perspective. Clin Infect Dis 35(8):921–928

Prakash A, Medcof JC, Tennant, AD (1971) Paralytic shellfish poisoning in eastern Canada. *Bulletin-Fisheries Resarch Board of Canada*, 177, 87 p.

Prapaiwong N, Wallace RK, Arias CR (2009) Bacterial loads and microbial composition in high pressure treated oysters during storage. Int J Food Microbiol 131:145–150

Rainbow PS, Phillips DH (1993) Cosmopolitan bioindicators of trace metals. Mar Pollut Bull 26:593–601

Ramos RJ, Miotto M, Lagreze Squella FJ, Cirolini A, Ferreira JF, Werneck Vieira CR (2012) Depuration of oysters (*Crassostrea gigas*) contaminated with *Vibrio parahaemolyticus* and *Vibrio vulnificus* with UV light and chlorinated seawater. J Food Prot 75:1501–1506

Rashid HO, Ito H, Ishigaki I (1992) Distribution of pathogenic vibrios and other bacteria in imported frozen shrimps and their decontamination by gamma-irradiation. World J Microbiol Biotechnol 8:494–498

Raszl SM, Froelich BA, Vieira CRW, Blackwood AD, Noble RT (2016) *Vibrio parahaemolyticus* and *Vibrio vulnificus* in South America: water, seafood and human infections. J Appl Microbiol 121(5):1201–1222

Reboreda A, Lago J, Chapela MJ, Vieites JM, Botana LM, Alfonso A, Cabado AG (2010) Decrease of marine toxin content in bivalves by industrial processes. Toxicon 55:235–243

Rees G, Karunasagar I, Santo Domingo J (2010) Driving forces and risk management. In: Rees G, Pond K, Kay D, Bartram J, Santo J (eds) Domingo safe management of shellfish and harvest waters. IWA Publishing, London, pp 39–51

Richards GP, McLeod C, Guyader FSL (2010) Processing strategies to inactivate enteric viruses in shellfish. Food Environ Virol 2:183–193

Rodrick GE, Schneider KR (1991) Vibrios in depuration. In: Rodrick WS, Martin GE, Otwell RE (eds) Molluscan shellfish depuration. CRC Press, Boca Raton, pp 115–125

Rubini S, Galletti G, D'Incau M, Govoni G, Boschetti L, Berardelli C, Barbieri S, Merialdi G, Formaglio A, Guidi E, Bergamini M, Piva S, Serraino A, Giacometti F (2018) Occurrence of Salmonella enterica subsp. enterica in bivalve molluscs and associations with *Escherichia coli* in molluscs and faecal coliforms in seawater. Food Control 84:429–435

Salam MA, Paul SC, Noor SNBM, Siddiqua SA, Aka TD, Wahab E, Aweng ER (2019) Contamination profile of heavy metals in marine fish and shellfish. Glob J Environ Sci Manag 5(2):225–236

Salem DMSA, El Sikaily A, Ahmed El Nemr A (2014) Organochlorines and their risk in marine shellfish collected from the Mediterranean coast, Egypt. Egypt J Aquat Res 40:93–101

Schroeder G, Bates SS, Spallino J (2015) Amnesic shellfish poisoning: emergency medical management. J Mar Sci Res Dev 6(1):179

Schwab KJ, Neill FH, Estes MK, Metcalf TG, Atmar RL (1998) Distribution of Norwalk virus within shellfish following bioaccumulation and subsequent depuration by detection using RTPCR. J Food Prot 61:1674–1680

Scoging A, Bahl M (1998) Diarrhetic shellfish poisoning in the UK. Lancet 352(11):117

SEAFISH (2017) Classification of shellfish harvesting areas. February 2017. https://www.seafish.org/media/publications/Classification-of-Harvesting-Areas-Feb-2017.pdf. Accesed 22 June 2020

Setala O, Lehtinen S, Kremp A, Hakanen P, Kankaanpaa H, Erler K, Suikkanen S (2014) Bioaccumulation of PSTs produced by Alexandrium ostenfeldii in the northern Baltic Sea. Hydrobiologia 726:143–154

Shabarinath S, Sanath Kumar H, Khushiramani R, Karunasagar I, Karunasagar I (2007) Detection and characterization of Salmonella associated with tropical seafood. Int J Food Microbiol 114(2):227–233

Sharma V, Parmar P, Kumari N (2016) Differential cadmiım stress tolerance in wheat genotypes under mycorrhizal association. J Plant Nutr 39:2025–2036

Shumway SE, Van Egmond HP, Hursti JW, Bean LL (1995) Management of shellfish resources. In: IOC, Manual of harmful marine microalgae. UNESCO, pp 433–461

Singh AS, Nagalakshimi K (2013) Depuration of bivalves for food quality and safety. INFOFISH Int 2:46–49

Soniat T (2009) Managing molluscan shellfish-borne microbial diseases. In: Shumway SE, Rodrick GE (eds) Shellfish safety and quality. Woodhead Publishing, Cambridge, pp 250–269

Szkoda J, Durkalec M, Nawrocka A, Michalski M (2015) Mercury concentration in bivalve molluscs. Bull Vet Inst Pulawy 59:357–360

Tamele IJ, Loureiro PV (2020) Lead, mercury and cadmium in fish and shellfish from the Indian ocean and red sea (African Countries): public health challenges. J Mar Sci Eng 8:344

Tayar M (2010) Gıda güvenliği. T.C. Marmara Belediyeler Birliği Yayınları, Istanbul, 325 p (in Turkish)

Thomas MA (2016) Rapid cooling of eastern oysters (*Crassostrea virginica*) using an on-board icing unit to slow the growth of *Vibrio vulnificus* and *Vibrio parahaemolyticus*. Louisiana State University Master's Theses, p 1618

Torres Z, Bernuy B, Zapata G, Vivanco M, Kahn G, Guzman E, Leon R (2001) Inactivation of Vibrio cholerae O1 El Tor inoculated into Peruvian "Choro" mussels (Aulacomya ater) and two species of clams (*Argopecten purpuratus* and *Gari solida*) using medium-dose irradiation. Irradiation to control vibrio infection from consumption of raw seafood and fresh produce. Results of a co-ordinated research project organized by the Pan American Health Organization and the Joint FAO/IAEA Division of Nuclear Techniques in Food and Agriculture, IAEA, Vienna, pp 37–47

Turnbull A, Harrison R, McKeow S (2013) Paralytic shellfish poisoning in south eastern Tasmania. Commun Dis Intell 37(1):E51–E54

Tzouros NE, Arvanitoyannis IS (2000) Implementation of hazard analysis critical control point (HACCP) system to the fish/seafood industry: a review. Food Rev Intl 16(3):273–325

Ueki Y, Shoji M, Suto A, Tanabe T, Okimura Y, Kikuchi Y, Saito N, Sano D, Omura T (2007) Persistence of caliciviruses in artificially contaminated oysters during depuration. Appl Environ Microbiol 73:5698–5701

Ujevic I, Ninčević-Gladan Z, Roje R, Skejić S, Arapov J, Marasović I (2010) Domoic acid: a new toxin in the Croatian adriatic shellfish toxin profile. Molecules 15:6835–6849

Usero J, GonzBlez-Regalado E, Gracia I (1997) Trace metals in the bivalve molluscs *Ruditapes decussatus* and *Ruditapes philippinarum* form the Atlantic coast of southern Spain. Environ Int 23(3):291–298

Vasconcelos GJ, Lee JS (1972) Microbial flora of Pacific oysters (*Crassostrea gigas*) subjected to ultraviolet-irradiated seawater. Appl Microbiol 23:11–16

Vazquez FG, Sharma VK, Alexander VH, Frausto CA (1995) Metals in some lagoons of Mexico. Environ Health Perspect 103(1):33–34

Venugopal V, Gopakumar K (2017) Shellfish: nutritive value, health benefits, and consumer safety. Compr Rev Food Sci Food Saf 16:1219–1242

Vidal A, Correa J, Blanco J (2009) Effect of some habitual cooking processes on the domoic acid concentration in the cockle (*Cerastoderma edule*) and Manila clam (*Ruditapes philippinarum*). J Food Addit Contam Part A 26(7):1089–1095

Vieites JM, Botana LM, Vieytes MR, Leira FJ (1999) Canning process that diminishes paralytic shellfish poison in naturally contaminated mussels (*Mytilus galloprovincialis*). J Food Prot 62:515–519

Wang WX, Hong Kong University of Science and Technology (2009) Metals and organic contaminants in bivalve molluscs. In: Shumway SE, Rodrick GE (eds) Shellfish safety and quality. Woodhead Publishing, Cambridge, pp 228–243

Wang W-X, Wong P (2006) Dynamics of trace metal concentrations in an intertidal rocky shore food chain. Mar Pollut Bull 52:332–337

Ward D, Hart K (1996) Hazards found in seafood, appendix 111. In HACCP: hazard analysis and critical control point training curriculum. North Carolina Sea Grant, North Carolina State University, Raleigh, pp 177–194

Wittman RJ, Flick GJ (1995) Microbial contamination of shellfish: prevalence, risk to human health, and control strategies. Annu Rev Public Health 16:123–140

Woody CA, O'Neal SL (2012) Effects of copper on fish and aquatic resources. The Natura Conversancy, 26 p. https://www.conservationgateway.org/ConservationByGeography/NorthAmerica/UnitedStates/alaska/sw/cpa/Documents/W2013ECopperF062012.pdf. Accessed 20June 2020

Wright AC, Danyluk MD, Otwell WS (2009) Pathogens in raw foods: what the salad bar can learn from the raw bar. Curr Opin Biotechnol 20:172–177

Yingliang J, Pei L, Yongning W, Jie M, Cannan W, Jinfang S, Yafei Z (2014) A systematic review on food lead concentration and dietary lead exposure in China. Chin Med J 127(15):2844–2849

Yulianto B, Oetari PS, Februhardi S, Putranto TWC, Soegianto A (2005) Heavy metals (Cd, Pb, Cu, Zn) concentrations in edible bivalves harvested from northern coast of central Java, Indonesia. In: International conference of science and technology for environment protection. IOP Conference Series: Earth and Environmental Science, vol 259, p 012005

Zaccaroni A, Scaravelli D (2008) Toxicity of sea algal toxins to humans and animals. In: Evangelista V, Barsanti L, Frassanito AM, Passarelli V, Gualtieri P (eds) Algal toxins: nature, occurrence, effect and detection. Springer, Dortrecth, pp 45–91

Index

N. Gökoğlu, *Shellfish Processing and Preservation*,
https://doi.org/10.1007/978-3-030-60303-8